dimension

construction

3D

coordinates

rendering

Introducing
AutoCAD® 2006

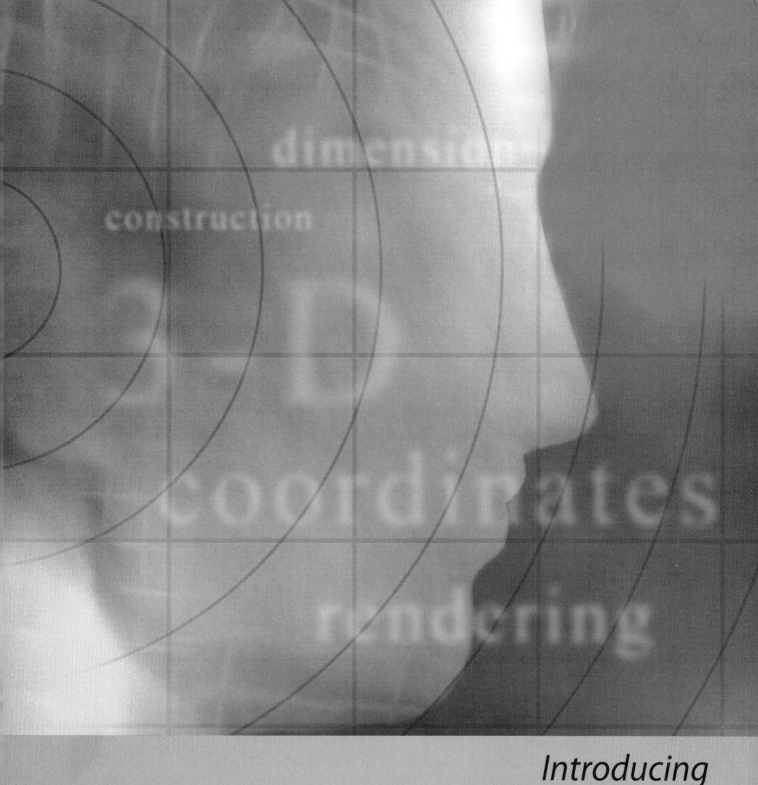

Introducing
AutoCAD® 2006

G.V. KRISHNAN
THOMAS A. STELLMAN

THOMSON
DELMAR LEARNING

Australia • Canada • Mexico • Singapore • Spain • United Kingdom • United States

Introducing AutoCAD 2006

G.V. Krishnan & Thomas A. Stellman

Autodesk Press Staff:

Vice President, Technology and Trades SBU:
Alar Elken

Editorial Director:
Sandy Clark

Senior Acquisitions Editor:
James DeVoe

Senior Development Editor:
John Fisher

Marketing Director:
Dave Garza

Channel Manager:
Dennis Williams

Marketing Coordinator:
Stacey Wiktorek

Production Director:
Mary Ellen Black

Production Manager:
Andrew Crouth

Production Editor:
Jennifer Hanley

Art & Design Specialist:
Mary Beth Vought

Editorial Assistant:
Tom Best

Cover Image:
PictureQuest

Library of Congress Cataloging-in-Publication Data:
Card Number:

ISBN: 1-4180-2033-8

NOTICE TO THE READER

contents

SECTION 2: WORLD POWER 129

CHAPTER 5 SELECTING, COPYING, AND CHANGING OBJECTS.........131

introduction

OPENING THE DOOR TO THE WORLD OF AUTOCAD 2006

Introducing AutoCAD 2006 presents an organized series of concise instructions for learning how to use today's leading desktop design/drafting software! Written specifically for first time users of AutoCAD, this up-to-date textbook contains easy-to-follow instructions and illustrations that clearly explain key AutoCAD commands and capabilities. It covers basic installation through 2D drawing and customization. AutoCAD 2006 state-of-the-art functionality is featured, by way of progressive examples of "prompt and response" sequences which are accompanied by illustrations that reinforce understanding. An extensive assortment of discipline-specific exercises and projects assist new users to become proficient in using AutoCAD 2006.

STEPS TO EXCELLENCE

This edition has been written to introduce AutoCAD to students or new users in an organized and logical sequence of lessons. The book is divided into five sections as follows:

- **Section One – In a Nutshell** introduces the program and how to use the main features, while creating fundamental drawing elements in AutoCAD.
- **Section Two – World Power** covers modifying objects, using Layers to manage objects and parts of a drawing, creating text (lettering) and accurate dimensions suitable for various disciplines.
- **Section Three – A Bigger World** explains how to use complex objects, that are a powerful part of the AutoCAD program, along with advanced modifying features to create new objects from existing objects.
- **Section Four – The Package is Complete** covers advanced elements and features of AutoCAD such as advanced dimensioning commands, using Paper Space for layouts, and plotting (producing paper copies).
- **Section Five – Icing on the Cake** explains how to combine simple objects into composite groups of objects for easier manipulation and redrawing, using already existing drawings in your new drawing, and the use of many other utility features that add speed and accuracy to the CADD drawing process.

STYLE CONVENTIONS

In order to make this text easier for you to use, we have adopted specific conventions that are used throughout the book:

Convention	Example
Command names are in small caps.	The MOVE command
Menu names appear with the first letter capitalized.	The Format menu
Toolbar menu names appear with the first letter capitalized.	The Standard toolbar
Command sequences are indented. User input is indicated by boldface type. Instructions are indicated by italics and are enclosed in parentheses.	Command: **move** Enter variable name or [?]: **snapmode** Enter group name: *(enter group name)*
When a command is invoked from a toolbar, it is shown as a screen capture with a tooltip. If it can be invoked from a menu, then it too is shown as a screen capture.	

HOW TO INVOKE COMMANDS

Like most Windows based programs, AutoCAD offers more that one method of invoking a command or accomplishing a particular task. As you progress through the different concepts and skill levels of using AutoCAD and as you use the program in your job, you will want to determine which method best suits your applications.

In early DOS-based versions, interfacing with AutoCAD usually involved typing command names at the "Command line", in the "Command Window" or using the cursor to navigate through the nested levels of the "Side Screen Menu". The Side Screen Menu became obsolete with the migration from DOS to Windows and its Menus and Toolbars for selecting commands. Now, with the new "Cursor-focused" interface, the Command line has taken a back seat as the prime method of interface and command entry.

Almost all commands offer options to the default sequence of prompts and responses. For example, the CIRCLE command's default method of drawing a circle is to specify the center and then the radius. You can override the default by using the center-diameter option or the tangent-tangent option. These options have previously been displayed in the Command Window and also available in the Shortcut menu accessed by right-clicking after invoking the command. This means you can turn off the display of the Command Window, but still access the list of options by right-clicking during a command.

For purposes of expediency, explanations and examples in *Harnessing AutoCAD 2006* assume that the user is entering commands at the On-Screen cursor, rather than entering them at the Command line.

DRAWING EXERCISES

Introducing AutoCAD 2006 offers comprehensive exercises associated with each chapter. These exercises are representative of drawing problems found in various disciplines of the design industry today. The CD in the back of this book contains PDF files of exercises that correlate with each chapter of *Introducing AutoCAD 2006*. These PDF files contain project exercises for the following disciplines: mechanical, architectural, civil, electrical, and piping. Icons identify the related discipline for each discipline (refer to the following table of exercise icons).

EXERCISE ICONS

Step-by-step Project Exercises are identified by the special icon shown in the following table. Exercises that give you practice with types of drawings often found in a particular discipline, are identified by the icons shown in the following table.

Type of Exercise	Icon	Type of Exercise	Icon
Project Exercises		Civil	
Electrical		Mechanical	
Piping		Architectural	

ONLINE COMPANION™

The Online Companion™ is your link to AutoCAD on the Internet. Updates are posted monthly, including a command of the month, tutorials, and FAQs. To access the Online Companions, go to the following URL:

http://www.autodeskpress.com/resources/olcs/index.html

E-RESOURCE

E-resource is an educational resource that creates a truly electronic Classroom. It is a CD-ROM containing tools and instructional resources that enrich your classroom and make your preparation time shorter. The elements of e-resource link directly to the text and tie together to provide a unified instructional system. Spend your time teaching, not preparing to teach.

Features contained in e-resource include:

Syllabus: Lesson plans created by chapter that list goals and discussion topics. You have the option of using these lesson plans with your own course information.

Chapter Hints: Objectives and teaching hints that provide direction on how to present the material and coordinate the subject matter with student projects.

Answers to Review Questions: These solutions enable you to grade and evaluate end of chapter tests.

PowerPoint® Presentation: These slides provide the basis for a lecture outline that helps you to present concepts and material. Key points and concepts can be graphically highlighted for student retention.

Exam View Computerized Test Bank: Over 800 questions of varying levels of difficulty are provided in true/false and multiple-choice formats. Exams can be generated to assess student comprehension, or questions can be made available to the student for self-evaluation.

Animations: These *.AVI* files graphically depict the execution of key concepts and commands in drafting, design, and AutoCAD and let you bring multi-media presentations into the classroom.

Spend your time teaching, not preparing to teach!

ISBN 1-4180-2034-6

ABOUT THE AUTHORS

Thomas A. Stellman received a B.A. degree in Architecture from Rice University and has over 20 years of experience in the architecture, engineering, and construction industry. He has taught at the college level for over ten years and has been teaching courses in AutoCAD since the introduction of version 2.0 in 1984. He has conducted seminars covering both introductory and advanced AutoLISP. In addition, he has developed and marketed third-party software for AutoCAD.

G.V. Krishnan is director of the Applied Business and Technology Center, University of Houston-Downtown, a Premier Autodesk Training Center. He has used AutoCAD since the introduction of version 1.4 and writes about AutoCAD from the standpoint of a user, instructor, and general CADD consultant to area industries. Since 1985 he has taught courses ranging from basic to advanced levels of AutoCAD, including customizing, 3D AutoCAD, solid modeling, and AutoLISP programming.

ACKNOWLEDGMENTS

We would like to thank and acknowledge the many professionals who reviewed the manuscript to help us publish this *Introducing AutoCAD 2006* text. Special thanks go to Phil Kreiker, Technical Editor and Gail Taylor, Copy Editor. Thanks to Jeff Cope, Central Georgia Technical College, Macon, Georgia for reviewing the content of the manuscript.

The authors would like to acknowledge and thank the following staff members of Thomson Delmar Learning:

Vice President, Technology and Trades SBU: Alar Elken

Senior Acquisitions Editor: James DeVoe

Production Manager: Andrew Crouth

Senior Developmental Editor: John Fisher

Production Editor: Jennifer Hanley

Art & Design Coordinator: Mary Beth Vought

The authors also would like to acknowledge and thank the following:

Composition: John Shanley and Phoenix Creative Graphics

SECTION

1

In a Nutshell

introduction

AutoCAD has grown from a fundamental PC-based computer-aided drafting program into one of the most innovative and powerful design tools available. It is very versatile, and its advanced features can be complex. It is possible to be overwhelmed by the many menus, toolbars, and dialog boxes full of commands, options, and system variable settings. Therefore, a well planned strategy is needed to learn AutoCAD.

If you are new to computer-aided drafting, specifically AutoCAD, the first step to competence is to become familiar with the parts of the screen layout. If you are new to computers, it is advisable to become skilled at computer file management. The purpose of this chapter is to guide you through the first steps toward becoming proficient in AutoCAD.

After completing this chapter, you will be able to do the following:

- Get to know the AutoCAD screen layout
- Use the AutoCAD Help
- Begin a new drawing and open an existing drawing
- Save a drawing to a given file name
- Save a drawing as a template
- Close a drawing and exit the AutoCAD program

GETTING STARTED

Design/drafting is what AutoCAD and this book are all about. There are three areas of expertise that should be developed in order for you to become a proficient AutoCAD user: design/drafting, Microsoft Windows, and AutoCAD itself.

Design/drafting is a field of exact communication, requiring the use of precise graphic and descriptive instructions. The conventions of drafting and the particular disciplines (architectural, mechanical, electrical, process, surveying, structural/civil, etc.) have carried over from traditional board drafting to computer-aided drafting. AutoCAD accommodates these conventions with its many optional styles. However, AutoCAD does not automatically apply the correct symbol, dimension, linetype, or other aspect of drafting to the drawing in progress. It is up to the user to know how the final product should look. For example, one discipline might require arrows at the end of the dimension lines, while another might call for cross marks. AutoCAD makes it easy to draw either type of symbol. But you, as the CAD drafter/designer, must be knowledgeable enough to know the correct conventions to apply. The power and speed of the computer is no substitute for knowing your trade or how the final drawing is supposed to appear.

The Microsoft Windows operating system, like AutoCAD, has grown into a complicated combination of application and data files. In addition to learning AutoCAD, it is helpful to become familiar with the basic functions of the operating system. A CAD station normally includes printers, plotters, network connections, Internet access, digitizers, and other peripherals that Windows manages, in addition to the drives and folders where files are stored.

This section teaches you how to start the AutoCAD program and describes the purposes of the features you will see on your startup screen.

Starting AutoCAD

How do you get into AutoCAD? Choose the Start button (Windows 2000/ME, or Windows XP operating system), select the AutoCAD 2006 program group, and then select the AutoCAD 2006 program.

The Startup Window

The window that appears when you start AutoCAD depends on whether you have selected the AutoCAD startup icon from one of its optional locations on one of the Windows features, such as the Desktop, Taskbar, and Explorer, or if you have selected a drawing file and opened it directly. The appearance of the startup window will also vary depending on how certain startup parameters are set up in the AutoCAD program and which of the various available template files you have used as the basis for starting your drawing. Instructions on how to modify program parameters and set up custom template files are covered in the advanced sections of this book.

First Time Startup

By default, when AutoCAD is started, it displays a blank drawing window surrounded by menus and toolbars, similar to Figure 1–1.

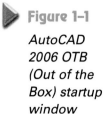

Figure 1-1

AutoCAD 2006 OTB (Out of the Box) startup window

The window shown in Figure 1-1 is one of the possible windows that might appear when AutoCAD is opened. This one appears when you start the program for the first time. The layout in the graphics area conforms to a particular set of drawing parameters. Other arrangements of the startup window and layout area are possible, depending on how AutoCAD has been configured. This arrangement along with other settings is referred to as the drawing environment. Individual configurations and arrangements surrounding the drawing area can be named and saved as unique workspaces and then recalled when needed. If you double-click on an AutoCAD drawing file icon in Windows Explorer, the layout and settings affecting the drawing area will conform to that drawing's environment within the current workspace arrangement. As covered later in this chapter, it is possible to configure AutoCAD to open with a Startup dialog box to assist you in setting up parameters different from the default opening layout.

Within the AutoCAD program window, you can create drawings for viewing, printing (referred to as plotting in the trade), solving geometry and engineering problems, accumulating data, creating three-dimensional views of objects, and various other design, graphics, and engineering applications. Whatever your objective is, you will very likely have to make changes in the layout and drawing parameters, or you can change the startup configuration to suit your needs.

note The difference between a workspace environment and a particular drawing environment is that the workspace controls the tools set up on the particular station that you use to make a drawing (menus, toolbars, palettes, etc.), while the drawing environment controls features that are configured for the drawing file itself (units, limits, layers, styles, etc.) and travel with the drawing from station to station.

The Drawing Layout

The initial AutoCAD startup graphics area (default) is a full view of a 12-unit wide by 9-unit high drawing sheet. The features and commands in AutoCAD permit you to move your view around the drawing area and zoom in for a closer look or zoom out to see a broader area.

Startup with an Existing Drawing

You can start the AutoCAD program by choosing a drawing file (one with the extension of *.dwg*) from the Windows Explorer window by double-clicking on its icon or file name. The AutoCAD program will be started. This is similar to the way other Windows-based programs are started by double-clicking on one of the types of files that are created and edited by that particular program. When AutoCAD is started in this manner (double-clicking on a *.dwg* file), the initial screen will display the drawing that was double-clicked using the view in which it was last saved.

AutoCAD Screen

The AutoCAD screen (see Figure 1–2) consists of the following elements: the Graphics window, Status bar and Tool Tray, Title bar, toolbars, menu bar, Model/Layout tabs, and the Command window. Other elements that appear from time to time are tool palettes, dialog boxes, on-screen prompts, the DesignCenter window, Command window, and graphic geometric values.

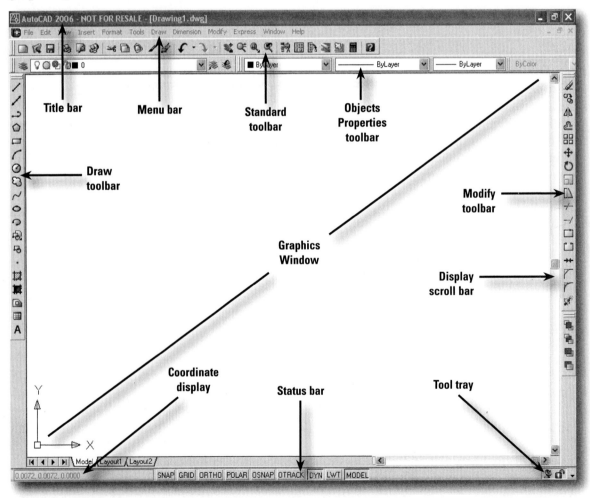

Figure 1–2

The AutoCAD screen

Graphics Window

The Graphics window is where you can create and view the objects for your design. In this window, AutoCAD displays the cursor, indicating your current working point. As you move your pointing device (usually a mouse or puck) around on a mouse pad, digitizing tablet, or other suitable surface, the cursor mimics your movements on the screen. Near the cursor is where on-screen prompts and graphic geometric values are displayed.

When AutoCAD prompts you to select a point, the cursor is in the form of crosshairs. It changes to a small pick box when you are being prompted to select an object on the screen. AutoCAD uses combinations of crosshairs, boxes, dashed rectangles, and arrows in various situations so you can quickly see what type of selection or pick mode is in effect.

note   It is possible to enter coordinates outside the viewing area for AutoCAD to use for creating or modifying objects. As you become more adept at AutoCAD, you may find a need to do this. Until then, working within the viewing area is recommended.

Status Bar and Tool Tray

The Status bar at the bottom of the screen displays the cursor's coordinates and important information about the status of various modes. On the right end of the Status bar is the Tool Tray, which contains icons for quick access to the Communications Center, Xref Manager, Locking Toolbar and Windows status, CAD Standards alert, and Digital Signature authenticator, as shown in Figure 1–3.

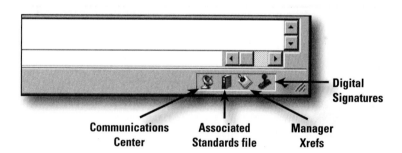

▷ **Figure 1–3**

Status bar with Tool Tray icons

Communications Center

The Communications Center icon, when selected, causes the Communication Center dialog box to be displayed. A Welcome message appears stating "The Communication Center is your direct connection to the latest software updates, product support announcements and more." From here you specify your country and preferred update frequency, connect to the Internet and download available information, and specify which information channels you wish to view.

External Reference Manager

The Xref Manager icon appears when your drawing has an external drawing attached. A message appears when an Xref needs to be reloaded or resolved. Chapter 18 explains using external references.

CAD Standards

The CAD Standards icon appears when there is a standards file associated with your drawing. A message appears when a standards violation occurs.

Digital Signatures

The Validate Digital Signatures icon appears when the drawing has a digital signature. Select the icon to validate a digital signature.

Toolbar and Window Locking

The Toolbar and Window Locking icon allows you to lock toolbars and windows to prevent accidentally changing their size or location. You can temporarily override the locking status by holding CTRL while moving a toolbar or window that has been locked.

Controlling Status Bar Display and Tray Settings

When selected, the down-arrow at the right end of the status bar tray, displays a shortcut menu. This menu provides a way to change the appearance of the status bar. Items with a check mark next to them are displayed on the status bar. Checking Cursor coordinate values causes the coordinate values to be displayed at the cursor location on the left end of the status bar. Checking any of the button names causes that button to be displayed. These include toggle buttons for snap, grid, ortho, polar tracking, object snap, object snap tracking, dynamic input, lineweight, and model. Selecting Tray Settings causes the Tray Settings dialog box to be displayed. From this dialog box you can select **Display icons from services**, which, when cleared, causes the tray to not be displayed. Checking **Display notifications from services** (under which you can select **Display time**, or **Display until closed**) displays notifications from services such as Communications Center, a service to AutoCAD users from Autodesk.

Title Bar

The Title bar displays the current drawing name, including the path where the drawing file is saved.

Toolbars

The toolbars contain tools, represented by icons, from which you can invoke commands. Select a toolbar button to invoke a command, and then select options from a dialog box or respond to the prompts on the command line. If you position your pointer over a toolbar button and wait a moment, the name of the tool is displayed, as shown in Figure 1–4. This is called the ToolTip. In addition to the ToolTip, AutoCAD displays a very brief description of the function of the command on the Status bar.

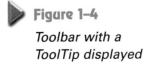

Figure 1–4

Toolbar with a ToolTip displayed

Some of the toolbar buttons have a small triangular symbol in the lower right corner of the button indicating that there are flyout buttons underneath that contain subcommands. Figure 1–5 shows the ZOOM command flyout located on the Standard toolbar. When you pick a flyout option, it remains on top to become the default option.

Figure 1-5

Display of the Zoom flyout located on the Standard toolbar

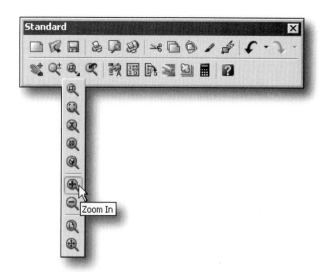

You can display multiple toolbars on screen at once, change their contents, resize them, and dock or float them. A docked toolbar attaches to any edge of the Graphics window. A floating toolbar can lie anywhere on the screen and can be resized.

Figure 1–6 shows the command icons available on the Standard toolbar; Figure 1–7 shows the commands available on the Properties toolbar. Appendix C lists the available toolbars.

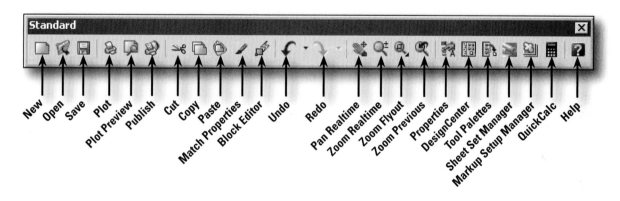

Figure 1-6

The Standard toolbar

Figure 1-7

The Properties toolbar

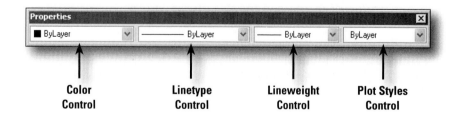

Docking and Undocking a Toolbar

To dock a toolbar, position the cursor on the caption, and press the pick button on the pointing device. While continuing to hold down the pick button, drag the toolbar to a docking location to the top, bottom, or either side of the Graphics window. When the outline of the toolbar appears in the docking area, release the pick button. To undock a toolbar, position the cursor on the left end (for horizontal toolbars) or the top end (for vertical toolbars) of the toolbar and drag and drop it outside the docking regions. To place a toolbar in a docking region without docking it, hold down CTRL as you drag. By default, the Standard toolbar and the Properties toolbar are docked at the top of the Graphics window (see Figure 1–2). Figure 1–8 shows several toolbars docked at the top of the Graphics window, the Draw toolbar docked on the left side, and the Modify and Draw Order toolbars docked on the right side.

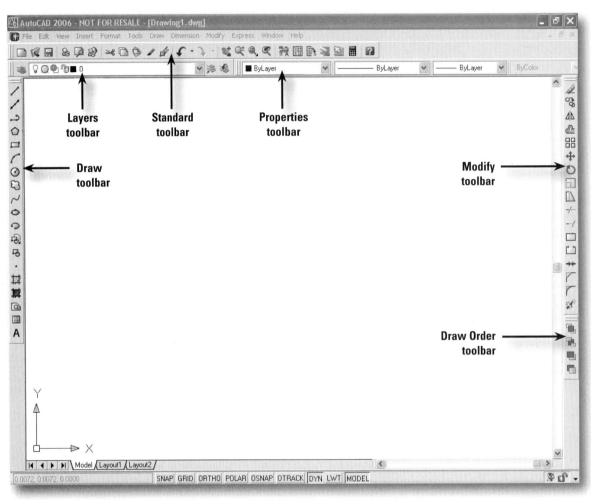

Figure 1–8

Docking of toolbars in the Graphics window

Resizing a Floating Toolbar

If necessary, you can resize a floating toolbar. To resize a floating toolbar, position the cursor anywhere on the border of the toolbar, and drag it in the direction you want to resize. Figure 1–9 shows different combinations of resizing the Draw toolbar.

 **Figure 1-9**

Draw toolbar in different resizing positions

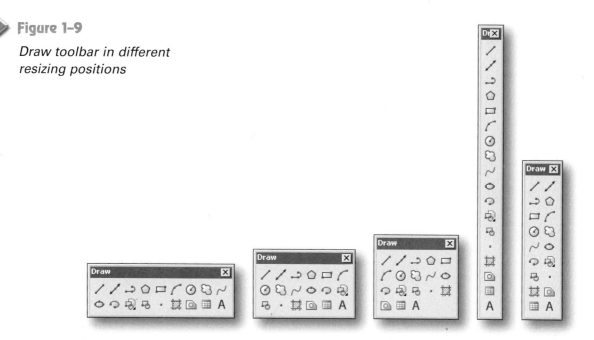

Closing a Floating Toolbar

To close a floating toolbar, position the cursor on the X located in the upper right corner of the toolbar, as shown in Figure 1–10, and press the pick button on your pointing device. The toolbar will disappear from the Graphics window.

 **Figure 1-10**

Positioning the cursor to close a toolbar

Opening a Toolbar

AutoCAD comes with 30 regular toolbars and 4 "ET" toolbars. To open any of the available toolbars, place your cursor anywhere on a toolbar that is displayed in the Graphics window, and right-click your pointing device. A Shortcut menu appears, listing all the available toolbars (see Figure 1–11). Select the toolbar you want to open. You can also close a toolbar by selecting it to remove the check mark next to the name of the toolbar. To open an "ET" toolbar, instead of right-clicking on a toolbar, right-click on the blank area behind where toolbars can be placed just off of the screen (not on the blank area to the right of the menus). Choose EXPRESS from the shortcut menu and then choose ET:Blocks, ET:Layers, ET:Standard or ET:Text.

Figure 1–11

Shortcut menu listing the available toolbars

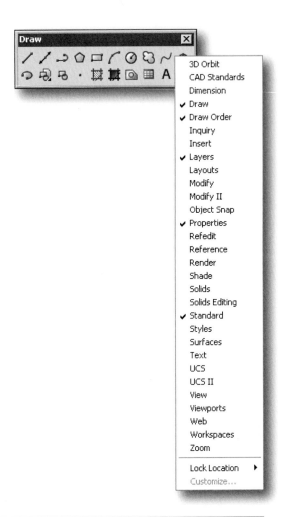

note It is advisable to always keep at least one toolbar visible on the screen. Having a toolbar visible makes it possible to right-click on it to cause the toolbar shortcut menu to appear. This provides easy access for turning other toolbars on and off. If all toolbars have been closed, you must use the TOOLBAR command to display the Customize User Interface dialog box to open at least one toolbar.

Locking and Unlocking Toolbars and Windows

When you right-click on a toolbar, it causes a shortcut menu to be displayed. Selecting Lock Location option at the bottom of the shortcut menu causes a flyout menu to appear (see Figure 1–12) with options to lock and unlock windows. You can lock and unlock floating and docked toolbars and windows. When one of the options is in the locked mode, a check will be displayed beside that option. Toolbars and windows that are locked cannot be repositioned or resized. This prevents inadvertently changing a toolbar or window position or size. The All option allows you to lock or unlock all floating and docked toolbars and windows with one selection.

Figure 1-12

Shortcut menu with Lock Location flyout displayed

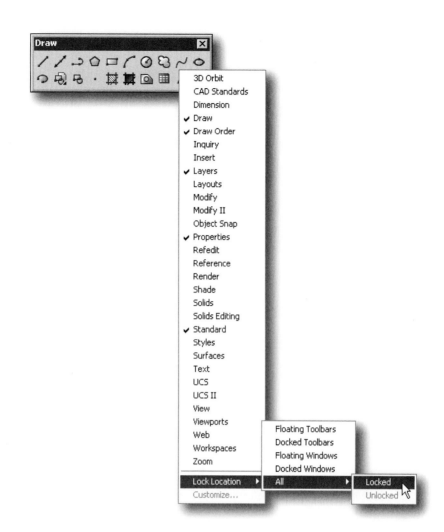

Tool Palettes Window

Tool palettes are tabbed areas within the Tool Palettes window that provide an efficient method for organizing, sharing, and placing blocks and hatches. Tool palettes can also contain custom tools provided by third-party developers. Blocks (see Chapter 17) and Hatch Patterns (see Chapter 12) are the primary tools that are managed with tool palettes. You can also create a tool on a tool palette that executes a single AutoCAD command or a string of commands. The Tool Palettes feature allows blocks and hatch patterns of similar usage and type to be grouped in their own tool palette. For example, one tool palette can be named Electrical and contain blocks representing electrical symbols, as shown in Figure 1–13.

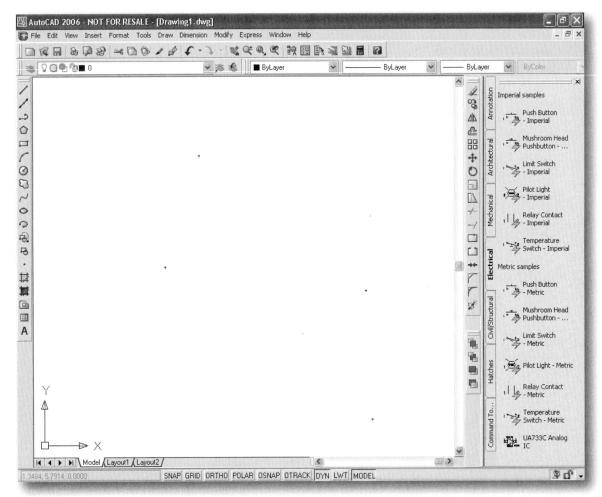

△ **Figure 1–13**

The Tool Palettes window in the docked position with the Electrical Tool palette displayed

Figure 1–13 shows the default Tool Palettes window that comes with AutoCAD 2006. Other Tool Palettes attached to the Tool Palettes window are Annotation, Architectural, Mechanical, Electrical, Civil/Structural, Hatches, and Command Tools. Each of these contain icons representing blocks, hatch patterns, or commands. The TOOLPALETTES command opens the Tool Palettes window.

The default position for the Tool Palettes window is docked on the right side of the screen. Its position can be changed by placing the cursor over the double line bar at the top of the window and either double-clicking or dragging the window into the screen area (or across to a docking position on the left side of the screen). Double-clicking causes the Tool Palettes window to become undocked and to float in the drawing area, as shown in Figure 1–14. When the Tool Palettes window is undocked, it can be docked by double-clicking in the title bar (which may be on the left or right side of the window) or by placing the cursor over the title bar and dragging the window all the way to the side where you wish to dock it.

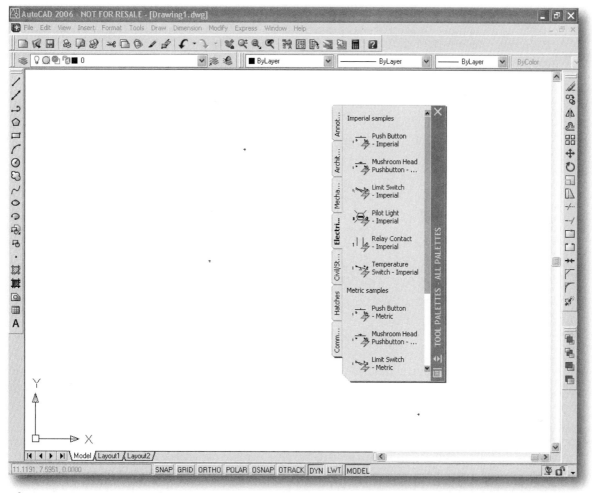

Figure 1-14

The Tool Palettes window in the floating position

Menu Bar

Pull-down menus are available from the menu bar at the top of the screen. To select any of the available commands, move the crosshairs cursor into the menu bar area and press the pick button on your pointing device, which pops that menu bar onto the screen (see Figure 1–15). Selecting from the list is a simple matter of moving the cursor until the desired item is highlighted and then pressing the designated pick button on the pointing device. If a menu item has an arrow to the right, it has a cascading submenu. To display the submenu, move the pointer over the item and it will automatically be displayed. Menu items that include ellipses (...) display dialog boxes. To select one of these, just pick that menu item.

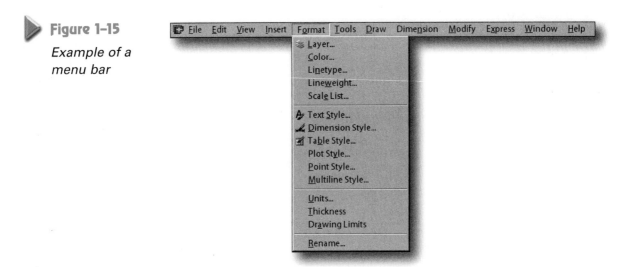

Figure 1-15

Example of a menu bar

Focusing on the Design – Dynamic Input

Dynamic Input allows you to "keep your eyes on the road" when working by providing a prompt and input interface where you work, on the screen, at the cursor. How to customize Dynamic Input is explained in Chapter 4 (Dynamic Input, On-Screen Prompts and Geometric Values Display section). Previous versions of AutoCAD emphasized the need to keep a constant vigil on the Command line in the Command Window (discussed next). This meant having to look back and forth between the Command Window and the point where you were working. Now, with the on-screen (near the cursor) interface, almost everything you need to know about what is going on with the program and your current work is right there, where you are working. Figure 1–16 shows the four stages of information that is displayed on the screen during the process of drawing a circle. The first view shows how the cursor appears when AutoCAD is ready for you to enter a command. The second view shows the text box that appears if you type in a command name from the keyboard (this step is skipped if you choose to initiate the command by another method like selecting it from a toolbar). The third view shows the prompt that appears, asking you to specify a center point for the circle. The "or" and the down arrow "⬇" indicates that options (other than just entering the center point) are available and that this is the time to choose one of them. The fourth view shows the graphics feedback that appears while you are being prompted to input the circle radius. This type of feedback varies depending on the command in effect. The prompt for the second point of a line being drawn might include both the distance and the angle.

note   Make sure **DYN,** located on the status bar is, set to ON to enable the dynamic input feature.

Figure 1-16

On-Screen Input, Prompts, and Graphics Feedback

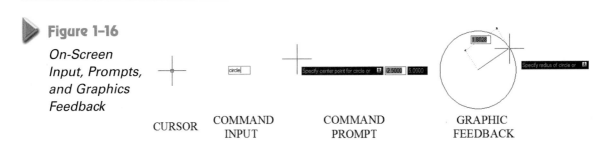

CURSOR   COMMAND INPUT   COMMAND PROMPT   GRAPHIC FEEDBACK

Command Window

The Command window is a window in which you enter commands and in which AutoCAD displays prompts and messages. The Command window can be a floating window with a caption and frame. You can move the floating Command window anywhere on the screen and resize its width and height by dragging a side, bottom, or corner of the window. With the introduction of the On-Screen input/prompt/graphics feedback feature in AutoCAD 2006, the Command Window has taken a "back seat" as the primary way to interface with the program. It is now possible to do most of the design/drafting work with the Command Window closed. However, if you close the window, you will need to enter CTRL+9 or invoke the COMMANDLINE command to reopen the Command window.

There are two components to the Command window: the single command line where AutoCAD prompts for input and you see your input echoed back, as shown in Figure 1–17, and the Command History area, which shows what has transpired in the current drawing session. One display of the single command line remains at the bottom of the screen.

Figure 1–17

Command window

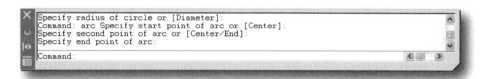

The Command History area can be enlarged like other windows by picking the top edge and dragging it to a new size. You can also scroll inside the enlarged area to see previous command activity by using the scroll bars (see Figure 1–18).

Figure 1–18

Command history

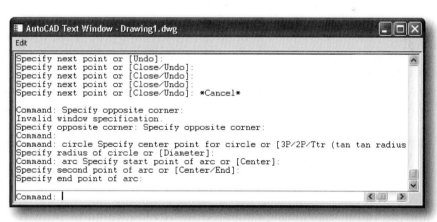

When you press F2, the Command History text window switches between being displayed and being hidden.

When you see "Command:" displayed in the Command window, it signals that AutoCAD is ready to accept a command. After you type a command name and press ENTER or select a command from one of the menus or toolbars, the prompt area continues to inform you of the type of response(s) that you must furnish, until the command is either completed or terminated. For example, when you pick the LINE command, the prompt displays "Specify first point:"; after selecting a starting point by appropriate means, you will see "Specify next point or [Undo]:" asking for the endpoint of the line.

Each command has its own series of prompts. The prompts that appear when a particular command is used in one situation may differ from the prompts or sequence of prompts when invoked in another situation. You will become familiar with the prompts as you learn to use each command.

When you enter the command name or give any other response by typing from the keyboard, be sure to press ENTER. Pressing ENTER sends the input to the program for processing. For example, after you type **line**, you must press ENTER in order for AutoCAD to start the part of the program that lets you draw lines. If you type **lin** and press ENTER, you will get an error message, unless someone has customized the program and created a command alias or command named "lin." Typing **lin** and pressing ENTER is not a standard AutoCAD command.

Pressing SPACEBAR has the same function as ENTER except when entering strings of words, letters, or numbers in response to TEXT, MTEXT or other command prompting for a text string.

To repeat the previous command, you can press ENTER or SPACEBAR at the Command: prompt. A few commands skip some of their normal prompts and assume default settings when repeated in this manner.

Terminating a Command

There are three ways to terminate a command:

- Complete the command sequence and return to the Command: prompt.
- Press ESC to terminate the command before it is completed.
- Invoke another command from one of the menus, which automatically cancels any command in progress.

---

note All the command explanations in this book are based on prompts displayed in Dynamic Input.

INTERFACING WITH AUTOCAD

It is important to be aware that there is almost always more than one way to tell AutoCAD to do something. After you have learned how to communicate with AutoCAD by the methods and examples shown in this book, you may wish to explore other methods that might better suit your needs. What is most important is to learn the commands and parameters that are necessary to produce the drawings and contract documents in an acceptable manner, whether a particular command is initiated from the On-screen prompt, Command: prompt, a menu, a toolbar, or a Shortcut menu.

This section introduces the methods available to initiate AutoCAD commands.

As much as possible, AutoCAD divides commands into related categories. For example, "Draw" is not a command, but a category of commands used for creating primary objects such as lines, circles, arcs, text (lettering), and other useful objects that are visible on the screen. Categories include Modify, View, and and Tools listing various commands and tools that will help in managing AutoCAD drawing. The commands under Format are also referred to as drawing aids and utility commands throughout the book. Learning the program can progress at a better pace if the concepts and commands are mentally grouped into their proper categories. This not only helps you find them when you need them, but also helps you grasp the fundamentals of computer-aided drafting more quickly.

Input Methods

There are several ways to start an AutoCAD command: the keyboard, toolbars, menu bars, dialog boxes, the Shortcut menu, or the digitizing tablet.

Keyboard

To invoke a command from the keyboard, simply type the command name at the On-screen prompt ("Command:" if you are using the Command window) and then press ENTER or SPACEBAR. To repeat a command you have just used, press ENTER, SPACEBAR, or right-click on your pointing device. Right-clicking causes a Shortcut menu to appear on the screen, from which you can choose the Repeat <last command> option. If you are using the Command window you can also repeat a command by using the UP ARROW and DOWN ARROW keys to display the commands you previously entered from the keyboard. Use the UP ARROW key to display the previous line in the command history; use the DOWN ARROW key to display the next line in the command history. Depending on the buffer size, AutoCAD stores all the information you entered from the keyboard in the current session.

AutoCAD also allows you to use certain commands transparently, which means they can be entered on the command line while you are using another command. Transparent commands are usually commands that change drawing settings or drawing tools, such as GRID, SNAP, and ZOOM. To invoke a command transparently, enter an apostrophe (') before the command name while you are using another command. After you complete the transparent command, the original command resumes.

Toolbars

The toolbars contain tools that represent commands. Click a toolbar button to invoke the command, and then select options from a dialog box or follow the On-screen prompts.

Menu

The menus are available from the menu bar at the top of the screen. You can invoke almost all of the available commands from the menu bar. You can choose menu options in one of the following ways:

- First select the menu name to display a list of available commands, and then select the appropriate command.
- Press and hold down ALT and then enter the underlined letter in the menu name. For example, to invoke the LINE command, first hold down ALT and then press **d** (that is, press ALT + D) to open the Draw menu, and then press **l**.

The default menu file (customize user interface) is *acad.cui*. You can load a different menu file by invoking the MENU command.

Dialog Boxes

Many commands, when invoked, cause a dialog box to appear, unless you prefix the command with a hyphen. For example, entering **insert** causes the dialog box to be displayed as shown in Figure 1–19, and entering **-insert** causes responses to be displayed in the On-screen prompt area. Dialog boxes display the lists and descriptions of options, long rectangles for receiving your input data, and, in general, are the more convenient and user-friendly method of communicating with the AutoCAD program for that particular command.

The commands listed in the menu bar that include ellipses (...), such as Plot... and Hatch..., display dialog boxes when selected. AutoCAD dialog boxes have features that are similar to Windows file handling dialog boxes.

 Figure 1–19

Dialog box invoked from the INSERT *command*

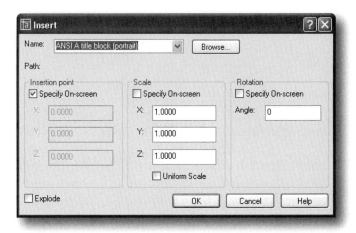

Shortcut Menu

The AutoCAD Shortcut menu appears at the location of the cursor when you press the right button (right-click) on the pointing device. The contents of the Shortcut menu depend on the situation at hand.

If you right-click in the Graphics window when there are no commands in effect, the Shortcut menu will include options to repeat the last command, a section for editing objects such as Cut and Copy, a section with Undo, Pan, and Zoom, and a section with Quick Select, Find, and Options, as shown in Figure 1–20. Selections that cannot be invoked under the current situation will appear in lighter text than those that can be invoked.

 Figure 1–20

Shortcut menu when no command is in effect

If you select one or more objects (system variable PICKFIRST set to ON), and no commands are in effect, and then right-click, the Shortcut menu will include some of the editing commands, as shown in Figure 1–21.

 Figure 1–21

Shortcut menu with one or more objects selected when no command is in effect

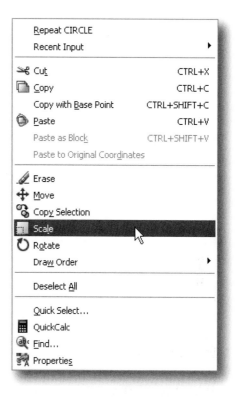

Whenever you have entered a command and do not wish to proceed with the default option, you can invoke the Shortcut menu and select the desired option with the mouse. For example, if instead of the default center-radius method of drawing a circle, you wished to use the TTR (tangent-tangent-radius), 2P (two-point), or 3P (three-point) option, you can select one of them from the Shortcut menu, as shown in Figure 1–22. You can also select the PAN and ZOOM commands (transparently) or cancel the command. If pressing ENTER is required, that is also available.

 Figure 1–22

Shortcut menu when the CIRCLE command is in effect

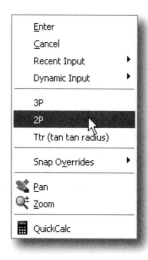

If you are using the Command window, right-click anywhere in the Command window, and the Shortcut menu provides access to the six most recently used commands (see Figure 1–23).

 Figure 1–23

Shortcut menu while the cursor is in the Command window

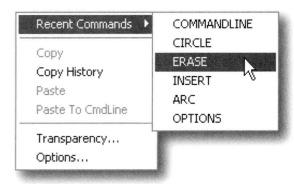

Right-click on any of the buttons in the Status bar, and the Shortcut menu provides toggle options for drawing tools and a means to modify their settings.

Right-click on the Model tab or Layout tabs of the drawing area, and the Shortcut menu provides display plotting, page setup manager, and various layout options.

Right-click on any of the open AutoCAD dialog boxes and windows, and the Shortcut menu provides context-specific options. Figure 1–24 shows an example for the Layer Properties Manager dialog box with a Shortcut menu.

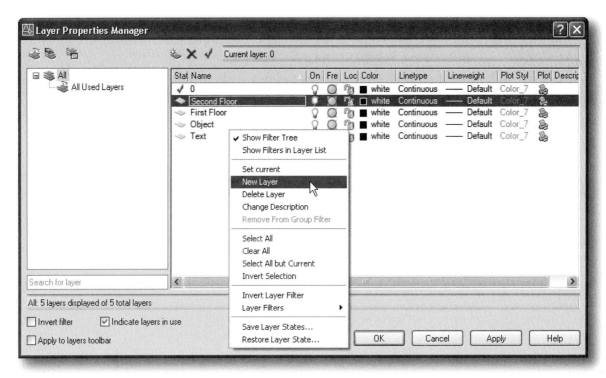

Figure 1–24

Shortcut menu in the Layer Properties Manager dialog box

Cursor Menu

The AutoCAD Cursor menu (see Figure 1–25) appears at the location of the cursor if you press the middle button on a mouse with three or more buttons. On a two-button mouse, you can invoke this feature by pressing SHIFT and right-clicking. On a two-button mouse, the right button usually causes the Shortcut menu to appear. The Cursor menu (different from the Shortcut menu) includes the handy Object Snap mode options along with the X,Y,Z filters. The importance of having the Object Snap modes in such ready access will become evident when you learn the significance of these functions.

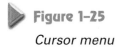

Figure 1–25

Cursor menu

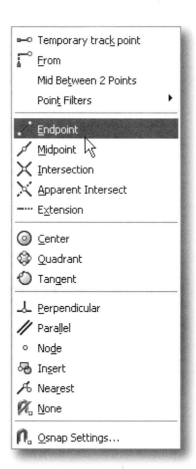

Digitizing Tablet

The most common input device after the mouse is the digitizing tablet. It combines the screen cursor control of a mouse with its own printed menu areas for selecting items. However, with the new Heads-up features and customizability in AutoCAD since release 2000, the tablet overlay with command entries is practically obsolete. One powerful feature of the digitizing tablet (not related to entering commands) is that it allows you to lay a map or other picture on the tablet and trace over it with the puck (the specific pointing device for a digitizing tablet), thereby transferring the objects to the AutoCAD drawing. The new interfaces with other platforms that allow you to insert a picture of an aerial photograph, for example, make tablet digitizing less of a demand.

GETTING HELP

Help is available through either a continuously resident help window, through a traditional Windows-type help interface, or through Online help on the Internet.

Info Palette

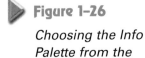 Figure 1–26

Choosing the Info Palette from the Help menu

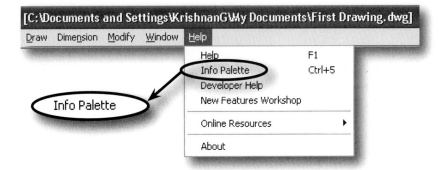

Invoking the ASSIST command or choosing the Info Palette (see Figure 1–26) provides automatic or on-demand context-sensitive quick help in the form of an Info Palette (see Figure 1–27).

 Figure 1–27

Info Palette

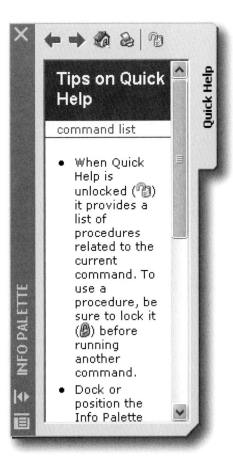

By dragging a side, bottom, or corner of the Info Palette, you can enlarge it to display more (sometimes all) of the information in the palette.

While the Info Palette is being displayed, and you invoke a command, the Active Assistance palette will display information about the command just invoked. For example, when you invoke the CIRCLE command, Help topics about the CIRCLE command will be displayed as shown in Figure 1–28. When you select one of the Help topics, the quick help is displayed with information about the topic selected. The procedure that is displayed in the palette can be locked by choosing the Lock from the toolbar located in the Info Palette. By locking the procedure, when you invoke another command, the procedure will not reflect the information of the active command.

Figure 1–28

The Info Palette automatically displaying Help topics about the CIRCLE command

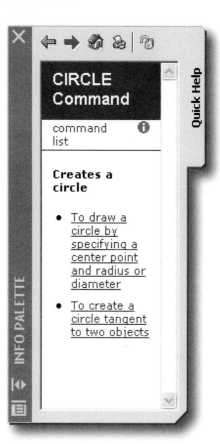

Traditional Help

Figure 1–29

Invoking the HELP command from the Standard toolbar

Choose Help from the Standard toolbar (see Figure 1–29). AutoCAD displays the AutoCAD Help: User Documentation window, as shown in Figure 1–30.

Figure 1-30

*AutoCAD Help: User
Documentation
dialog box*

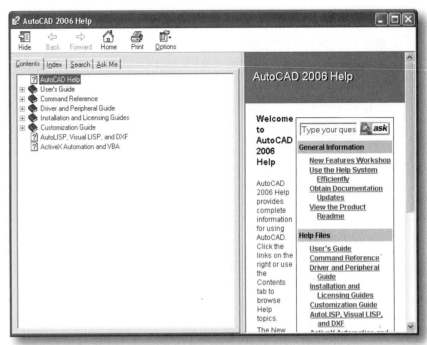

The Help window provides a context-sensitive help facility to list commands and what they do. The Help window provides online assistance within AutoCAD. When an invalid command is entered, AutoCAD displays a message to remind you of the availability of the Help facility.

The Help window can be opened while you are in the middle of a command. This is referred to as a transparent command. To invoke a command transparently (if it is one of those that can be used that way), simply prefix the command name with an apostrophe. For example, to use Help transparently, enter **'help** or **'?** in response to any prompt that is not asking for a text string. AutoCAD displays help for the current command. Often the help is general in nature, but sometimes it is specific to the command's current prompt.

As an alternative, press the function key F1 to open the Help window. Or you can open the Help window by choosing the Help menu from the menu bar at the top of your screen.

Help switches to an independent window, so that when you are through with the help utility, you will need to switch to the AutoCAD program window to continue drawing. You do not have to close the AutoCAD Help: User Documentation window to work in an AutoCAD drawing session.

At the top of the AutoCAD Help: User Documentation window is a button that will let you hide the tab (left) side, therefore shrinking the size of the window with the **Hide** button or, if it has been shrunk, enlarge the window and show the tabs with the **Show** button. The **Back** button returns you to the previous screen when possible. The **Forward** button reverses the previous action of the **Back** button. The information (right) side is the instruction or description area where information about the selected subject or command is displayed. The tab side has text boxes and list boxes to aid in getting the help you need on all subjects and commands in AutoCAD. The **Home** button will take you to the home page of the User documentation. The **Print** button sends the contents of the information area to the printer.

The **Contents** tab of the AutoCAD Help: User Documentation dialog box presents an overview of the available documentation in a list of topics and subtopics. It allows you to browse by selecting and expanding topics.

The **Index** tab displays an alphabetical list of keywords related to the topics listed on the **Contents** tab. You can access information quickly when you already know the name of a feature, command, or operation, or when you know what action you want AutoCAD to perform.

The **Search** tab provides full-text search of all the topics listed on the **Contents** tab. It allows you to perform an exhaustive search for a specific word or phrase. It displays a ranked list of topics that contain the word or words entered in the keyword field.

The **Favorites** tab provides an area where you can save "bookmarks" to important topics.

The **Ask Me** tab allows you to find information using a question phrased in everyday language. It displays a ranked list of topics that correspond to the word or phrase entered in the question field.

Developer Help

The Developer Help option of the Help menu, when selected, causes the AutoCAD 2006 Help: Developer Documentation dialog box to be displayed. The topics in this dialog box have to do with subjects important to third-party developers such as customization, AutoLISP, DXF, ActiveX, and VBA applications.

New Features Workshop

The New Features Workshop option of the Help menu, when selected, causes the New Features Workshop dialog box to be displayed. New features introduced in AutoCAD 2006 are described and explained.

Additional Resources

The Additional Resources option of the Help menu, when selected, displays a submenu with options for Product Support, Training, Customization, and Autodesk User Group International. Each of these options, when selected, launches your Internet browser and links to the Autodesk address associated with the option selected.

DRAWINGS AS COMPUTER FILES

AutoCAD keeps a running record of the objects created, changes made to objects, and system variable settings during a drawing session. These records comprise the data that are stored as a computer data file when it is saved to its selected destination on a computer storage device. This computer data file can be stored, transferred, copied, emailed, opened and edited (by using AutoCAD), viewed (with an appropriate program), and otherwise managed like most other computer data files. It will automatically be given a file name extension of *.dwg*, which designates it as an AutoCAD drawing file.

Beginning a Drawing

You can start AutoCAD in a new drawing file by just starting the program. You can also start a new drawing after you have started AutoCAD and are in an ongoing drawing session. In many cases, the new AutoCAD drawing is like a clean sheet of drawing paper, velum, Mylar, or other drawing medium with nothing drawn on it except perhaps a border and blank title block that can be filled in as necessary.

Opening AutoCAD in a New Drawing: Drawing1.dwg

When the first new drawing is started in an AutoCAD drawing session, it is given the temporary name *Drawing1.dwg*. It will not be saved with a name of your choice until you use a form of the SAVE command. You can begin working immediately and save the drawing to a file name later, using the SAVE or SAVEAS command. The second new drawing in a session is given the temporary name *Drawing2.dwg* (and so on).

The setting of the system variable STARTUP affects what you see on the screen when an AutoCAD drawing session is begun. It also controls the type of dialog box that is displayed when the NEW command is invoked. Here we discuss using the NEW command when AutoCAD is configured with the system variable STARTUP set to 0 (default setting).

Beginning a Drawing with the New Command

Figure 1–31

Invoking the QNEW (Quick NEW) command from the Standard toolbar

The NEW command (or QNew as it is called on the Standard toolbar) allows you to begin a new drawing (see Figure 1–31). When you started AutoCAD initially, it automatically invoked the NEW command once. If you invoke the NEW command in this situation, it is really for the second time in the session. AutoCAD prompts for the selection of a template from Select Template dialog box (Figure 1–32) and will utilize the temporary name *drawing2.dwt*.

Figure 1–32

Select Template dialog box

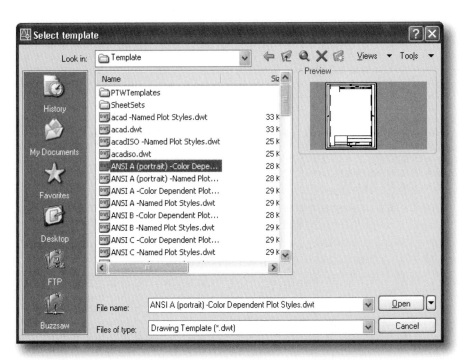

The Select Template dialog box operates in a manner similar to Windows file management dialog boxes. It contains a **Preview** window that will show a thumbnail sketch of the template file selected if a thumbnail sketch is available.

---

note The NEW/QNEW command causes the Select Template dialog box to be displayed when the STARTUP system variable is set to 0 (default) and the Create New Drawing dialog box to be displayed when STARTUP is set to 1.

Drawing Template File

A Drawing Template file is a drawing file with selected parameters already preset to meet certain requirements, so that you do not have to go through the process of setting them up each time you wish to begin drawing with those parameters. A template drawing might have the imaginary drawing sheet dimensions (LIMITS) preset, and the units of measurement preset, or it could contain objects already drawn on the drawing. In many cases a blank title block has already been created on a standard sheet size. Or the type of coordinate system that is needed to make the drawing could already be set up with the origin (X,Y,Z coordinates of 0,0,0) located where needed relative to the edges of the envisioned drawing sheet.

The AutoCAD program contains over 60 templates for drawings of various standard sizes containing predrawn title blocks conforming to standards such as ANSI, DIN, ISO, and JIS. You can create templates by making a drawing with the desired preset parameters and predrawn objects and then saving the drawing as a template file with the extension of *.dwt*. Unless the startup parameters have been changed, the default drawing template file that AutoCAD uses when you start the program is *acad.dwt*.

Existing Drawings

You can open an existing drawing by selecting and opening the desired drawing file (double-click the file name or select Open from the Shortcut menu that appears when you select the file name in the Windows Explorer). This will automatically open the AutoCAD program. You can also open an existing drawing after you have started AutoCAD and are in an ongoing drawing session.

Opening an Existing Drawing with the Open Command

 Figure 1–33

Invoking the OPEN command from the Standard toolbar

The OPEN command allows you to open an existing drawing (see Figure 1–33). AutoCAD displays the Select File dialog box, as shown in Figure 1–34. The dialog box is similar to the standard file selection dialog box, except that it includes options for selecting an initial view and for setting Open Read-Only, Partial Open, and Partial Open Read-Only modes. In addition, when you select the file name, AutoCAD displays a bitmap image in the **Preview** section.

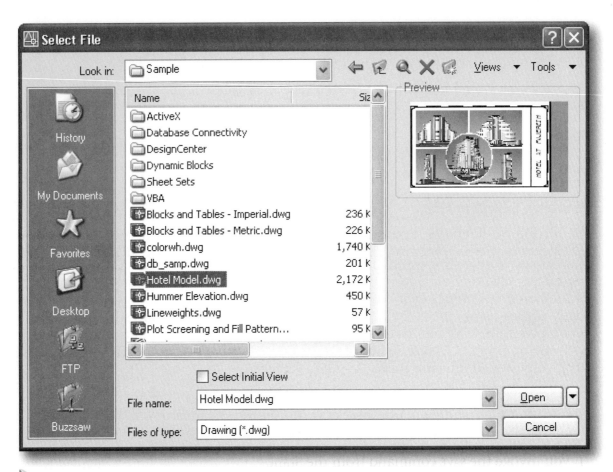

Figure 1-34

Select File dialog box

The **Select Initial View** check box permits you to specify a view name in the named drawing to be the startup view. If there are named views in the drawing, an M or P beside their name will tell if the view is model or paper space, respectively.

The **Views** button displays a menu with the options of list or details to determine how folders and files are displayed and a preview option that opens the **Preview** window to show a thumbnail sketch of drawing selected.

note You can open and edit an AutoCAD drawing created in any previous release. If necessary, you can save the drawing in other formats by using the SAVEAS command. Possible formats include 2006 Drawing (*.dwg), 2004 and 2000/LT 2000 Drawing (*.dwg), Drawing Standards (*.dws), Template (*.dwt), 2006, 2004 DXF (*.dxf), 2000/LT 2000 DXF (*.dxf), and R12/LT 2 DXF (*.dxf). Certain limitations apply when doing this, which are explained in the "Saving with Save As" section later in this chapter.

A menu is displayed when you select the down arrow to the right of the **Open** button. From this menu you may open a drawing in the read-only mode, which permits you to view the drawing but not save it with its current name. You can open a drawing with the **Partial Open** option, which loads a portion of a drawing, including geometry on a specific view or layer.

Closing a Drawing

You can end a drawing session by using the CLOSE command to close and save it with its current name or save it with a specified name if it has not been previously saved. You can also use the SAVE command to save the drawing periodically. You can use the SAVEAS command to save the drawing with a new name. This action closes the drawing under its current name at the status that it was last saved. Or you can exit the session and abandon all changes since the last saved status.

Saving a Drawing

While working in AutoCAD, you should save your drawing once every 10 to 15 minutes without exiting AutoCAD. By saving your work periodically, you are protecting your work from possible power failures, editing errors, and other disasters. This can be done automatically by setting the SAVETIME system variable to a specific interval (in minutes), saving to a temporary name. In addition, you can also manually save by using the SAVE, SAVEAS, or QSAVE commands.

Saving with Save

 Figure 1–35

Invoking the SAVE command from the Standard toolbar

If you invoke the SAVE command from the Standard toolbar for the first time while you are working in a drawing (see Figure 1–35), AutoCAD displays a standard dialog box prompting for a drawing file name. Select the appropriate folder in which to save the file, and type the name of the file in the **File name** box.

The file name can contain up to 255 characters including embedded spaces and punctuation. File names cannot include any of the following characters: forward slash (/), backslash (\), greater than sign (>), less than sign (<), asterisk (*), question mark (?), quotation mark ("), pipe symbol (|), colon (:), or semicolon (;). The following are examples of valid file names:

This-is-my-first-drawing

First_house

machine part one

AutoCAD automatically appends *.dwg* as a file extension. If you save it as a template file, then AutoCAD appends *.dwt* as a file extension. If you invoke the SAVE command in a drawing session with a drawing that has previously been saved, the SAVE command will save the drawing without prompting for a file name.

Saving with Save As

 Figure 1-36

Invoking the SAVEAS command from the File Menu

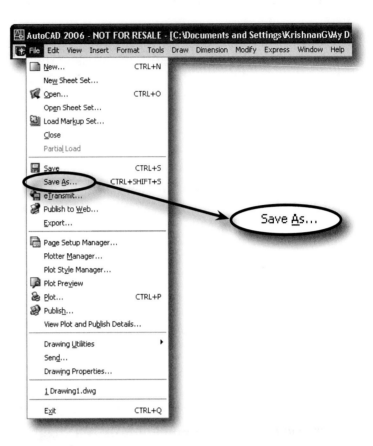

If you invoke the SAVEAS command (see Figure1–36), AutoCAD displays the Save Drawing As dialog box.

The SAVEAS command saves an unnamed drawing with a file name that you specify or renames the current drawing. If the current drawing is already named, AutoCAD prompts for a new file name and sets the current drawing to the new file name you specify. If the current drawing is already named and you accept the current default file name, AutoCAD saves the current drawing and continues to work on the updated drawing. If you specify a file name that already exists in the current folder, AutoCAD displays a warning that you are about to overwrite another drawing file. If you do not want to overwrite it, specify a different file name. The SAVEAS command also allows you to save in various formats. Possible formats include 2006 Drawing (*.*dwg*), 2004 and 2000/LT 2000 Drawing (*.*dwg*), Drawing Standards (*.*dws*), Template (*.*dwt*), 2006, 2004 DXF (*.*dxf*), 2000/LT 2000 DXF (*.*dxf*), and R12/LT 2 DXF (*.*dxf*).

Closing the Current Drawing

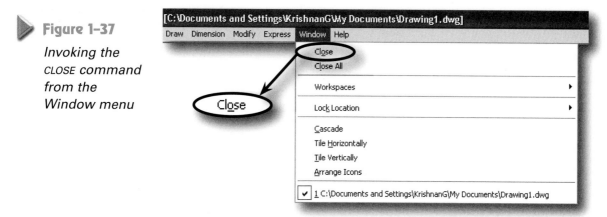

Figure 1–37

Invoking the CLOSE command from the Window menu

The CLOSE command closes the active drawing (see Figure 1–37). If you have not saved the drawing, since the last change, AutoCAD displays the AutoCAD alert box – Save changes to *filename.dwg*. If you choose **No**, then AutoCAD closes the drawing. If you choose **Yes**, then AutoCAD saves the drawing to the given file name and closes the drawing.

Closing All the Open Drawings

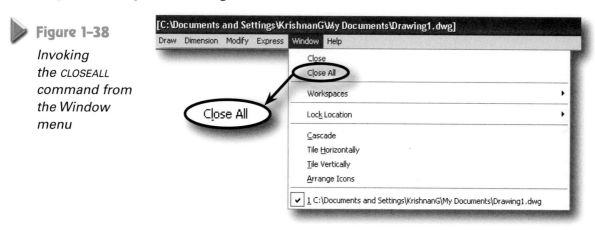

Figure 1–38

Invoking the CLOSEALL command from the Window menu

If you are working in multiple drawings, the CLOSEALL command closes all the open drawings (see Figure 1–38). AutoCAD displays a message for any unsaved drawing, in which you can save any changes (since the last SAVE) to the drawing before closing it.

Exiting AutoCAD

Figure 1–39

Invoking the EXIT command from the File menu

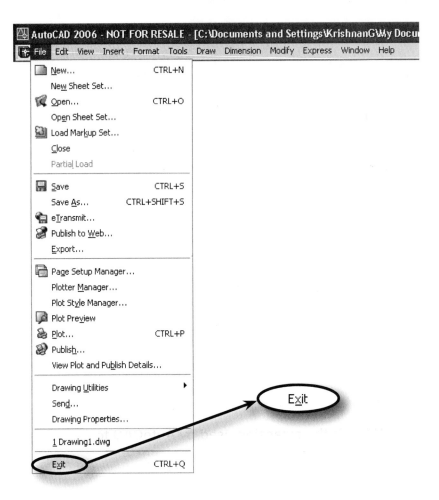

The EXIT or QUIT command allows you to exit the AutoCAD program (see Figure 1–39). The EXIT or QUIT command exits the AutoCAD program if there have been no changes since the drawing(s) was last saved. If the drawing has been modified, AutoCAD displays the Drawing Modification dialog box to prompt you to save or discard the changes before quitting.

review questions

1. If you executed the following commands in order: LINE, CIRCLE, ARC, and ERASE, what would you need to do to re-execute the CIRCLE command?

   a. *Press* PAGE UP, PAGE UP, PAGE UP, ENTER
   b. *Press* DOWN ARROW, DOWN ARROW, DOWN ARROW, ENTER
   c. *Press* UP ARROW, UP ARROW, UP ARROW, ENTER
   d. *Press* PAGE DOWN, PAGE DOWN, PAGE DOWN, ENTER
   e. *Press* LEFT ARROW, LEFT ARROW, LEFT ARROW, ENTER

2. In a dialog box, when there is a set of mutually exclusive options (a list of several from which you must select exactly one), these are called:

   a. *Text boxes.*
   b. *Check boxes.*
   c. *Option buttons.*
   d. *Scroll bars.*
   e. *List boxes.*

3. If you lost all pull-down menus, what command could you use to load the standard menu?

   a. *LOAD*   d. *MENU*
   b. *OPEN*   e. *PULL*
   c. *NEW*

4. What is the extension used by AutoCAD for template drawing file?

   a. *.DWG*   d. *.TEM*
   b. *.DWT*   e. *.WIZ*
   c. *.DWK*

5. What is an external tablet called that is used to input absolute coordinate addresses to AutoCAD by means of a puck or stylus?

   a. *Digitizer*
   b. *Input pad*
   c. *Coordinate tablet*
   d. *Touch screen*

6. The menu that can be made to appear at the location of the crosshairs is called:

   a. *Mouse menu.*
   b. *Cross-hair menu.*
   c. *Cursor menu.*
   d. *None of the above, there is no such menu.*

7. In order to save basic setup parameters (such as snap, grid, etc.) for future drawings, you should:

   a. *Create an AutoCAD macro.*
   b. *Create a prototype drawing template file.*
   c. *Create a new configuration file.*
   d. *Modify the ACAD.INI file.*

8. To cancel an AutoCAD command, press:

   a. *CTRL + A.*
   b. *CTRL + X.*
   c. *ALT + A.*
   d. *ESC.*
   e. *CTRL + ENTER.*

9. The SAVE command:

   a. *Saves your work.*
   b. *Does not exit AutoCAD.*
   c. *Is a valuable feature for periodically storing information to disk.*
   d. *All of the above.*

10. Whenever you begin a new drawing, AutoCAD creates a new drawing called:

    a. *DRAWING1.DWG.*   **c.** *DRAWING.DXF.*
    b. *DRAWING.BMP.*   **d.** *DRAWING.CAD.*

11. The AutoCAD screen consists of the following:

    a. *Graphics window.*   **c.** *Command window.*
    b. *Toolbars.*   **d.** *All of the above.*

12. Toolbars can:

    a. *Be resized.*   **c.** *Change their contents.*
    b. *Be docked or floating.*   **d.** *All of the above.*

13. To repeat the previous command, you can press:

    a. ENTER.   **c.** *Either a or b.*
    b. SPACEBAR.   **d.** *None of the above.*

14. When you use a template to create a new drawing, AutoCAD copies all the information from the template drawing to the new drawing.

    a. *TRUE*   **b.** *FALSE*

15. It is possible to enter coordinates outside of the viewing area for AutoCAD to use for creating objects.

    a. *TRUE*   **b.** *FALSE*

16. The Status bar at the bottom of the screen displays the cursor's coordinates and important information on the status of various modes.

    a. *TRUE*   **b.** *FALSE*

17. The Title bar displays the current drawing name for the AutoCAD application window.

    a. *TRUE*   **b.** *FALSE*

18. A floating toolbar can lie anywhere on the screen and can be resized.

    a. *TRUE*   **b.** *FALSE*

19. Toolbars can only be docked at the top and left of the graphics screen.

    a. *TRUE*   **b.** *FALSE*

20. By default, the Standard toolbar and the Properties toolbar are docked at the top of the graphics window.
    a. *TRUE*   **b.** *FALSE*

21. The command history area can be enlarged like other windows.
    a. *TRUE*   **b.** *FALSE*

22. AutoCAD allows you to use certain commands transparently.
    a. *TRUE*   **b.** *FALSE*

23. When a child dialog box appears, you must respond to the options in the child dialog box before the underlying dialog box can continue.
    a. *TRUE*   **b.** *FALSE*

24. Selecting one option button will de-activate any other in the group.
    a. *TRUE*   **b.** *FALSE*

25. A check box acts as a toggle.
    a. *TRUE*   **b.** *FALSE*

26. The HELP command is a transparent command.
    a. *TRUE*   **b.** *FALSE*

a world for objects

introduction

The first chapter explained the AutoCAD program interface and how to use menus, dialog boxes, and toolbars. This chapter explains how to find your way around in that computer-simulated space.

After completing this chapter, you will be able to do the following:

- Use coordinate systems in AutoCAD

- Use Units of Measurement

- Set Units and Drawing Limits

- Set Snap, Grid, and Ortho

THE ELECTRONIC DRAWING SHEET

For all practical purposes, the computer-simulated surface that AutoCAD provides for you to create your drawing is limitless in dimensions. To illustrate this, early versions of AutoCAD came with a scale drawing of our solar system. Within this drawing, you could zoom out to view the orbit path of the largest planet and then zoom in to view a scaled detail of a six-inch high plaque on the lunar lander. The size ratio of the smallest to largest distances that could practically be dimensioned was fourteen significant figures. Quite adequate for ballpark work: meaning you can draw, to scale, a particle of sand that is one one-thousandth of an inch, within a drawing of a ballpark that is a billion feet long (almost eight times around the earth). So, when you start to lay out your drawing, the question is not "Is there enough space?" but "Where should I put it, so that I can keep track of it?" There will always be adequate space combined with exceptional accuracy.

The Plane Facts

A comparison can be made between the Drawing Area that appears on your AutoCAD screen and the board drafter's drawing sheet. For instance, the AutoCAD Drawing Area and the board drafter's sheet both represent flat planes. Sometimes, the approach you take to create objects with a computer is similar to the approach taken by the board drafter. However, there are other times, when the approach is notably different.

Board Drafting – the Drawing Sheet

Board drafters use lines, circles, arcs, and freehand sketches on the flat plane of the drawing sheet to communicate outlines and edges of actual three-dimensional solid objects, as they project onto the plane. For example, a circle can be drawn to represent a ball or the end of a cylinder, a triangle to represent a cone or pyramid, and an elongated rectangle might represent the side view of a piece of pipe or the edge of a 2 x 4 wood stud (see Figure 2–1).

▶ **Figure 2–1**

How various objects are represented in a 2D drawing: sphere, cylinder, cone, pyramid, long thin cylinder, long thin bar.

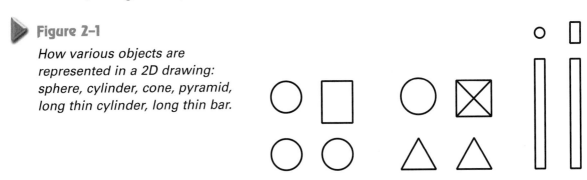

When the board drafter draws an object on paper, the points on that object are normally located in terms of their distance and direction from a reference point. That reference point may be on the object itself, or on another object in the drawing. In a drawing of a piece of property, the corners of the lot are usually given by their relationship to other corners. A surveyor's drawing of the lot would specify points by using the "meets and bounds" method, that is, giving the direction and distance of one point to another (see Figure 2–2).

▶ **Figure 2–2**

Using "meets and bounds" or directions and distances

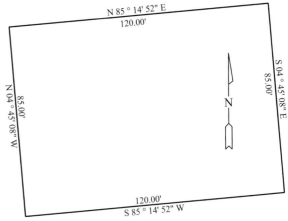

AutoCAD Drafting – the Drawing Window

The basic drawing elements created in the computer are similar to those made by board drafters. The concepts and skills covered by this book are, for the most part, the ones needed for creating two-dimensional computer-generated elements for the purpose of plotting on paper sheets, even though the elements usually represent three-dimensional objects.

In AutoCAD, the Drawing Area displayed on the screen is a view of a specific area on a plane in computer-simulated space. This plane is the customary 2D drawing surface in

AutoCAD. If you envision your point of view as an "eye-in-the-sky" and you wish to depict a fence, a tree, and a building, you could draw a line, a circle, and a rectangle on this plane to represent these three objects. This is the fundamental method of communicating, to any person who will read your drawing, the relative sizes and positions of solid objects. Like an artist about to paint a picture, you use the plane on the AutoCAD drawing screen as your canvas.

---

 note AutoCAD does not limit you to drawing in a single plane. AutoCAD is capable of generating three-dimensional objects in a computer-simulated three-dimensional space (see Figure 2–3). This allows you to be not only a painter, but a sculptor as well. But you should first learn to make drawings in 2D before graduating to the third dimension.

▶ **Figure 2–3**

2D and 3D drawings of three-dimensional object in AutoCAD

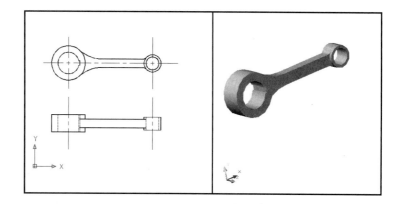

Coordinate Systems

The purpose of a coordinate system is to specify the locations of points in space or on a plane. Any time you are working in AutoCAD, there will be only one current drawing plane in effect. All the dimensions and points on the current drawing plane can be expressed in terms of the particular coordinate system currently in effect (see Figure 2–4).

▶ **Figure 2–4**

AutoCAD screen showing the Drawing Area

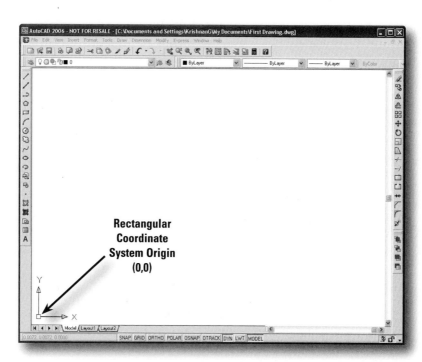

As mentioned earlier, the boundary of a piece of property can be drawn from point to point, using the "meets and bounds" method. If a house is located on the lot, a point representing one corner of the house may be specified on the drawing by its direction and distance from a point that represents a corner of the lot. Usually the location of this point is specified by its distances from property lines.

If two property lines are perpendicular to each other, they could be considered a pair of axes. Points representing other corners of the house or points on other features such as driveways, sidewalks, and trees could be placed in positions relative to the first established corner of the house. Or these points could be placed according to their distances from the perpendicular property lines, the axes (see Figure 2–5).

 Figure 2–5

House on a lot located by reference to property lines

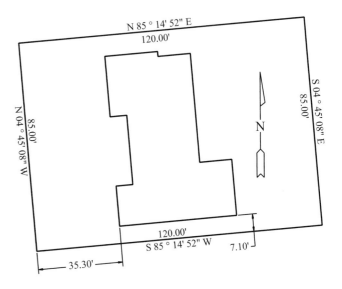

What's the Point?

A child's picture puzzle book guides you to create a "connect-the-dots" picture by drawing lines between points that are numbered 1, 2, 3 and so on. These numbers have nothing to do with distances from one dot to another or their location in the picture. They signify only the sequence of starting and ending a series of lines. However, if the picture had been generated in AutoCAD, these points would have other numbers associated with them, their coordinates. That is because the AutoCAD drawing plane has a built-in coordinate system. Whether you elect to use of them or not, all points in an AutoCAD drawing have pairs (triplets, if you advance to the concept of 3D) of numbers associated with them (see Figure 2–6).

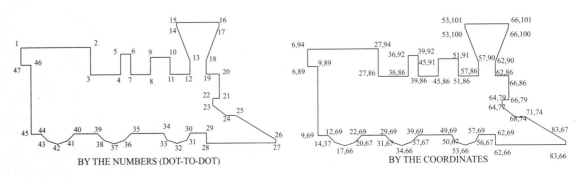

 Figure 2–6

"Connect-the-dots" versus coordinates

The Fifth Point on the Compass

Scouting teaches that there are five points on a compass: North, East, South, West, and the point where you are. If you don't know where you are on the map, then knowing directions is not very helpful. An important use of drafting is to create maps. In fact, almost every drawing you produce is some sort of map. The layout of a circuit board is a sort of map, showing the routes of the various circuits on the board. The front elevation of a building is a map locating the doors, windows, and other features on the front plane of the structure, relative to each other.

Most board drafters' drawings do not utilize a coordinate system, especially architectural and mechanical drawings. And electrical schematics and piping flow diagrams aren't even concerned with dimensions and spatial relationships. However, mapping by surveyors and layouts of petrochemical and manufacturing plants, often associate objects in their drawings to some basic coordinate system. When a large petrochemical plant is situated on several hundred acres, those who design the original layout will establish two major axes (usually one east-west and the other north-south). These imaginary lines are perpendicular to each other (see Figure 2–7). Where they cross is called the origin. If the origin is in the middle of the plant, then depending on which quadrant a point is situated, one or both coordinates might be negative.

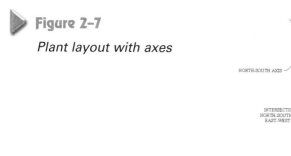

Figure 2–7

Plant layout with axes

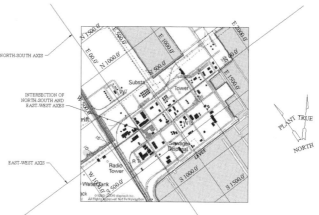

Whether you utilize AutoCAD's coordinate system or choose not to, *it is there*. If you wish to make a plot of the drawing, you must understand how the coordinate system functions and how it affects the arrangement and location of objects that you have created. Understanding the coordinate system is essential in determining how objects will appear on the plotted sheet.

The board drafter can start with the plotted sheet size and fit the objects on it without being concerned if either axis of a coordinate system (if it even exists) or the origin is somewhere on the drawing. The AutoCAD drafter has more difficulty ignoring the coordinate system. You can arbitrarily draw a line or other entity on the screen. You can even complete a drawing without regard to the coordinate system. This might happen when you draw an electrical schematic or flow diagram. But every point on every entity will still have a pair of numbers (coordinates) associated with it.

One *big* advantage of working in AutoCAD is that when entities (along with all of their associated points) are drawn correctly but not located as they should be in relation to the coordinate system, they can be moved en masse to the proper locations. It's just easier if you understand how the coordinate system works and you place the objects in their proper locations in the first place, if feasible.

A later discussion in this section explains how to apply the concept of a scale factor. The board drafter makes large entities (lines, circles, arcs, etc.) proportionately small enough to fit a certain size sheet. The AutoCAD drafter does just the opposite by drawing entities at their true sizes and then placing an imaginary rectangle, that represents the paper sheet, proportionally large enough to include the objects. Then, a plot configuration can be set up so that as the information defining the objects is sent by AutoCAD to the plotter, everything is reduced to fit the actual sheet size. How to proportion the imaginary rectangle and where to locate it, relative to the objects, is where an understanding of the coordinate system is important.

Three types of Coordinate Systems

In three-dimensional space, there are three commonly used coordinate systems. Each system uses three numbers.

The Spherical Coordinate System

The Spherical coordinate system is used for specifying points on a sphere. It is the basis for navigation on the surface of the earth (latitude and longitude). The first number of this coordinate system is the radius of the sphere (**r** in Figure 2–8). The second number (Θ in Figure 2–8) is the angle between a line through a zero point on the equator and a line that goes through where the point is projected onto the equatorial plane of the sphere. This number is given as the longitude in navigational terms. The third number (Φ in Figure 2–8) is the angle between a line through the point and the equatorial plane. This number is given as the latitude in navigational terms. AutoCAD expects Spherical coordinates to be entered as **r** (distance from the origin)<Φ (angle from X axis)<Θ (angle from XY plane). In Figure 2–8, the illustration on the right shows how AutoCAD calculates the Spherical Coordinates 10<75<60.

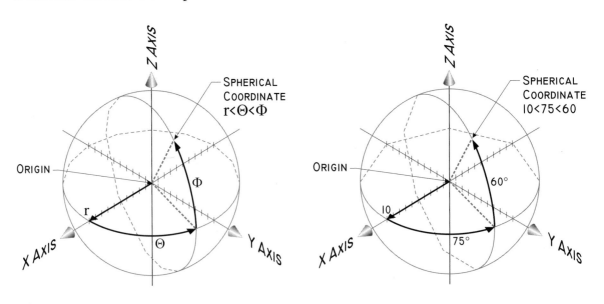

▲ **Figure 2–8**

Spherical coordinate system: mathematical model (left); AutoCAD's calculation of point 10<75<60 (right)

The Cylindrical Coordinate System

The Cylindrical coordinate system is used for specifying a point on a cylinder. It has as its base a horizontal plane that is perpendicular to the centerline of the cylinder. On the base plane there is also a zero base line from the center of the cylinder that extends to the surface of the cylinder. The first number (**r** in Figure 2–9) is the radius of the cylinder. The second number (Θ in Figure 2–9) is the angle of rotation from the zero baseline. The third point (**z** in Figure 2–9) is the distance along the Z axis from the base plane. In Figure 2–9, the illustration on the right shows how AutoCAD calculates the Cylindrical coordinate 5<75,9.

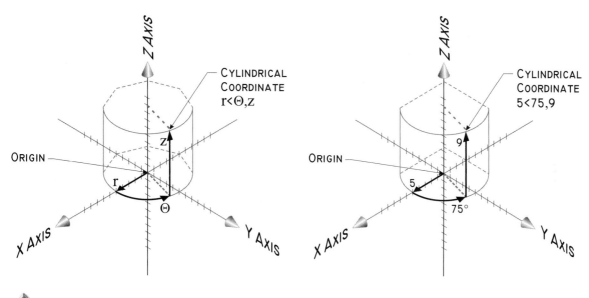

▲ Figure 2–9

Cylindrical coordinate system: mathematical model (left); AutoCAD's calculation of point 5<75,9 (right)

The Cartesian Coordinate System – AutoCAD's Default System

AutoCAD adheres to the conventions of the Cartesian coordinate system. From the origin (0,0,0) distances along the X axis (to the right) increase in value. On the Y axis, points above the origin have positive values. On the Z axis, points closer to the viewer than the origin, have a positive value; this provides a sense of depth. These axes define the Word Coordinate System (WCS).

AutoCAD comes with a rectangular coordinate system that does not need curved surfaces, circles, arcs, or angles to describe points in 2D and 3D space. It consists of three mutually perpendicular planes. One plane is considered horizontal, which means that the other two are vertical. The three lines created by the intersections of the three pairs of planes are called the axes (see Figure 2–10). Where the three axes intersect is known as the origin, with the coordinates 0,0,0.

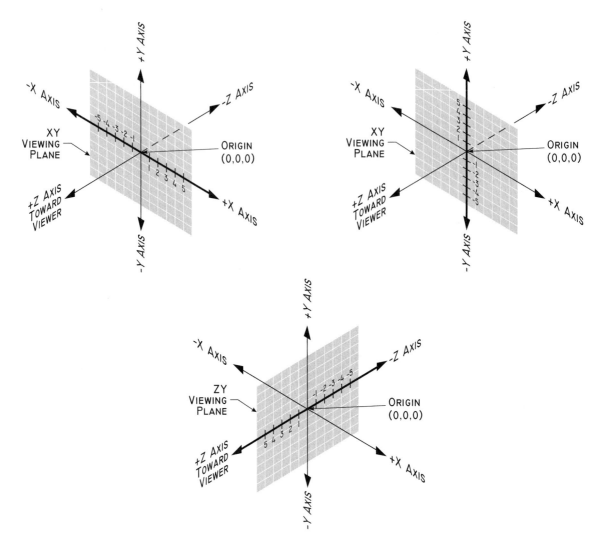

 Figure 2-10

Cartesian coordinate system showing the Origin, XY and ZY viewing planes

The significance of the WCS is that it is always in your drawing; it cannot be altered. An infinite number of other coordinate systems can be established relative to it. These others are called *User Coordinate Systems* (UCS) and can be created with the UCS command. Even though the WCS is fixed, you can view it from any angle, side, or rotation without changing to another coordinate system.

AutoCAD provides what is called a coordinate system icon to help you keep your bearings while working in different coordinate systems in a drawing. The icon will show you the orientation of your current UCS by indicating the positive directions of the X and Y axes. Figure 2–11 shows some examples of coordinate system icons.

 Figure 2-11

Examples of the UCS icons

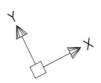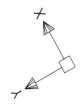

Computer-aided drafting permits you to draw an object at its true size and then make the border, title block, and other non-object associated features fit on and around the object. The completed combination is reduced (or increased) to fit the sheet size you require when you plot.

A more complicated situation arises when you wish to draw objects at different scales on the same drawing. This can be handled easily by one of several methods with the more advanced features and commands provided by AutoCAD.

Drawing a schematic that is not to scale is one situation where the graphics and computing power are hardly used to their potential. But even though the symbols and the distances between them have no relationship to any real-life dimensions, the sheet size, text size, line widths, and other visible characteristics of the drawing must be considered in order to give your schematic the readability you desire. Some planning, including sizing, needs to be applied to all drawings.

Methods to Specify Points

When AutoCAD prompts for the location of a point, you can use one of several available point entry methods, including Spherical coordinates, Cylindrical coordinates, and Cartesian coordinates: absolute rectangular, relative rectangular, and relative polar coordinates.

Absolute Rectangular Coordinates

The rectangular coordinates method is based on specifying the location of a point by providing its distances from two intersecting perpendicular axes in a 2D plane or from three intersecting perpendicular planes for 3D space. Each point's distance is measured parallel to the X axis (left-to-right on the drawing plane), Y axis (up-and-down on the drawing plane), and Z axis (toward or away from the viewer). The intersection of the axes, called the origin (X,Y,Z = 0,0,0) divides the coordinates into four quadrants for 2D or eight sections for 3D (see Figure 2–10).

Points are located by absolute rectangular coordinates in relation to the axes. You specify the reference to the WCS origin or UCS origin. In AutoCAD, by default the origin (0,0) is located at the lower left corner of the Grid display, as shown in Figure 2–12.

Figure 2–12

Default location of the AutoCAD origin

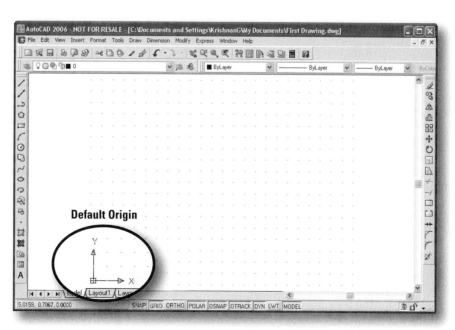

As mentioned earlier, the left-to-right distance on the drawing plane increases in the positive X direction from the origin, and the up-and-down distance on the drawing plane increases in the positive Y direction from the origin. You specify a point by entering its X and Y coordinates in decimal, architectural, fractional, or scientific notation separated by commas. AutoCAD automatically assigns the current elevation as the Z coordinate. Unless it has been changed, the default value is zero (0). Advanced three-dimensional drafting involves specifying X,Y, and Z coordinates when appropriate.

Relative Rectangular Coordinates

Points are located by relative rectangular coordinates in relation to the last specified position or point, rather than the origin. This is like specifying a point as an offset from the last point you entered. In AutoCAD, whenever you specify relative coordinates, the @ ("at" symbol) must precede your entry. This symbol is entered by holding down the SHIFT key and simultaneously pressing the key for the number 2 at the top of the keyboard. The following table shows examples of relative rectangular coordinate keyboard input when prompted by AutoCAD to specify a point, the absolute coordinates of the last point specified (from which the newly specified point is offset), and the absolute coordinates of the point resulting from applying the relative rectangular coordinates with the @ prefix.

Absolute Coordinates of Last Specified Point	Relative Rectangular Coordinates Keyboard Input	Resulting Absolute Coordinates of Point Specified by Keyboard Input
3,4	@2,2	5,6
5,5	@-7,0	-2,5
3.25,8.0	@0,12.5	3.25,20.5

If you are working in a UCS (User Coordinate System) and would like to enter points with reference to the WCS (World Coordinate System), prefix the coordinates with an asterisk (*). For example, to specify the point with an X coordinate of 3.5 and a Y coordinate of 2.57 with reference to the WCS, regardless of the current UCS, enter:

*3.5,2.57

In the case of relative coordinates, the asterisk will be preceded by the @ symbol. For example:

@*4,5

This represents an offset of 4,5 from the previous point with reference to the WCS.

Relative Polar Coordinates

Polar coordinates are based on a distance from a fixed point at a given angle. In AutoCAD, a polar coordinate point is determined by the distance from a previous point and angle measured from the zero degree, rad, or gradient. By default the angle is measured in the counterclockwise direction. It is important to remember that for points located using relative polar coordinates they are to be positioned relative to the previous point and not the origin (0,0). You can specify a point by entering its distance from the previous point and its direction in the XY plane, separated by < (not a comma). This symbol is selected by holding the SHIFT key and simultaneously pressing the comma (",") key at the bottom of the keyboard. Failure to use the @ symbol will cause the point to be located relative to the origin (0,0). The following table shows examples of relative polar coordinate keyboard input when prompted by AutoCAD to specify a point, the absolute coordinates of the last point specified (from which the newly speci-fied point is offset), and the absolute coordinates of the point resulting from applying the relative polar coordinates with the "@" prefix.

Absolute Coordinates of Last Specified Point	Relative Polar Coordinates Keyboard Input	Resulting Absolute Coordinates of Point Specified by Keyboard Input
3,4	@2<0	5,4
5,5	@4<180	1,5
2.00,2.00	@ 1.4142135623<45	3.00,3.00

Coordinates Display

 Figure 2–13

Coordinates display in the Status bar

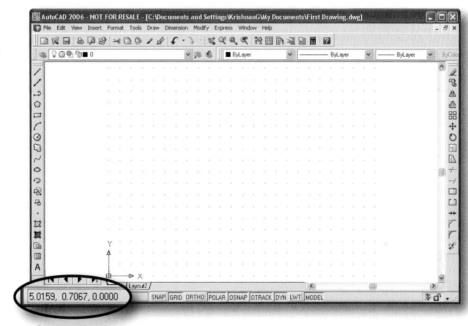

The Coordinates display (see Figure 2–13) is a report of the cursor coordinates in the Status bar at the bottom of the screen. It has three settings. On most systems the F6 function key toggles between the three settings. The three settings are as follows:

1. This setting causes the display to report the location of the cursor when the prompt is in the "Command:" status or when you are being prompted to select the first point selection of a command. It then changes to a relative polar mode when you are prompted for a second point that could be specified relative to the previous point. In this case, the report is in the form of the direction/distance. The direction is given in terms of the current angular units setting and the distance in terms of the current linear units setting.

2. This setting is similar to the previous one, except that the display for the second location is given in terms of its absolute coordinates, rather than relative to the previous point.

3. This setting is used to save either the location in the display at the time you toggle to this setting or the last point entered. It does not change dynamically with the movement of the cursor.

Overrides for Absolute/Relative Coordinate Entry

Under certain conditions, when you specify a point by typing in the coordinates, you might be specifying relative coordinates when you are intending to enter absolute coordinates. This can occur if you are using the dynamic input (described in Chapter 4) and by default, set for relative coordinates input mode. In the dynamic input (prompt/responses displayed near the cursor) mode, AutoCAD automatically prefixes the coordinates with the @ (last point) symbol whether you type it in or not for second and subsequent points. For example, if you start a line at 1,1 and specify the next point as 3,3, the line will be drawn to coordinates 4,4 (not 3,3). This is because the coordinate input you specified will be applied as a displacement from the last point (relative) rather than the point at 3,3 (absolute). To change the pointer input to absolute format, set the DYNPICOORDS system variable to 1 (default is set to 0). This prevents the coordinates from being relative (even though you haven't put in the @ symbol). This rule does not apply if you are using the Command window instead of dynamic input. See Chapter 4 (Dynamic Input, On-Screen Prompts and Geometric Values Display section) for detailed explanation on dynamic input mode settings.

To change the DYNPICOORDS system variable setting, type in **dynpicoords** at On-Screen prompt and AutoCAD prompts:

> Enter new value for DYNPICOORDS <0>: (specify 1 and press ENTER)

AutoCAD will now use the input as absolute and not automatically prefix the coordinates entered in dynamic input with the @ symbol.

---

note   The prompts and responses shown in examples throughout this book and drawing exercises assume that you are using the new dynamic input introduced in AutoCAD 2006. Where they call for you to enter coordinates without the @ prefix, they are based on the input being absolute coordinates. Because the default setting (0) of the DYNPICOORDS system variable causes the @ prefix to be added (without your seeing it on the screen), you should change the setting to 1 if you wish to try out the examples on your system. Or, if you do not wish to change the DYNPICOORDS setting to 1, you can use an override by prefixing the relative coordinates with the # symbol. This forces the input to be absolute and only works in dynamic input, not at the Command line in the Command window.

UNITS OF MEASUREMENT

Just as AutoCAD's computer-simulated space comes with a coordinate system, the coordinate system comes with units of measurement, both linear and angular. If you set up the drawing in Engineering or Architectural formats, linear measurement will be in feet and inches. These formats assume one drawing unit represents one inch. The other formats (scientific, decimal, and fractional) make no such assumption, and they can represent whatever real-world units you like.

The Drawing Area and Scale

The LIMITS command allows you to place an imaginary rectangular drawing sheet in the CAD drawing space. But, unlike the limitations of the drawing sheet of the board drafter, you can move or change the size of the CAD electronic sheet (the limits) after you have started your drawing. The LIMITS command does not affect the current display on the screen. The area defined by the limits governs the portion of the drawing where the visible grid is displayed (for more on the GRID command, see the section on GRID). Limits are also factors that determine how much of the drawing is displayed by the ZOOM ALL command (for more on the ZOOM ALL command, see the section on ZOOM).

The limits are expressed as a pair of 2D points in the World Coordinate System, a lower left and an upper right limit. For example, to set limits for an A-size sheet in landscape orientation, set lower left as 0,0 and upper right as 11,8.5 or 12,9; for a B-size sheet set lower left as 0,0 and upper right as 17,11 or 18,12 and so on. An example of using the limits to simulate the final plotted sheet size at a scale of 1/4"=1'-0" would be if you drew a building that is 100' long on a drawing sheet that is 36" wide. A board drafter would use the 1/4"= 1'-0" (a ratio of 1:48) scale and draw the building. The 100' building is 1200" long. So at the scale of 1:48 it would be drawn 25" long on the 36" wide sheet. It would fit without a problem with 11" to spare. There would even be room on each end of the drawing to draw the dimensions and notes.

As an AutoCAD designer you would do just the opposite. You would draw the building to its actual length of 100'. Then, if the drawing is to be plotted on a 36" (or 3') by 24" (or 2') sheet to a scale of 1/4"= 1'-0" (a ratio of 1:48) you would set your limits in the drawing to 48 times the sheet size or 48 times 3' (for the 36" wide sheet) by 48 times 2' (for the 24" high sheet). This would be 144' by 96' on the drawing. Countless combinations of lower left corners and upper right corners of the limits would achieve this; 0,0 and 144',96'; -144',-96' and 0,0; -72',-48' and 72',48'; 100',100' and 244',196'; and so on. So long as the width of the limits (upper right X-coordinate minus lower left X-coordinate) equals 144' and the height of the limits (upper right Y-coordinate minus lower left Y-coordinate) equals 96', when you plot the drawing to 1/4"=1'0" (1:48), the limits will be plotted 36" wide by 24" high.

Scale Factors

Unless you are drawing a schematic diagram, objects in AutoCAD usually have a dimension. And as noted earlier, you can (and it is highly recommended that you do) draw objects to their true size. So how can you draw a twenty-four foot long object on a screen that isn't even twenty-four inches across? The same way you can see all of an eagle with a wing span of six feet through a zoom telescope with only a two-inch diameter lens at its big end.

note   It is advisable not to use an architect's or an engineer's scale to determine the size of objects on the plotted sheet when they are drawn to their true size and plotted to a standard scale factor. This practice is subject to errors. It is better if accurate, clear, and sufficient dimensions have been included. Also, having objects drawn to their true size is a great advantage when drawing dimensions. When drawing dimensions, AutoCAD can automatically write in the dimension of the object. Thus, if the object is drawn at its true size, and you select the proper points on the object when dimensions are drawn, the correct dimension text will be automatically written.

The use of a scale factor normally does not come into effect in an AutoCAD drawing until you are ready to plot the drawing.

By using the appropriate display commands, discussed later in this section, you can select the area in the plane that you wish to view by panning back and forth and up and down. It is like looking through the lens of a camera with a zoom lens at a board draw-ing. You can also determine how large or small that view will be by zooming in and out. However, even though you pan and zoom around the computer-simulated plane, the relationships of points and objects to each other all remain unchanged. A circle that is 2 units in diameter stays 2 units in diameter. Two parallel lines that are 0.75 units apart remain 0.75 units apart even though you might zoom in so close that only one of them can be seen on the screen or you zoom out so far that they appear to be just one line.

Hard Copy, Easy as Pi

One objective has not changed much in the transition from board drafting to CAD: producing a hard copy. The term "hard copy" describes a tangible reproduction of a screen image. The hard copy is usually a reproducible medium from which prints are made and can take many forms, including slides, videotape, prints, and plots. In manu-al drafting, if you need objects in your drawing to be drawn to two different scales, you physically draw objects in two different scales. In AutoCAD, with minor modifications, you plot or print the same drawing to different scale factors on different sizes of paper. You can even compose your drawing in paper space (explained later) with limits that equal the sheet size, and plot it at 1:1 scale.

Planning the Plotted Sheet

Planning ahead is still required in laying out the objects to be drawn on the final sheet. The objects drawn on the plotted sheet must be arranged. At least in AutoCAD, with its true-size capability, an object can be started without first laying out a plotted sheet. But eventually, limits, or at least a displayed area, must be determined. For schemat-ics, diagrams, and graphs, plotted scale is of little concern. But for architectural, civil, and mechanical drawings, plotting to a conventional scale is a professionally accepted practice that should not be abandoned just because it can be circumvented.

When setting up the drawing limits, you must take the plotted sheet into consideration to get the entire view of the object(s) on the sheet. So, even with all the power of the AutoCAD system, some thought must still be given to the concept of scale, which is the ratio of true size to the size plotted. In other words, before you start drawing, you should have an idea about what scale the final drawing will be plotted at or printed on a given size of paper.

The limits should correspond to some factor of the plotted sheet. If the objects will fit on a 24" x 18" sheet at full size with room for a border, title block, bill of materials, dimensioning, and general notes, then set up your limits to (0,0) (lower left corner) and (24,18) (upper right corner). This can be plotted or printed at 1:1 scale, that is, one object unit equals one plotted unit.

Plot scales can be expressed in several formats. Each of the following five plot scales is exactly the same; only the display formats differ.

1/4" = 1'-0"

1" = 4'

1 = 48

1:48

1/48

A plot scale of 1:48 means that a line 48 units long in AutoCAD will plot with a length of 1 unit. The units can be any measurement system, including inches, feet, millimeters, nautical miles, chains, angstroms, and light-years, but by default, plotting units in AutoCAD are inches.

There are four variables that control the relationship between the size of objects in the AutoCAD drawing and their sizes on a sheet of paper produced by an AutoCAD plot:

- Size of the object in AutoCAD. For simplification it will be referred to as ACAD_size.
- Size of the object on the plot. For simplification it will be referred to as ACAD_plot.
- Maximum available plot area for a given sheet of paper. For simplification it will be referred to as ACAD_max_plot.
- Plot scale. For simplification it will be referred as to ACAD_scale.

The relationship between the variables can be described by the following three algebraic formulas:

$$ACAD\_scale = ACAD\_plot / ACAD\_size$$

$$ACAD\_plot = ACAD\_size \times ACAD\_scale$$

$$ACAD\_size = ACAD\_plot / ACAD\_scale$$

Example of Computing Plot Scale, Plot Size, and Limits

An architectural elevation of a building 48' wide and 24' high must be plotted on a 36" x 24" sheet. First, you determine the plotter's maximum available plot area for the given sheet size. This depends on the model of plotter you use.

In the case of an HP plotter, the available area for 36" x 24" is 33.5" x 21.5". Next, you determine the area needed for the title block, general notes, and other items, such as an area for revision notes and a list of reference drawings. For the given example, let's say that an area of 27" x 16" is available for the drawing.

The objective is to arrive at one of the standard architectural scales in the form of x in. = 1 ft. The usual range is from 1/16" = 1'-0" for plans of large structures to 3" = 1'0" for small details. To determine the plot scale, substitute these values for the appropriate variables in the formula:

$$ACAD\_scale = ACAD\_plot/ACAD\_size$$

$$ACAD\_scale = 27"/48' \text{ for X axis}$$

$$= 0.5625"/1'\text{-}0" \text{ or } 0.5625"=1'\text{-}0"$$

The closest standard architectural scale that can be used in the given situation is 1/2" = 1'-0" (0.5" = 1'-0", 1/24 or 1:24).

To determine the size of the object on the plot, substitute these values for the appropriate variables in the formula:

$$ACAD\_plot = ACAD\_size \times ACAD\_scale$$

$$ACAD\_plot = 48' \times (0.5"/1') \text{ for X axis}$$

$$= 24" \text{ (less than the 27" maximum allowable space on the paper)}$$

$$ACAD\_plot = 24' \times (0.5"/1') \text{ for Y axis}$$

$$= 12" \text{ (less than the 16" maximum allowable space on the paper)}$$

If, instead of 1/2" = 1'-0" scale, you wish to use a scale of 3/4" = 1'-0", then the size of the object on the plot will be 48' x (0.75"/1') = 36" for the X axis. This is more than the available space on the given paper, so the drawing will not fit on the given paper size. You must select a larger paper size.

Once the plot scale is determined and you have verified that the drawing fits on the given paper size, you can then determine the drawing limits for the plotted sheet size of 33.5" x 21.5".

To determine the limits for the X and Y axes, substitute the appropriate values in the formula:

$$ACAD\_limits \text{ (X axis)} = ACAD\_max\_plot/ACAD\_scale$$
$$= 33.5"/(0.5"/1'-0")$$
$$= 67'$$
$$ACAD\_limits \text{ (Y axis)} = 21.5"/(0.5"/1'-0")$$
$$= 43'$$

Appropriate limits settings in AutoCAD for a 36" x 24" sheet with a maximum available plot area of 33.5" x 21.5" at a plot scale of 0.5" = 1'-0" would be 0,0 for lower left corner and 67',43' for upper right corner.

Another consideration in setting up a drawing for user convenience is to have the (0,0) coordinates at some point other than the lower left corner of the drawing sheet. Many objects have a reference point from which other parts of the object are dimensioned. Being able to set that reference point to (0,0) is very helpful. In many cases, the location of (0,0) is optional. In other cases, the coordinates should coincide with real coordinates, such as those on an industrial plant area block. In still other cases, only one set of coordinates might be a governing factor.

In this example, the 48' wide x 24' high front elevation of the building is to be plotted on a 36" x 24" sheet at a scale of 1/2" = 1'-0". It has been determined that (0,0) should be at the lower left corner of the front elevation view, as shown in Figure 2–14.

Centering the view on the sheet requires a few minutes of layout time. Several approaches allow the drafter to arrive at the location of (0,0) relative to the lower left corner of the plotted sheet or limits. Having computed the limits to be 67' wide x 43' high, the half-width and half-height (dimensions from the center) of the sheet are 33.5' and 21.5' to scale, respectively. Subtracting the half-width of the building from the half-width of the limits will set the X coordinate of the lower left corner at –9.5' (from the equation, 24' – 33.5'). The same is done for the Y coordinate –9.5'(12' – 21.5'). Therefore, the lower left corner of the limits is at (–9.5',–9.5').

 Figure 2–14

Setting the reference point to the origin (0,0) in a location other than the lower left corner

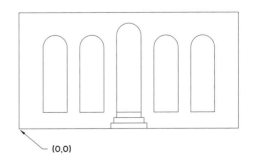

(0,0)

Appropriate limits settings in AutoCAD for a 36" x 24" sheet with a maximum available plot area of 33.5" x 21.5" by centering the view at a plot scale of 0.5" = 1'0" (see Figure 2–15) is –9.5',–9.5' for lower left corner and 57.5', 33.5' for upper right corner.

---

note The absolute X coordinate values, when added (57.5' + 9.5',) equal 67', which is the width of the limits, and the absolute Y coordinate values (33.5' + 9.5') equal 43', which is the height of the limits.

Figure 2-15

Setting the limits to the maximum available plot area by centering the view

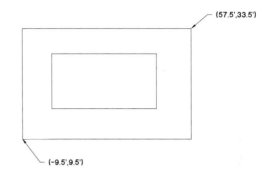

(57.5',33.5')

(-9.5',9.5')

SIZE, SHAPE, AND DIRECTIONS

This section covers the commands and features used to communicate the physical appearance of things. A rectangle might represent a very small computer chip on a printed circuit, a building, or the state of Colorado on a map. Whichever object is depicted; the appropriate type of units (metric, architectural, surveyor's) should be utilized. The type of units includes both linear and angular measurement. What shape and how much of the Drawing Area is to be set aside needs to be determined. The UNITS and LIMITS commands are used to accomplish these tasks.

Setting Units

Figure 2-16

Invoking the UNITS command from the Format menu

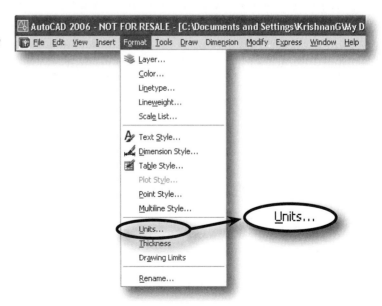

Invoking the UNITS command (see Figure 2–16) lets you change the linear and angular units by means of the Drawing Units dialog box (see Figure 2–17). In addition, it lets you set the display format measurement and precision of your drawing units. You can change any or all of the following:

Unit display format	Angle display precision
Unit display precision	Angle base
Angle display format	Angle direction

▶ **Figure 2–17**

*The Drawing Units
dialog box*

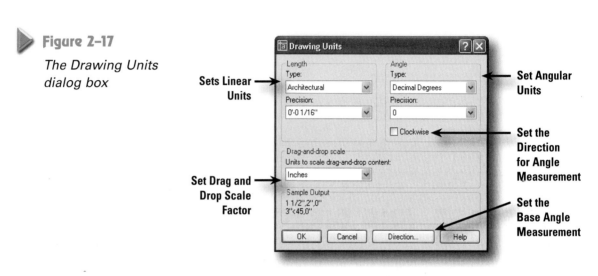

Setting Linear Units

The **Length** section of the Drawing Units dialog box allows you to change the type of units of linear measurement. From the **Type** box select one of the five types of report formats you prefer. For the selected report format, select a precision from the **Precision** box.

For engineering and architectural units, you work in feet and inches, and each drawing unit represents 1 inch. Scientific, decimal, and fractional units can be whatever units you choose to call them.

Drawing a 150-ft long object might differ, however, depending on the units chosen. For example, if you use decimal units and decide that 1 unit = 1 foot, then the 150-ft long object will be 150 units long. If you decide that 1 unit = 1 inch, then the 150-ft long object will be drawn 150 x 12 = 1,800 units long. In architectural and engineering unit types, the unit automatically equals 1 inch. You can then give the length of the 150-ft long object as 150' or 1800" or simply 1800.

Setting Angular Measurement

The **Angle** section of the Drawing Units dialog box allows you to set the drawing's angle measurement. From the **Type** box select one of the five types of report formats you prefer. For the selected format, select a precision from the **Precision** list.

Select the direction in which the angles are measured, clockwise or counterclockwise. If the **Clockwise** check box is set to ON, then the angles will increase in value in the clockwise direction. If it is set to OFF, the angles will increase in value in the counter-clockwise direction (see Figure 2–18).

 Figure 2-18

The default, counterclockwise direction of angle measurement

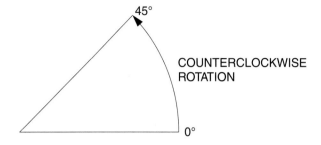

 note   The default of 0 degrees being *East* and angle values increasing in the *counterclockwise* direction is used for the angular prompts and responses throughout this book, unless otherwise noted.

Setting Insertion Scale Units

The units setting that you select from the **Insertion Scale** list box determines the unit of measure used for block insertions from AutoCAD DesignCenter, Tool Palettes, or i-drop. If a block is created in units different from the units specified in the list box, they will be inserted and scaled in the specified units. If you select **Unitless**, the block will be inserted as is, and the scale will not be adjusted to match the specified units. See Chapter 17 for detailed explanation on creating and modifying blocks.

Setting Base Angle for Angle Measurement

To set the base angle for angle measurement, choose **Direction**; the Direction Control dialog box appears, as shown in Figure 2–19.

AutoCAD, by its default setting, assumes 0 degrees is to the right (east, or 3 o'clock) (see Figure 2–20), and angles increase in the counterclockwise direction.

 Figure 2-19

Direction Control dialog box

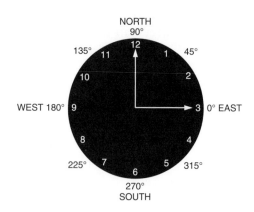

Figure 2–20

Default angle setting direction

You can change angle measurement to start at any compass point by selecting one of the five available options.

You can also show AutoCAD the direction you want for angle 0 by specifying two points. This can be done by selecting **Other** and choosing **Angle**. AutoCAD prompts you for two points and sets the direction for angle 0. Choose **OK** to close the Direction Control dialog box.

Once you are satisfied with all of the settings in the Drawing Units dialog box, choose **OK** to set the appropriate settings to the current working drawing and close the dialog box.

note   When AutoCAD prompts for a distance, displacement, spacing, or coordinates, you can always reply with numbers in integer, decimal, scientific, or fractional format. If the engineering or architectural report format is in effect, you can also enter feet, inches, or a combination of feet and inches. However, feet-and-inches input format differs slightly from the report format because it cannot contain a blank space. For example, a distance of 75.5 inches can be entered in the feet/inches/fractions format as 6'3-1/2". Note the absence of spaces and the hyphen in the unconventional location between the inches and the fraction. Normally, it will be displayed in the status area as 6'-3 1/2.

If you wish, you can use the SETVAR command to set the UNITMODE system variable to 1 (default UNITMODE setting is 0) to display feet-and-inches output in the accepted format. For example, if you set UNITMODE to 1, AutoCAD displays the fractional value of 45 1/4 as you enter it: 45-1/4. The feet input should be followed by an apostrophe (') and inches with a trailing double quote (").

When engineering or architectural report format is in effect, the drawing unit equals 1 inch, so you can omit the trailing double quote (") if you like. When you enter feet-and-inches values combined, the inch values should immediately follow the apostrophe, without an intervening space. Distance input does not permit spaces because, except when entering text, pressing the SPACEBAR is the same as pressing ENTER.

Setting Limits

As mentioned earlier, in the "Drawing Area and Scale" section, the LIMITS command (see Figure 2–21) allows you to place an imaginary rectangular drawing sheet in the CAD drawing space. The limits are expressed as a pair of *2D* points in the World Coordinate System, a lower left and an upper right limit.

Figure 2-21

Invoking the LIMITS command from the Format menu

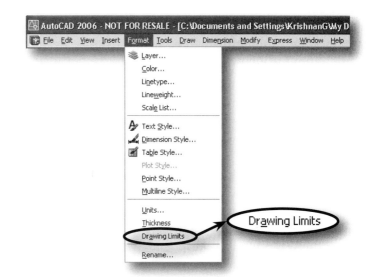

AutoCAD prompts:

limits (ENTER)

Specify lower left corner or ⬇ <current>: *(press ENTER to accept the current setting, specify the lower left corner, or right-click for the Shortcut menu and choose one of the available options)*

Specify upper right corner <current>: *(press ENTER to accept the current setting, or specify the upper right corner)*

zoom (ENTER)

Specify corner of window, enter a scale factor (nX or nXP) or ⬇: all (ENTER)

The response you give for the upper right corner gives the location of the upper right corner of the imaginary rectangular drawing sheet.

There are two additional options available for the LIMITS command. When AutoCAD prompts for the lower left corner, you can respond to the ON or OFF options. The ON/OFF options determine whether or not you can specify a point outside the limits when prompted to do so.

When you select the ON option, limits checking is turned on, and you cannot start or end an object outside the limits, nor can you specify displacement points required by the MOVE or COPY command outside the limits. You can, however, specify two points (center and point on circle) that draw a circle, part of which might be outside the limits. The limits check is simply an aid to help you avoid drawing off the imaginary rectangular drawing sheet. Leaving the limits checking ON is a sort of safety net to keep you from inadvertently specifying a point outside the limits. On the other hand, limits checking is a hindrance if you need to specify such a point.

When you select the OFF option (default), AutoCAD disables limits checking, allowing you to draw objects and specify points outside the limits.

Whenever you change the limits, you will not see any change on the screen unless you use the All option of the ZOOM command. ZOOM ALL lets you see entire newly set limits on the screen. For example, if your current limits are 12 by 9 (lower left corner 0,0 and upper right corner 12,9) and you change the limits to 42 by 36 (lower left corner 0,0 and upper right corner 42,36), you still see the 12 by 9 area. You can draw the objects anywhere on the limits 42 by 36 area, but you will see on the screen only the objects that are drawn in the 12 by 9 area. To see only the entire limits, invoke the ZOOM command using the All option (see Figure 2-22).

Figure 2-22

Invoking the ZOOM ALL command from the View menu

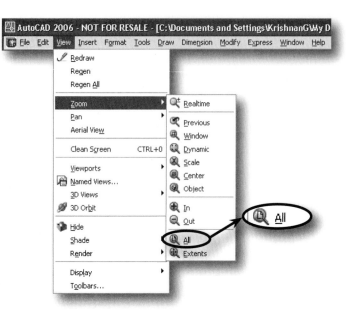

You see the entire limits or current extents (whichever is greater) on the screen. If objects are drawn outside the limits, ZOOM ALL displays all objects. (For a detailed explanation see Chapter 4 for an explanation on the ZOOM command).

Whenever you change the limits, you should always invoke ZOOM ALL to see the entire limits or current extents on the screen.

SETTING SNAP, GRID, AND ORTHO

The SNAP, GRID, and ORTHO commands do not create objects. However, they make it possible to create and modify them more easily and accurately. Each of these Drafting Settings commands can be readily toggled ON when needed and OFF when not. These commands, when turned ON, operate according to settings that can also be changed easily. When used appropriately, these commands provide the power, speed, and accuracy associated with Computer Aided Design/Drafting.

To change the Snap or Grid settings, right-click the SNAP or GRID button on the Status bar at the bottom of the screen and choose Settings from the Shortcut menu. AutoCAD will display the Drafting Settings dialog box with the Snap and Grid tab selected (see Figure 2–23).

 Figure 2-23

Changing the Snap or Grid settings from the Snap and Grid tab of the Drafting Settings dialog box

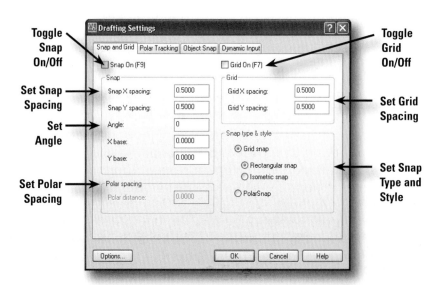

Locking to an Invisible Grid

The SNAP command provides an invisible reference grid in the Drawing Area. When set to ON, the Snap feature forces the cursor to lock onto the nearest point on the specified Snap Grid. Using the SNAP command, you can specify points quickly, letting AutoCAD ensure that they are placed precisely. You can always override the Snap spacing by entering absolute or relative coordinate points by means of the keyboard, or by simply turning the Snap mode OFF. Locking of the cursor to one of the Snap grid points can also be overridden by an Object Snap mode. When the Snap mode is set to OFF, it has no effect on the cursor. When it is set to ON, you cannot place the cursor on a point that is not on one of the specified Snap Grid locations.

Setting Snap ON and OFF

Snap can be toggled ON and OFF by choosing SNAP on the Status bar or by pressing the function key F9.

Changing Snap Spacing

From the **Snap** section in the **Snap and Grid** tab on the Drafting Settings dialog box (see Figure 2–24), you can change the settings of the X and Y spacing by entering the desired values in the **Snap X Spacing** and **Snap Y Spacing** boxes.

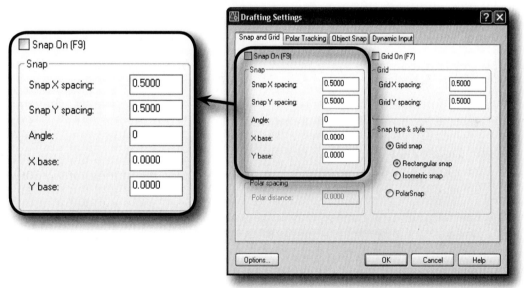

 Figure 2–24

Setting the Snap settings from the Snap and Grid tab of the Drafting Settings dialog box

Setting Aspect Ratio and Rotation Angle

AutoCAD allows you to set the Y Snap Spacing different from the X Snap Spacing (aspect ratio). This is done by entering a value in the **Snap Y Spacing** box that is different from the value in the **Snap X Spacing** box.

AutoCAD allows you to specify an angle to rotate both the visible Grid and the invisible Snap grid. It is a simple version of the more complicated User Coordinate System. It permits you to set up a Snap grid with an origin (X coordinate, Y coordinate of 0,0) and an angle of rotation specified with respect to the default origin and Zero-East system of direction. In conjunction with the X and Y spacing of the Snap grid, the Rotate option can make it easier to draw certain shapes.

From the **Snap** section in the **Snap and Grid** tab of the Drafting Settings dialog box (see Figure 2–24), you can set the angle to rotate the Snap grid by entering a value in the **Angle** box. If you wish to offset the origin, this is done by entering values in the **X base** box and **Y base** box for the X and Y coordinates of the new Origin respectively.

Setting Style formats

The Snap type and style section controls Snap mode settings (see Figure 2–25). The Grid snap selection sets the snap type to Grid and chooses the Rectangular Snap option be default. The Rectangular Snap option refers to the normal rectangular grid and the Isometric option refers to a Grid and Snap designed for Isometric drafting purposes (see Figure 2–26).

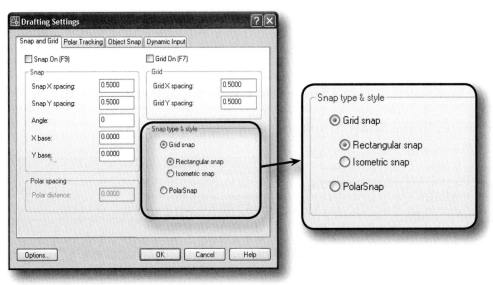

 Figure 2–25

Setting the Style Format from the Snap and Grid tab of the Drafting Settings dialog box

 Figure 2–26

Setting the Snap for isometric drafting

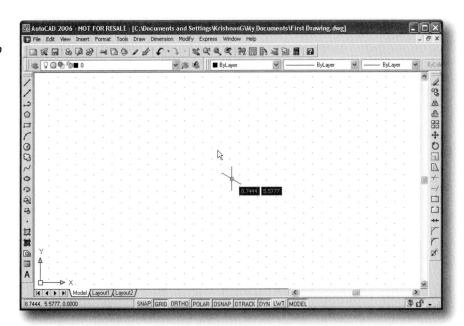

You can switch the Isoplanes between Left (90- and 150-degree angles), Top (30- and 150-degree angles), and Right (30- and 90-degree angles) by CTRL + E (the combination key strokes of holding down CTRL and then pressing E) or by simply press the function key F5.

The PolarSnap selection sets the Snap to Polar Tracking angles. See the explanations about Polar Tracking in Chapter 4.

Displaying the Visible Grid

The GRID command is used to display a visible array of dots with row and column spacings that you specify. AutoCAD creates a grid that is similar to a sheet of graph paper. You can set the Grid display ON and OFF, and you can change the dot spacing. The Grid is a drawing tool and is not part of the drawing; it is for visual reference and is never plotted. In the World Coordinate System, the Grid fills the area defined by the limits.

The Grid has several uses within AutoCAD. First, it shows the extent of the drawing limits. For example, if you set the limits to 42 by 36 units and Grid spacing is set to 0.5 units, then each row will have 85 dots and each column will have 73 dots. This will give you a better sense of the drawing's size relative to the limits than if it were on a blank background.

Second, using the GRID command with the SNAP command is helpful when you create a design in terms of evenly spaced units. For example, if your design is in multiples of 0.5 units, then you can set Grid spacing as 0.5 to facilitate point entry. You could check your drawing visually by comparing the locations of the grid dots and the crosshairs. Figure 2–27 shows a drawing with a Grid spacing of 0.5 units, with limits set to 0,0 and 17,11.

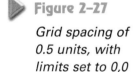

Figure 2–27

Grid spacing of 0.5 units, with limits set to 0,0 and 17,11

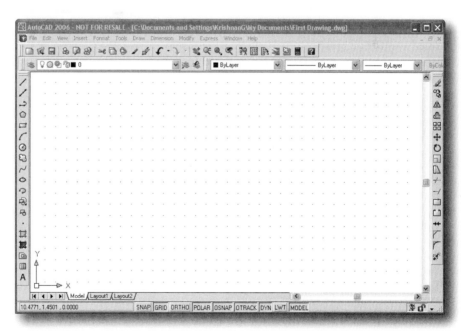

Setting Grid ON and OFF

The Grid can be toggled ON and OFF by choosing the GRID button on the Status bar or by pressing the function key F7.

Changing Grid Spacing

From the **Grid** section in the **Snap and Grid** tab on the Drafting Settings dialog box (see Figure 2–28), you can change the settings of the X and Y spacing by entering the desired values in the **Grid X Spacing** and **Grid Y Spacing** boxes.

Figure 2-28

Setting the Grid settings from the Snap and Grid section of the Drafting Settings dialog box

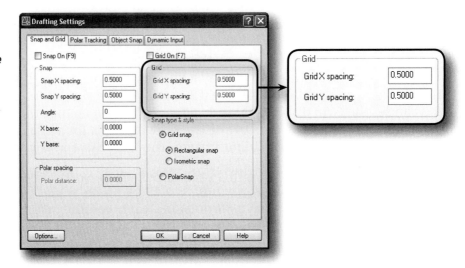

If the spacing of the visible grid is set too small, AutoCAD displays the following message and does not show the dots on the screen:

    Grid too dense to display

To display the Grid, specify a larger grid spacing.

Setting Aspect Ratio

AutoCAD allows you to set the Y Grid Spacing different from the X Grid Spacing (aspect ratio). From the **Grid** section in the **Snap and Grid** tab of the Drafting Settings dialog box (See 2–28), you can set the Y spacing different from the X spacing. This is done by entering a value in the **Grid Y Spacing** box that is different from the value in the **Grid X Spacing** box.

An example of applying the Aspect option, setting the Grid X spacing to **0.5** and the Grid Y spacing to **0.25** provides the Grid dot spacing shown in Figure 2–29.

Figure 2-29

Display after setting the Grid aspect to 0.5 for horizontal and 0.25 for vertical spacing

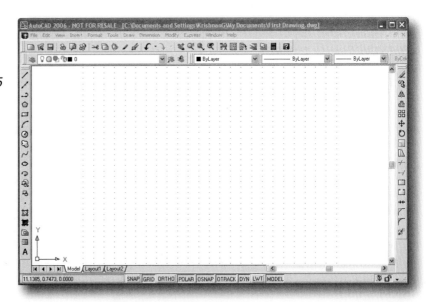

Relationship to Snap Setting

It is often useful to set the Grid spacing equal to the Snap resolution, or make it a multiple of it.

To specify the Grid spacing the same as the Snap value, invoke the GRID command at the Command: prompt:

> **grid** (ENTER)
>
> Specify grid spacing(X) or [ON/OFF/Snap/Aspect] <current>: **s** (ENTER)

To specify the Grid spacing as a multiple of the Snap value, enter **x** after the value. For example, to set up the Grid value as three times the current Snap value (snap = 0.5 units), enter **3x** for the prompt, as shown below:

> **grid** (ENTER)
>
> Specify grid spacing(X) or ⊞ <current>: **3x** (ENTER)

---

note The relationship between the Grid setting and the Snap setting, when established as described in the previous section, is based on the current Snap setting. If the Snap setting is subsequently changed, the Grid setting does not change accordingly. For example, if the Snap setting is 1.00 and you enter *s* in response to the "Specify grid spacing(X) or [ON/OFF/Snap/Aspect] <current>:" prompt, the Grid setting becomes 1.00 and remains 1.00 even if the Snap setting is later set to something else. Likewise, if you set the Grid setting to *3x*, it becomes 3.00 and will not change with a subsequent change in the Snap setting. To have the Grid follow the Snap setting, set the Grid X and Y spacing to 0.

Constraining Cursor Movement

The ORTHO command constrains the cursor movement and lets you draw lines and specify point displacements that are parallel to either the X or Y axis. Lines drawn with the Ortho mode set to ON are therefore either parallel or perpendicular to each other. This mode is helpful when you need to draw lines that are exactly horizontal or vertical. Also, when the Snap style is set to Isometric, it forces lines to be parallel to one of the three isometric axes.

Setting Ortho ON and OFF

Ortho can be toggled ON and OFF by choosing ORTHO button on the Status bar or by pressing the function key F8.

---

note The Ortho and Polar Tracking modes (explained in Chapter 4) cannot both be set to ON at the same time. They can both be set to OFF, or either one can be set to ON.

When the Ortho mode is active, you can draw lines and specify displacements only in the horizontal or vertical directions, regardless of the cursor's on-screen position. The direction in which you draw is determined by the change in the X value of the cursor movement compared to the change in the cursor's distance to the Y axis. AutoCAD allows you to draw horizontally if the distance in the X direction is greater than the distance in the Y direction; conversely, if the change in the Y direction is greater than the change in the X direction, then it forces you to draw vertically. The Ortho mode does not affect keyboard entry of points.

drawing exercises

Open the Exercise Manual PDF file for Chapter 2, on the accompanying CD, for discipline-specific exercises.

review questions

1. If you performed a ZOOM ALL and the object in the drawing shrunk to a small portion of the screen, one possible problem might be:
   a. *Out of computer memory.*
   b. *Misplaced drawing object.*
   c. *Grid and Snap set incorrectly.*
   d. *This should never happen in AutoCAD.*
   e. *Limits are set much larger than current drawing objects.*

2. Points are located by relative rectangular coordinates in relation to:
   a. *The last specified point or position.*
   b. *The global origin.*
   c. *The lower left corner of the screen.*
   d. *All of the above.*

3. Relative Polar coordinates are based on a distance from:
   a. *The global origin.*
   b. *The last specified position at a given angle.*
   c. *The center of the display.*
   d. *All of the above.*

4. Which of the following coordinates will define a point at the screen default origin?
   a. *000*
   b. *00*
   c. *0.0,0.0*
   d. *112*
   e. *@0,0*

5. To draw a line a length of eight feet, four and five-eights inches in the 12 o'clock direction from the last point selected, enter:
   a. *@8'4-5/8<90.*
   b. *8'-4-5/8<90.*
   c. *@8-45/8<90.*
   d. *8'-45/8<90.*
   e. *None of the above.*

6. The default settings of AutoCAD:
   a. *Assume that 0 degrees is to the right.*
   b. *Assume that angles increase in the counterclockwise direction.*
   c. *Both a and b.*
   d. *None of the above.*

7. By default, what direction does a positive number indicate when specifying angles in degrees?
    a. *Clockwise*
    b. *Counterclockwise*
    c. *Has no impact when specifying angles in degrees*
    d. *None of the above*

8. An absolute rectangular coordinate is based on the last specified point.
    a. *True*
    b. *False*

9. The invisible grid, which the cursor locks onto, is called:
    a. *Snap.*   **c.** *Ortho.*
    b. *Grid.*   **d.** *Cursor lock.*

10. Under which menu on the menu bar can you find Drafting Settings?
    a. *Edit*   **c.** *Format*
    b. *Insert*   **d.** *Tools*

11. Which of the following cannot be modified in the Drafting Settings dialog box?
    a. *Snap*
    b. *Grid*
    c. *Snap type and style*
    d. *Limits*
    e. *All of the above*

12. If the spacing of the visible grid is set too small, AutoCAD responds as follows:
    a. *Does not accept the command.*
    b. *Produces a "Grid too dense to display" message.*
    c. *Produces a display that is distorted.*
    d. *Automatically adjusts the size of the grid so it will display.*
    e. *Displays the grid anyway.*

13. Which command constrains cursor movements to the horizontal or vertical directions (relative to the UCS)?
    a. *GRID*   **c.** *ORTHO*
    b. *MODE*   **d.** *SNAP*

14. Which function key toggles the display of the GRID on and off?
    a. *F6*   **c.** *F8*
    b. *F7*   **d.** *F9*

15. The smallest number that can be displayed in the denominator when setting units to architectural units is:
    a. *8.*   **d.** *128.*
    b. *16.*   **e.** *None of the above.*
    c. *64.*

16. It is possible to enter coordinates outside of the viewing area for AutoCAD to use for creating objects.
   a. *True*
   b. *False*

17. In architectural and engineering unit modes, the unit automatically equals one foot.
   a. *True*
   b. *False*

18. Which of the following is an absolute coordinate entry?
   a. *2,1*   **c.** *@2<90*
   b. *@2,1*   **d.** *None of the above*

creating objects

introduction

This chapter introduces some of the basic commands and concepts in AutoCAD that can be used to create geometric objects. When you learn how to access and use the commands, how to find your way around the screen, and how AutoCAD applies coordinate geometry to the entities that make up a drawing, you can apply these skills to the chapters containing more advanced concepts.

After completing this chapter, you will be able to do the following:

- Construct geometric figures using the LINE, RECTANGLE, POLYGON, POINT, CIRCLE, and ARC commands

- Use the UNDO and REDO commands

You will progress at a better rate if you make an effort to learn as much as possible as soon as possible about the descriptive properties of the individual objects. When you become familiar with how AutoCAD creates, manipulates, and stores the data that describes the objects, you are then able to create drawings more effectively.

STRAIGHT OBJECTS

This section covers objects that can be defined by the coordinates of their endpoints alone. The objects that are included in this section are Line, Rectangle, Polygon, and Point. These objects are comprised of straight line segments. Even though the Point object is defined by only one set of coordinates, it is included in this section.

Drawing Lines

 Figure 3-1

Invoking the LINE
command from the
Draw toolbar

The primary drawing object is the line. A series of connected straight line segments can be drawn by invoking the LINE command (see Figure 3–1) and then selecting the proper sequence of endpoints. AutoCAD connects the points with a series of lines.

You can specify the endpoints using either two dimensional (x,y) or three dimensional (x,y,z) coordinates, or a combination of the two. If you enter 2D coordinates, AutoCAD uses the current elevation as the Z element of the point (zero is the default). This chapter is concerned only with 2D points whose elevation is zero.

If you are placing points with a cursor instead of specifying coordinates, a rubber-band preview line is displayed between the starting point and the crosshairs. This helps you see where the resulting line will go. In Figure 3–2 the dotted lines represent previous cursor positions.

AutoCAD prompts:

Specify first point: *(specify Point 1 as the starting point of the line)*
Specify next point or ⬇: *(specify Point 2 as the ending point of the line)*

 Figure 3-2

Placing points with the cursor rather
than with keyboard coordinates input

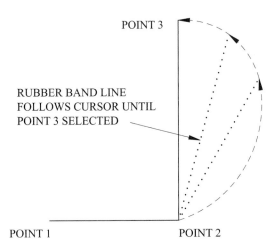

POINT 3

RUBBER BAND LINE
FOLLOWS CURSOR UNTIL
POINT 3 SELECTED

POINT 1   POINT 2

Specify next point or ⬇: *(specify Point 3 as the ending point of the second line)*

After two line segments have been drawn with one LINE command, the prompt will include the Close option.

The LINE command is one of the few AutoCAD commands that automatically repeat. It uses the ending point of one line as the starting point of the next, continuing to prompt you for each subsequent ending point. To terminate this continuing feature you must give a null response (press ENTER or right-click and choose Enter from the Shortcut menu). Even though a series of lines is drawn using a single LINE command, each line is a separate object, as though it has been drawn with a separate LINE command.

Most of the AutoCAD commands have a variety of options. They can be chosen from the Shortcut menu that appears when you right-click on your pointing device after invoking the command. For the LINE command, three options are available: Continue, Close, and Undo.

Continue Option

When you invoke the LINE command, if instead of specifying a starting point, you respond to the "Specify first point:" prompt by pressing ENTER, AutoCAD automatically sets the starting point of the new line at the endpoint of the most recently drawn line or arc. This provides a simple method for constructing a tangentially connected line in an arc-line continuation.

The subsequent prompt sequence depends on whether a line or arc was more recently drawn. If the line is more recent, the starting point of the new line will be set as the ending point of that most recent line, and the "Specify next point:" prompt appears as usual. For example, following the sequence shown above that resulted in the two line segments shown in Figure 3–2, the next three line segments can be drawn, as shown in Figure 3–3, with the Continue option using the following sequence:

Specify first point: *(to continue the next line from Point 3, press* ENTER *or the* SPACEBAR*)*
Specify next point or ⬇: *(specify Point 4)*
Specify next point or ⬇: *(specify Point 5)*
Specify next point or ⬇: *(specify Point 6)*
Specify next point or ⬇: (ENTER)

▶ **Figure 3–3**

Using the LINE *command's Continue option*

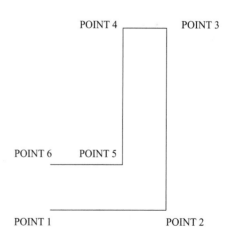

If an arc is more recent, its end defines the starting point and the direction of the new line. AutoCAD prompts:
Length of line: *(specify the length of the line to be drawn)*

AutoCAD continues with the normal "Specify next point:" prompt.

Close Option

If you are drawing a sequence of lines to form a polygon, you can use the Close option to join the last and first points automatically. AutoCAD draws the closing line segment if you respond to the "Specify next point:" prompt by right-clicking and choosing Close from the Shortcut menu.

AutoCAD performs two steps when you choose the Close option. The first step closes the polygon, and the second step terminates the LINE command (equivalent to a null response) and returns you to the On-screen prompt.

The following command sequence shows an example of using the Close option (see Figure 3–4).

Specify first point: *(specify Point 1)*

Specify next point or ⤓: *(specify Point 2)*

Specify next point or ⤓: *(specify Point 3)*

Specify next point or ⤓: *(specify Point 4)*

Specify next point or ⤓: *(specify Point 5)*

Specify next point or ⤓: *(specify Point 6)*

Specify next point or ⤓: *(choose Close from the Shortcut menu)*

▶ **Figure 3–4**

Using the LINE command's Close option

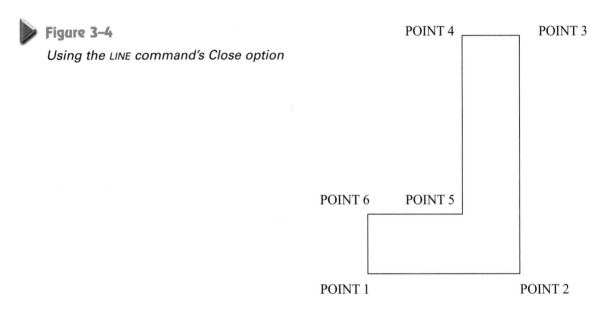

POINT 4   POINT 3

POINT 6   POINT 5

POINT 1   POINT 2

---

note Using the Close option causes the LINE command to automatically draw the last line of a sequence from the endpoint of the current line to the starting point of the first line of the current sequence. If the series of lines in Figure 3–4 had been drawn in two sequences, one from Point 1 to Point 3 and then using the Continue option to draw the second sequence from Point 3 to Point 6, then the Close option would cause the last line to be drawn from Point 6 to Point 3, not to Point 1.

Undo Option

While drawing a series of connected lines, you may wish to erase the most recent line segment and continue from the end of the previous line segment. You can do so and remain in the LINE command without exiting by using the Undo option. Whenever you wish to erase the most recent line segment, at the "Specify next point:" prompt, choose the Undo option from the Shortcut menu. If necessary, you can select multiple Undos; this will erase the most recent line segment one at a time. Once you are out of the LINE command, it is too late to use the Undo option of the LINE command to erase the most recent line segment.

The following command sequence shows an example using the Undo option (see Figure 3–5).

Specify first point: *(specify Point 1)*

Specify next point or : *(specify Point 2)*

Specify next point or : *(specify Point 3)*

Specify next point or : *(specify Point 4)*

Specify next point or : *(specify Point 5)*

Specify next point or : *(choose Undo option from the Shortcut menu)*

Specify next point or : *(choose Undo option from the Shortcut menu)*

Specify next point or : *(specify New Point 4)*

Specify next point or : *(specify New Point 5)*

Specify next point or : (ENTER)

▶ **Figure 3–5**

Using the LINE command's Undo option

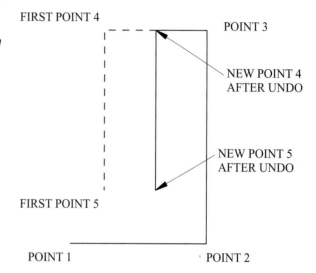

Drawing Rectangles

▶ **Figure 3–6**

Invoking the RECTANGLE command from the Draw toolbar

When it is necessary to create a closed rectangular box whose sides are parallel to the X and Y axes, you can use the RECTANGLE command (see Figure 3–6).

AutoCAD prompts:

Specify first corner point or : *(specify first corner point to define the start of the rectangle or choose one of the available options from the Shortcut menu)*

Specify other corner point or : *(specify a point to define the opposite corner of the rectangle or choose dimensions from the Shortcut menu)*

Chamfer Option

The Chamfer option sets the chamfers to be drawn at the corners of the rectangle to be drawn. Refer to Chapter 11 for a detailed explanation on the usage of the CHAMFER command and its available settings.

Elevation Option

The Elevation option specifies the elevation of the rectangle to be drawn.

Fillet Option

The Fillet option sets fillet radii to be drawn at the cornerss of the rectangle to be drawn. Refer to Chapter 11 for a detailed explanation on the usage of the FILLET command and its available settings.

Thickness Option

The Thickness option specifies the thickness of the rectangle to be drawn. Thickness is actually linewidth applied to the line in the verticle plane, giving it a three demensional effect. The use of the thickness option has been almost outmoded by the 3D capabilities of AutoCAD.

Width Option

The Width option allows you set the line width for the rectangle to be drawn. The default width is set to 0.0.

Area Option

The Area option allows you to create a rectangle of a specified area. You also specify either the length or the width value.

Dimensions

The Dimensions option allows you create a rectangle by specifying length and width values. When you choose the Dimensions option, AutoCAD prompts:

Specify length for rectangles <default>: *(specify length of the rectangle)*

Specify width for rectangles <default>: *(specify width of the rectangle)*

Specify other corner point or ⏷: *(specify a point by moving the cursor to one of the four possible locations for the diagonally opposite corner of rectangle)*

Rotation Option

The Rotation option allows you to create a non-orthogonal rectangle. AutoCAD draws a rectangle using a specified angle for the base and second corner.

Drawing Polygons

 Figure 3-7

Invoking the POLYGON command from the Draw toolbar

The POLYGON command (see Figure 3–7) creates an equilateral (edges with equal length) closed polyline. It offers three different methods for drawing 2D polygons: Inscribed in Circle, Circumscribed about Circle, and Edge. The number of sides can vary from 3 (which forms an equilateral triangle) to 1024.

Inscribed in Circle

The Inscribed in Circle option draws the polygon of equal length for all sides inscribed inside an imaginary circle having the same diameter as the distance across opposite polygon corners (for an even number of sides) as shown in the following example:

Enter number of sides<default>: **6** (ENTER)

Specify center of polygon or ⊻: **3,3** (ENTER)

Enter an option ⊻: *(choose Inscribed in Circle from the Shortcut menu)*

Specify radius of circle: **2** (ENTER)

AutoCAD draws a polygon with six sides (see Figure 3–8), centered at 3,3, whose edge vertices are 2 units from the center of the polygon.

Figure 3–8

Polygon drawn with six sides by selecting the Inscribed in Circle option

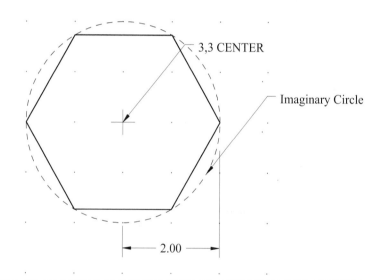

3,3 CENTER

Imaginary Circle

2.00

note Specifying the radius with a specific value draws the bottom edge of the polygon at the current snap rotation angle. If instead, you specify the radius with your pointing device or by means of coordinates, AutoCAD places one apex of the polygon on the specified point, which determines the rotation and size of the polygon.

Circumscribed about Circle

The Circumscribed about Circle option draws a polygon circumscribed around the outside of an imaginary circle having the same diameter as the distance across the opposite polygon sides (for an even number of sides) as shown in the following example:.

Enter number of sides<default>: **8** (ENTER)

Specify center of polygon or ⊻: **3,3** (ENTER)

Enter an option ⊻: *(choose Circumscribed about Circle from the Shortcut menu)*

Specify radius of circle: **2** (ENTER)

AutoCAD draws a polygon with eight sides (see Figure 3–9), centered at 3,3, whose edge midpoints are 2 units from the center of the polygon.

 Figure 3-9

Polygon drawn with eight sides by selecting the Circumscribed about Circle option

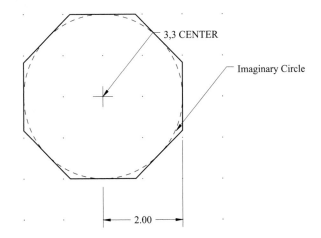

note Specifying the radius draws the bottom edge of the polygon at the current snap rotation angle. If instead, you specify the radius with your pointing device or by means of coordinates, AutoCAD places the midpoint of one edge of the polygon ate the specified point, which determines the rotation and size of the polygon.

Edge

The Edge option allows you to draw a polygon by specifying the endpoints of the first edge as shown in the following example:.

    Enter number of sides<default>: **7** (ENTER)

    Enter an option ⊡: *(choose Edge from the Shortcut menu)*

    Specify first endpoint of edge: **1,1** (ENTER)

    Specify second endpoint of edge: **3,1** (ENTER)

AutoCAD draws a polygon with seven sides (see Figure 3–10) for the specified endpoints of one of the sides of the polygon.

 Figure 3-10

Polygon drawn with seven sides by selecting the Edge option

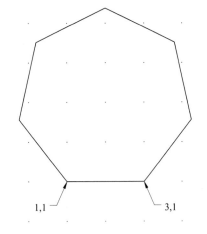

Drawing Point Objects

Figure 3-11

Invoking the POINT command from the Draw toolbar

The POINT command (see Figure 3–11) draws points on the drawing, and these points are drawn on the plotted drawing sheet with a single "pen down." Points can be used as reference points for object snapping when necessary. See Chapter 4 for an explanation of the Object Snap feature.

Points are entered by specifying 2D or 3D coordinates or with the pointing device.

Point: *(specify a point)*

Point Modes

When you draw the point, it appears on the display as a blip (+) if the BLIPMODE system variable is set to ON (default is OFF). After a REDRAW command, it appears as a dot (.). You can make the point appear as a +, x, 0, or any of the available symbols by changing the PDMODE system variable. This can be done by entering **pdmode** at the On-screen prompt and entering the appropriate value. You can also change the PDMODE value by using the Point Style icon menu, as shown in Figure 3–12, invoked by choosing Point Style from the Format menu. The default value of PDMODE is zero, which means the point appears as a dot. If PDMODE is changed, all previous points drawn are replaced with the current setting.

Figure 3-12

The Point Style icon menu

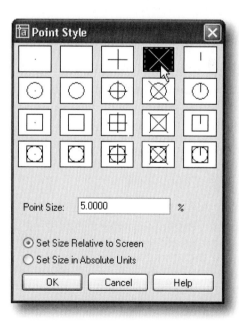

When PDMODE is set to a value other than zero, the size of the point that appears on the screen depends on the value to which the PDSIZE system variable is set. If necessary, you can change the size from the Point Style icon menu. The default for PDSIZE is zero (one pixel in size). Any positive value larger than this will increase the size of the point accordingly.

CURVED OBJECTS

This section covers curved objects that can be drawn with the CIRCLE and ARC commands.

Drawing Circles

▶ **Figure 3–13**

Invoking the CIRCLE command from the Draw toolbar

The CIRCLE command (see Figure 3–13) offers five different methods for drawing circles: Center-Radius (default), Center-Diameter, Three-Point, Two-Point, and Tangent, Tangent, Radius (TTR).

Center-Radius

The Center-Radius option draws a circle based on a center point and a radius. A circle is drawn with the default Center-Radius option by invoking the CIRCLE command and proceeding with the command sequence as shown in the following example:

> Specify center point for circle or ⊡: **2,2** (ENTER)
> Specify radius of circle or ⊡: **1** (ENTER)

AutoCAD draws a circle with 2,2 as the center point with a radius of 1 unit (see Figure 3–14).

The same circle can be drawn as follows (see Figure 3–15):

> Specify center point for circle or ⊡: **2,2** (ENTER)
> Specify radius of circle or ⊡: **3,2** (ENTER)

AutoCAD draws the circle with 2,2 as the center point and the distance between the center point and the second point specified as the value for the radius of the circle.

▶ **Figure 3–14**

A circle drawn with the CIRCLE command's default option: Center-Radius

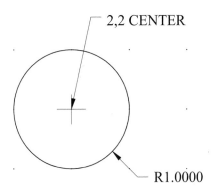

2,2 CENTER

R1.0000

 Figure 3-15

A circle drawn with the Center-Radius option by specifying the coordinates

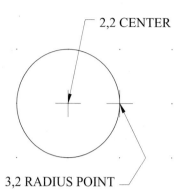

2,2 CENTER

3,2 RADIUS POINT

Center-Diameter

The Center-Diameter option draws a circle based on a center point and a specified diameter as shown in the following example;

Specify center point for circle or ⊡: **2,2** (ENTER)

Specify radius of circle or ⊡: *(choose Diameter from the Shortcut menu)*

Specify diameter of circle: **2** (ENTER)

AutoCAD draws a circle with 2,2 as the center point with a diameter of 2 units as shown in Figure 3–16.

 Figure 3-16

A circle drawn using the Center-Diameter option

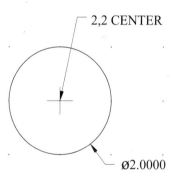

2,2 CENTER

Ø2.0000

Three-Point

The Three-Point circle option draws a circle based on three points on the circumference as shown in the following example:

Specify center point for circle or ⊡: *(choose 3P from the Shortcut menu)*

Specify first point on circle: **2,1** (ENTER)

Specify second point on circle: **3,2** (ENTER)

Specify third point on circle: **2,3** (ENTER)

 note The 3P response allows you to override the Center-Point default.

AutoCAD draws a circle based on three coordinates 2,1 3,2 and 2,3 as shown in Figure 3–17.

 Figure 3-17

A circle drawn with the Three-Point option

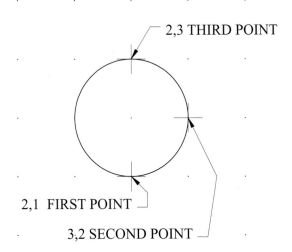

2,3 THIRD POINT

2,1 FIRST POINT

3,2 SECOND POINT

Two-Point

The Two-Point circle option draws a circle based on two endpoints of the diameter as shown in the following example:

Specify center point for circle or : *(choose 2P from the Shortcut menu)*
Specify first end point of circle's diameter: **1,2** (ENTER)
Specify second end point of circle's diameter: **3,2** (ENTER)

note   The 2P response overrides the Center-Point default.

AutoCAD draws a circle with 1,2 and 3,2 as endpoint of a of the diameter as shown in Figure 3–18.

 Figure 3-18

A circle drawn with the Two-Point option

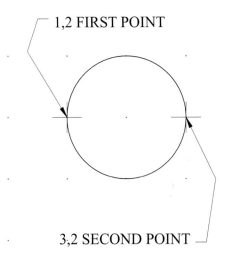

1,2 FIRST POINT

3,2 SECOND POINT

Tangent, Tangent, Radius (TTR)

The Tangent, Tangent, Radius option draws a circle tangent to two objects (either lines, arcs, or circles) with a specified radius as shown in the following example:

Specify center point for circle or ⊻: *(choose Ttr (tan tan radius) from the Shortcut menu)*

Specify point on object for first tangent of circle: *(specify an object for first tangent of circle)*

Specify point on object for second tangent of circle: *(specify an object for second tangent of circle)*

Specify radius of circle: **2** (ENTER)

AutoCAD draws a circle tangent to two objects with a radius of 2 as shown in Figure 3–19.

  Figure 3–19

A circle drawn with the Tangent, Tangent, Radius option

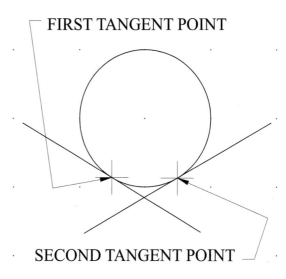

FIRST TANGENT POINT

SECOND TANGENT POINT

For specifying the "tangent-to" objects, normally it does not matter where on the objects you make your selection. However, if more than one circle can be drawn tangent to the object selected, AutoCAD will draw the one whose tangent point is nearest to the selection point.

---

note Until it is changed, the radius/diameter you specify in any one of the options becomes the default setting for subsequent circles to be drawn.

Drawing Arcs

Figure 3–20

Invoking the ARC command from the Draw toolbar

The ARC command (see Figure 3–20) offers four types of methods for drawing arcs:

- Combination of three points
- Combination of two points and an included angle or starting direction
- Combination of two points and a length of chord or radius
- Continuation from line or arc

Three Points (Start, Point-on-Circumference or Center, and End)

There are three ways to use a Three Points method of drawing arcs:

- Three-Point
- Start, Center, End (S,C,E)
- Center, Start, End (C,S,E)

Three-Point

The Three-Point option (default) draws an arc using three specified points on the arc's circumference. The first point specifies the start point, the second point specifies a point on the circumference of the arc, and the third point is the arc endpoint. You can specify a three-point arc either clockwise or counterclockwise as shown in the following command sequence:

Specify start point of arc or ⊡: **1,2** (ENTER)
Specify second point of arc or ⊡: **2,1** (ENTER)
Specify end point of arc: **3,2** (ENTER)

AutoCAD draws an arc based on the coordinates 1,2 2,1 and 3,2 as shown in Figure 3–21.

Figure 3–21

*An arc drawn with the **ARC** command's default option: Three Points*

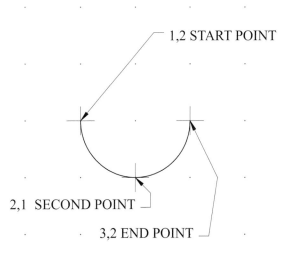

1,2 START POINT

2,1 SECOND POINT

3,2 END POINT

Start, Center, End (S,C,E)

The Start, Center, End option draws an arc using three specified points. The first point specifies the start point, the second point specifies the center point of the arc to be drawn, and the third point is the arc endpoint as shown in the following example:

Specify start point of arc or ⬇: **1,2** (ENTER)

Specify second point of arc or ⬇: *(choose Center from the Shortcut menu)*

Specify center point of arc: **2,2** (ENTER)

Specify end point of arc or ⬇: **2,3** (ENTER)

AutoCAD draws an arc based on a starting point of 1,2 with the center point 2,2 and an end point of 2,3 as shown in Figure 3–22.

Figure 3–22

An arc drawn with the Start, Center, End (S,C,E) option

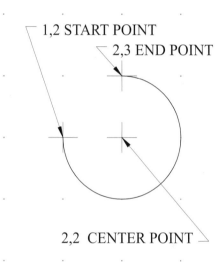

1,2 START POINT

2,3 END POINT

2,2 CENTER POINT

> **note**   Arcs drawn by this method are always drawn counterclockwise from the starting point. The distance between the center point and the starting point determines the radius. Therefore, the point specified in response to "end point" needs only to be on the same radial line of the desired endpoint.

Center, Start, End (C,S,E)

The Center, Start, End option is similar to the Start, Center, End (S,C,E) method, except that in this option, the first point selected is the center point of the arc rather than the start point.

Two Points and an Included Angle or Starting Direction

There are four ways to use a Two Points and an Included Angle or Starting Direction method of drawing arcs:

- Start, Center, Angle (S,C,A)
- Center, Start, Angle (C,S,A)
- Start, End, Angle (S,E,A)
- Start, End, Direction (S,E,D)

Start, Center, Angle (S,C,A)

The Start, Center, Angle option draws an arc similar to the Start, Center, End option method, but it places the endpoint on a radial line at the specified angle from the line between the center point and the start point. If you specify a positive angle as the included angle, an arc is drawn counterclockwise as shown in the following example; for a negative angle, the arc is drawn clockwise.

> Specify start point of arc or ⬇: **1,2** (ENTER)
>
> Specify second point of arc or ⬇: *(choose Center from the Shortcut menu)*
>
> Specify center point of arc: **2,2** (ENTER)
>
> Specify end point of arc or ⬇: *(choose Angle from the Shortcut menu)*
>
> Specify Included Angle: **270** (ENTER)

AutoCAD draws an arc based on a starting point of 1,2 with the center point 2,2 and an included angle of 270 degrees as shown in Figure 3–23.

 Figure 3–23

An arc drawn with the Start, Center, Angle (S,C,A) option

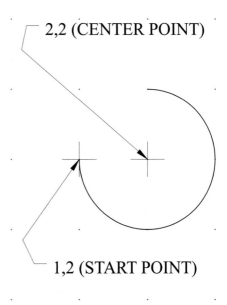

2,2 (CENTER POINT)

1,2 (START POINT)

---

note   If a point directly below the specified center were selected (in the previous example) in response to the "Included angle: prompt", AutoCAD would read the angle of the line (270 degrees from zero) as the included angle for the arc.

Center, Start, Angle (C,S,A)

The Center, Start, Angle option is similar to the Start, Center, Angle (S,C,A) method, except that in this option, the first point selected is the center point of the arc rather than the start point.

Start, End, Angle (S,E,A)

The Start, End, Angle option draws an arc similar to the Start, Center, Angle option method and places the endpoint on a radial line at the specified angle from the line between the center point and the start point. If you specify a positive angle for the included angle, an arc is drawn counterclockwise as shown in the following example; for a negative angle the arc is drawn clockwise.

Specify start point of arc or : **3,2** (ENTER)

Specify second point of arc or : *(choose End from the Shortcut menu)*

Specify end point of arc: **2,3** (ENTER)

Specify center point of arc or [Angle/Direction/Radius]: *(choose Angle from the Shortcut menu)*

Specify included angle: **90** (ENTER)

AutoCAD draws an arc based on a starting point of 3,2 with an end point of 2,3 and an included angle of 90 degrees as shown in Figure 3–24.

 Figure 3-24

An arc drawn counterclockwise with the Start, End, Included Angle (S,E,A) option

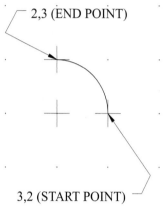

The arc shown in Figure 3–25 is drawn with a negative angle using the following example:

Specify start point of arc or : **3,2** (ENTER)

Specify second point of arc or : *(choose End from the Shortcut menu)*

Specify end point of arc: **2,3** (ENTER)

Specify center point of arc or [Angle/Direction/Radius]: *(choose Angle from the Shortcut menu)*

Specify included angle: **–270** (ENTER)

 Figure 3-25

An arc drawn clockwise with the Start, End, Included Angle (S,E,A) option

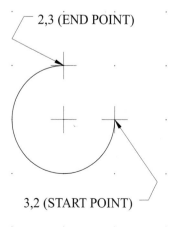

Start, End, Direction (S,E,D)

The Start, End, Direction option allows you to draw an arc between selected points by specifying a direction in which the arc will start from the selected start point. Either the direction can be keyed in as shown in the following example, or you can select a point on the screen with your pointing device. If you select a point on the screen, AutoCAD uses the angle from the start point to the selected point as the starting direction.

Specify start point of arc or ⬇: **3,2** (ENTER)

Specify second point of arc or ⬇: *(choose End from the Shortcut menu)*

Specify end point of arc: **2,3** (ENTER)

Specify center point of arc or [Angle/Direction/Radius]: *(choose Direction from the Shortcut menu)*

Direction from start point: **90** (ENTER)

AutoCAD draws an arc based on a starting point of 3,2 with an end point of 2,3 and direction set to 90 degrees as shown in Figure 3–26.

Figure 3–26

An arc drawn with the Start, End, Direction (S,E,D) option

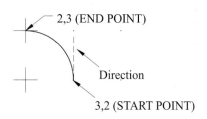

2,3 (END POINT)

Direction

3,2 (START POINT)

Two Points and a Length of Chord or Radius

There are three ways to use a Two Points and a Length of Chord or Radius method of drawing arcs:

- Start, Center, Length of Chord (S,C,L)
- Center, Start, Length of Chord (C,S,L)
- Start, End, Radius (S,E,R)

Start, Center, Length of Chord (S,C,L)

The Start, Center, Length of Chord option uses the specified chord length as the straight-line distance from the start point to the endpoint. With any chord length (equal to or less than the diameter length), there are four possible arcs that can be drawn: a major arc in either direction and a minor arc in either direction. Therefore, all arcs drawn by this method are counterclockwise from the start point. A positive value for the length of chord will cause AutoCAD to draw the minor arc as shown in the following example; a negative value will result in the major arc.

Specify start point of arc or ⬇: **1,2** (ENTER)

Specify second point of arc or ⬇: *(choose Center from the Shortcut menu)*

Specify center point of arc: **2,2** (ENTER)

Specify end point of arc or : *(choose chord Length from the Shortcut menu)*

Specify length of chord: **1.4142** (ENTER)

AutoCAD draws an arc based on a starting point of 1,2 with the center point of 2,2 and a length of chord set to 1.4142 as shown in Figure 3–27.

Figure 3-27

A minor arc drawn with the Start, Center, Length of chord (S,C,L) option.

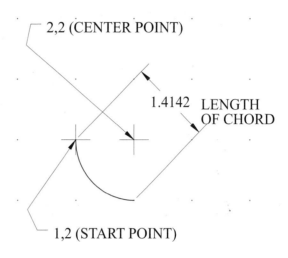

2,2 (CENTER POINT)

1.4142   LENGTH OF CHORD

1,2 (START POINT)

The following example shows drawing a major arc.

Specify start point of arc or : **1,2** (ENTER)

Specify second point of arc or : *(choose Center from the Shortcut menu)*

Specify center point of arc: **2,2** (ENTER)

Specify end point of arc or : *(choose chord Length from the Shortcut menu)*

Specify length of chord: **–1.414** (ENTER)

AutoCAD draws an arc based on a starting point of 1,2 with the center point of 2,2 and a length of chord set to –1.414 as shown in Figure 3–28.

Figure 3-28

A major arc drawn with the Start, Center, Length of chord (S,C,L) option

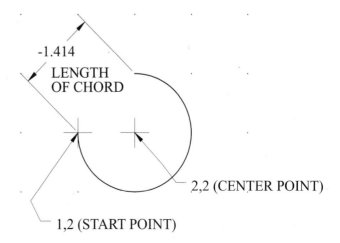

-1.414

LENGTH OF CHORD

2,2 (CENTER POINT)

1,2 (START POINT)

Center, Start, Length (C,S,L)

The Center, Start, Length option is similar to the Start, Center, Length (S,C,L) method, except that the first point selected is the center point of the arc rather than the start point.

Start, End, Radius (S,E,R)

The Start, End, Radius option allows you to specify a radius after selecting the two endpoints of the arc. As with the Chord Length method, there are four possible arcs that can be drawn: a major arc in either direction and a minor arc in either direction. Therefore, all arcs drawn by this method are counterclockwise from the start point. A positive value for the radius causes AutoCAD to draw the minor arc; a negative value results in the major arc as shown in the following example:

> Specify start point of arc or 🔽: **1,2** (ENTER)
>
> Specify second point of arc or 🔽: *(choose End from the Shortcut menu)*
>
> Specify end point of arc: **2,3** (ENTER)
>
> Specify center point of arc or [Angle/Direction/Radius]: *(choose Radius from the Shortcut menu)*
>
> Specify radius of arc: **–1** (ENTER)

AutoCAD draws an arc based on a starting point of 1,2 with an end point of 2,3 and a radius of –1 unit as shown in Figure 3–29.

▶ **Figure 3–29**

A major arc drawn with the Start, End, Radius (S,E,R) option

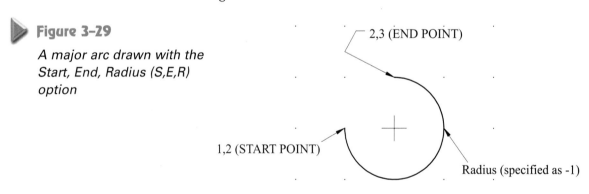

The following example shows drawing a minor arc.

> Specify start point of arc or 🔽: **2,3** (ENTER)
>
> Specify second point of arc or 🔽: *(choose End from the Shortcut menu)*
>
> Specify end point of arc: **1,2** (ENTER)
>
> Specify center point of arc or [Angle/Direction/Radius]: *(choose Radius from the Shortcut menu)*
>
> Specify radius of arc: **1** (ENTER)

AutoCAD draws an arc based on a starting point of 2,3 with an end point of 1,2 and a radius of 1 unit as shown in Figure 3–30.

Figure 3-30

A minor arc drawn with the Start, End, Radius (S,E,R) option

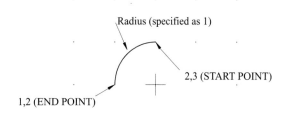

Radius (specified as 1)

2,3 (START POINT)

1,2 (END POINT)

Line-Arc and Arc-Arc Continuation

You can use an automatic Start Point, Endpoint, Starting Direction method to draw an arc by pressing ENTER as a response to the first prompt of the ARC command. After you press ENTER, the only other input is to select or specify the endpoint of the arc you wish to draw. AutoCAD uses the endpoint of the previous line or arc (whichever was drawn last) as the start point of the new arc. AutoCAD then uses the ending direction of that last drawn object as the starting direction of the arc. Examples are shown in the following sequences and figures.

The start point of the existing arc is 2,1 and the endpoint is 3,2 with a radius of 1. This makes the ending direction of the existing arc 90 degrees, as shown in Figure 3–31.

Figure 3-31

An arc drawn with a start point (2,1), an endpoint (3,2), and a radius of 1.0

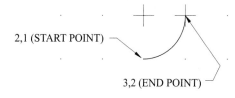

2,1 (START POINT)

3,2 (END POINT)

The following example continues drawing an arc from the last-drawn arc (Figure 3–31), as shown in Figure 3–32 (Arc-Arc-Continuation).

Specify start point of arc or ⬇: (ENTER)
Specify end point of arc: **2,3** (ENTER)

Figure 3-32

An arc drawn by means of the Arc-Arc Continuation method

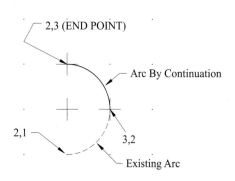

2,3 (END POINT)

Arc By Continuation

2,1

3,2

Existing Arc

The arc, as shown in Figure 3–31, is drawn clockwise instead, with its start point at 3,2 to an endpoint of 2,1 (see Figure 3–33).

The following example will draw the automatic start point, endpoint, starting direction arc, as shown in Figure 3–34.

> Specify start point of arc or ⏷: (ENTER)
> Specify end point of arc: **2,3** (ENTER)

▶ **Figure 3–33**

An arc drawn clockwise with start point (3,2)
and endpoint (2,1)

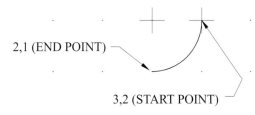

2,1 (END POINT)

3,2 (START POINT)

▶ **Figure 3–34**

An arc drawn by means of the
automatic Start point–Endpoint-
Starting direction method

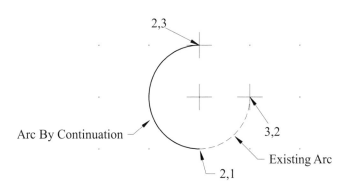

2,3

Arc By Continuation

3,2

Existing Arc

2,1

In the last case, the direction used is 180 degrees. The same arc would have been drawn if the last "line-or-arc" drawn were a line starting at 4,1 and ending at 2,1.

---

note This method uses the last drawn arc or a line. If you draw an arc, draw a line, draw a circle, and then use this continuation method, AutoCAD will use the line as the basis for the start point and direction. This is because the line was the last of the "line-or-arc" objects drawn.

REVERSE AND FORWARD

The UNDO command undoes the effects of the previous command or group of commands, depending on the option employed. The REDO command reverses the effects of the previous UNDO commands.

Undo

Figure 3–35

Invoking the UNDO command from the Standard toolbar

To undo the most recent action, choose Undo from the Standard toolbar as shown in Figure 3–35.

To undo a specific number of actions, click the Undo list arrow on the Standard toolbar, as shown in Figure 3–36. A list of actions that you can undo, starting with the most recent action, is displayed. Drag to select the actions to undo.

Figure 3–36

Displaying the UNDO list on the Standard toolbar

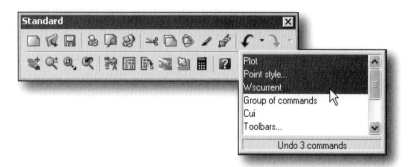

Redo

Figure 3–37

Invoking the REDO command from the Standard toolbar

To redo an action, choose Redo from the Standard toolbar as shown in Figure 3–37. Only the action immediately preceding an UNDO command can be reversed with REDO.

To redo a specific number of actions, click the Redo list arrow on the Standard toolbar as shown in Figure 3–38. A list of undo actions that you can redo, starting with the most recent action, is displayed. Drag to select the actions to redo.

Figure 3–38

Displaying the REDO list on the Standard toolbar

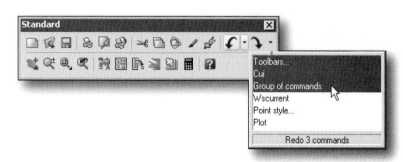

note REDO can reverse the operation only when it is invoked immediately after an undo or multiple sequential undo commands are performed.

drawing exercises

Open the Exercise Manual PDF file for Chapter 3 on the accompanying CD for discipline-specific exercises.

review questions

1. To draw multiple connected line segments, you must invoke the LINE command multiple times.
   a. *True*
   b. *False*

2. The "C" option used in the LINE command at the "From Point:" prompt:
   a. *Continues the line from the last line or arc that was drawn.*
   b. *Closes the previous set of line segments.*
   c. *Displays an error message.*

3. What happens when you select the "close" option while in the LINE command?
   a. *Terminates the line command*
   b. *Closes the polygon*
   c. *Closes the polygon and terminates the line command*
   d. *None of the above*

4. While creating a series of line segments, you can enter _____ to delete the last segment without exiting the LINE command.
   a. *ERASE*   c. *REDO*
   b. *U*   d. *OOPS*

5. A rectangle generated by the RECTANGLE command will always have horizontal and vertical sides.
   a. *True*
   b. *False*

6. The RECTANGLE command requests what information when drawing a rectangle?
   a. *An initial corner, the width, and the height*
   b. *The coordinates of the four corners of the rectangle*
   c. *The coordinates of diagonally opposite corners of the rectangle*
   d. *The coordinates of three adjacent corners of the rectangle*

7. The number of different methods by which a circle can be drawn is:
   a. *1.*
   b. *3.*
   c. *4.*
   d. *7.*
   e. *None of the above.*

8. When drawing a circle with the two-point potion, the distance between the two points is equal to:

   a. *The circumference.*
   b. *The perimeter.*
   c. *The shortest chord.*
   d. *The radius.*
   e. *The diameter.*

9. A circle may be created by any of the following options, except:

   a. *2P.*
   b. *3P.*
   c. *4P.*
   d. *Cen,Rad.*
   e. *TTR.*

10. The 3 point circle option draws a circle based on two points on the circumference and the center point.

    a. *True*
    b. *False*

11. Regarding the ARC options, what does "S,C,E" mean?

    a. *Start, Center, End*
    b. *Second, Continue, Extents*
    c. *Second, Center, End*
    d. *Start, Continue, End*

12. The ARC command has how many options?

    a. *7*   **c.** *11*
    b. *9*   **d.** *5*

13. To reverse the effect of the last 11 commands, you could:

    a. *Use the UNDO command.*
    b. *Use the U command multiple times.*
    c. *Either A or B.*
    d. *It is not possible, AutoCAD only retains the last 10 commands.*

exploration and discovery

introduction

This chapter introduces some of the basic commands and concepts in AutoCAD that can be used to maximize precision in drawing, get around in the world of object creation, change the drawing environment parameters, and gather information about objects in the drawing.

After completing this chapter, you will be able to do the following:

- Use commands that enhance the way you locate points and position the cursor

- Control the view of the drawing area that is displayed on the screen

- Control the number and arrangement of views of the drawing area

- Update the visibility of the objects in the viewing area

- Change the values of variables that affect almost every aspect of the way AutoCAD operates, including how you interact with AutoCAD, how text and number entry is accepted and displayed, the style of dimensions and how their elements appear, and countless other settings

- Gather data about objects such as distances between specified points, angles, layers, location (coordinates), and areas of closed shapes

FROM HERE TO THERE AND EXACTLY WHERE

This section covers the Object Snap, Direct Distance, and Tracking features of AutoCAD, which are essential tools in creating and modifying objects easily and accurately.

Object Snap

Figure 4-1

Toggling Object Snap from the Status bar

The Object Snap (Osnap, for short) feature lets you specify points based on existing objects in the drawing. For example, if you need to draw a line to an endpoint of an existing line or arc, you can apply the Object Snap mode called ENDpoint. In response to the "Specify next point or ⬇:" prompt, place the cursor so that it touches the line or arc nearer the desired endpoint. AutoCAD will lock onto the endpoint of the existing line or arc when you press the pick button of your pointing device. The endpoint becomes the ending point of the new line. This feature is similar to the basic SNAP command, which locks to a point on an invisible reference grid. Failure to use a snap or object snap method will result in a drawing with gaps or crosses at the endpoint. This is unsuitable for Numeric Control applications, and looks sloppy for others.

Invoking Object Snap

You can invoke an Object Snap mode whenever AutoCAD prompts for a point. Object Snap modes can be invoked while executing an AutoCAD command that prompts for a point, such as the LINE, CIRCLE, MOVE, and COPY commands.

Figure 4-2

Choosing Object Snap Settings from the Object Snap toolbar

Object Snap modes can be applied in either of two ways:

1. **Running Object Snap:** From the Object Snap toolbar, choose Object Snap Settings (see Figure 4–2), or on the Status bar at the bottom of the screen, right-click on OSNAP and choose Settings. The Drafting Settings dialog box will be displayed with the **Object Snap** tab selected, as shown in Figure 4–3. An Object Snap mode has been chosen if there is a check in its associated check box. Choose the desired Object Snap mode(s). This makes it possible to use the selected mode(s) any time you are prompted to specify a point. *Object Snaps* can be toggled ON and OFF in the same manner as the GRID and SNAP commands. The most common method is to use the button on the Status bar at the bottom of the screen (see Figure 4–1). When the OSNAP button appears to be pressed in, Object Snap is ON. When the OSNAP button appears to be out, Object Snap is OFF. To change its setting, simply place the cursor over the OSNAP button and press the pointing device pick button. You can also use the function key F3 to toggle the Osnap setting.

2. **Object Snap Overrides:** When prompted to specify a point, select one of the available Object Snap modes from the Object Snap toolbar before specifying the point or enter the first three letters of the name of the desired Object Snap mode and press enter. This is a *one-time-only* Osnap method. This will override the Running Object Snap mode for this one point selection only.

Figure 4-3

Drafting Settings dialog box displaying the Object Snap tab

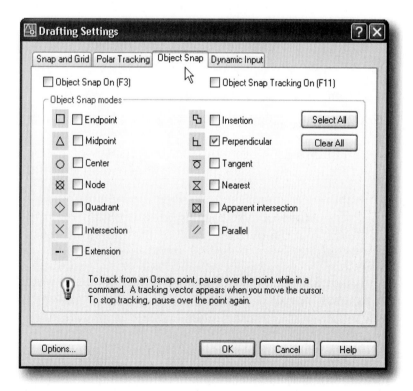

Object Snap Markers and Tooltips

Whenever one or more Object Snap modes are activated and you move the cursor target box over a snap point, AutoCAD displays a geometric shape (marker) and tooltip. By displaying a marker on the active Osnap points with a tooltip, you can see the point that will be selected and the Object Snap mode in effect. AutoCAD displays the marker depending on the Object Snap mode in effect. On the **Object Snap** tab of the Drafting

Settings dialog box, each marker is displayed next to the name of its associated Object Snap mode.

On the **Object Snap** tab of the Drafting Settings dialog box, choose **Options**. AutoCAD displays the Options dialog box with the **Drafting** tab selected. This is where the settings for displaying the markers and tooltips can be changed by checking or clearing the appropriate check boxes in the **AutoSnap Settings** section (see Figure 4–4).

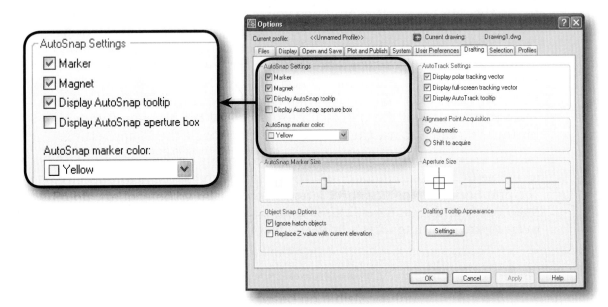

 Figure 4–4

Options dialog box with the Drafting tab selected

In the **AutoSnap Settings** section:

> The **Marker** box controls the display of the AutoSnap marker. As mentioned earlier, the marker is the geometric shape that is displayed when the cursor moves over a snap point.

> The **Magnet** box toggles the AutoSnap magnet ON or OFF. The magnet is an automatic movement of the crosshairs that locks the crosshairs onto the nearest snap point.

> The **Display AutoSnap tooltip** check box controls the display of the AutoSnap tooltip. The tooltip is a label that describes which Osnap mode is being applied.

> The **Display AutoSnap aperture box** setting controls the display of the AutoSnap aperture box. The aperture box is a box that appears at the crosshairs when you snap to an object.

> To change the color of the marker, select the desired color from the **AutoSnap marker color** box.

note   The default color of the marker is yellow, which shows up well against the default black background. If you change the background to white, you may wish to change the color of the marker to a contrasting color like dark red, blue, or green.

To change the size of the marker, press and hold the pick button of the pointing device while the cursor is over the slide bar in the **Autosnap Marker Size** section, and move it to the right to make the marker larger and to the left to make it smaller. The image tile shows the current size of the marker.

To change the size of the aperture, press and hold the pick button of the pointing device while the cursor is over the slide bar in the **Aperture Size** section, and move it to the right to make the aperture larger and to the left to make it smaller. The image tile shows the current size of the aperture.

After making the necessary changes to the AutoSnap settings, choose **OK** to close the Options dialog box and then choose **OK** again to close the Drafting Settings dialog box.

Object Snap Modes

 **Figure 4-5**

The Object Snap toolbar

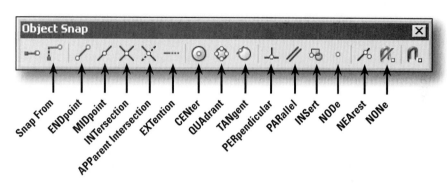

There are sixteen available Object Snap overrides: Endpoint, Midpoint, Intersection, Apparent Intersection, Extension, Center, Quadrant, Tangent, Perpendicular, Parallel, Insert, Node, Nearest, From, None, and MTP (snapping between specified points) (see Figure 4–5).

---

note It takes practice! And you must be alert whenever you need to specify a point near objects on the screen when one or more Running Osnap modes are in effect. For example, if you are trying to specify an ENDpoint on a line that is near the circumference of a circle while the CENter Osnap mode is active and the cursor pick box touches the circle, AutoCAD may select the center of the circle as the specified point. This kind of error is usually obvious, and you have the opportunity to correct it. But if AutoCAD places a point slightly off its intended location due to a Running Osnap mode taking over unintentionally, it could result in an undetected error in your drawing.

Endpoint, Intersection, Midpoint, and Perpendicular

The Endpoint mode allows you to snap to the closest endpoint of a line, arc, elliptical arc, multiline, polyline segment, spline, region or ray, or to the closest corner of a trace, solid, or 3Dface. As shown in Figure 4–6, LINE B is drawn from the indicated starting point to the end of the LINE A by using Endpoint mode.

The Intersection mode allows you to snap to the intersection of two objects that can include arcs, circles, ellipses, elliptical arcs, lines, multilines, polylines, rays, splines, or xlines. As shown in Figure 4–6, LINE C is drawn from the indicated starting point to the intersection of the CIRCLE C and LINE A by using the Intersection mode.

The Midpoint mode allows you to snap to the midpoint of a line, arc, elliptical arc, broken ellipse, multiline, polyline segment, xline, solid, or spline. As shown in Figure 4–6, LINE D is drawn from the indicated starting point to the midpoint of LINE A by using the Midpoint mode.

The Perpendicular mode allows you to snap to a point perpendicular to a line, arc, circle, elliptical arc, multiline, polyline, ray, solid, spline, or xline. Deferred Perpendicular snap mode is automatically turned on when more than one perpendicular snap is required by the object being drawn. As shown in Figure 4–6, LINE E is drawn from the indicated starting point to a point on and perpendicular to LINE A by using the Perpendicular mode. The location of the cursor may have to be adjusted to ensure that the LINE A is selected; otherwise, if CIRCLE C is selected, then the result will be LINE F drawn perpendicular to the circle.

 Figure 4–6

Lines drawn to designated points of another line using the Endpoint, Intersection, Midpoint, and Perpendicular Object Snap modes

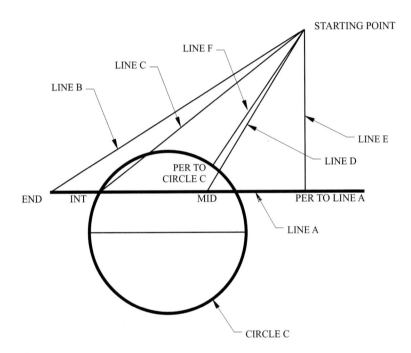

Quadrant, Tangent, and Center

The Quadrant mode allows you to snap to one of the quadrant points of a circle, arc, ellipse, or elliptical arc. The quadrant points are located at 0°, 90°, 180°, and 270° from the center of the circle or arc. The quadrant points are determined by the zero degree direction of the current coordinate system. As shown in Figure 4–7, LINE B is drawn from the indicated starting point to one of the quadrants of the CIRCLE C by using Quadrant mode.

The Tangent mode allows you to snap to the tangent of an arc, circle, ellipse, or elliptical arc. Deferred Tangent snap mode is automatically turned on when more than one tangent is required by the object being drawn. As shown in Figure 4–7, LINE C is drawn from the indicated starting point to tangent of CIRCLE C by using Tangent mode.

The Center mode allows you to snap to the center of an arc, circle, ellipse, or elliptical arc. As shown in Figure 4–7, LINE D is drawn from the indicated starting point to the center of the CIRCLE C by using Center mode.

Figure 4–7

Lines drawn to designated points of a circle using the Quadrant, Tangent, and Center Object Snap modes.

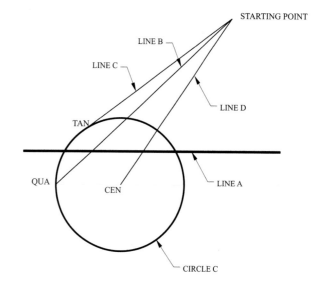

note The Quadrant point selected is one of four possible points on the circle that include points at 0°, 90°, 180°, and 270°.

Apparent Intersection, Extension, and Parallel

The Apparent Intersection mode allows you to snap to the apparent intersection of two objects that can include an arc, circle, ellipse, elliptical arc, line, multiline, polyline, ray, spline, or xline. These objects may or may not actually intersect, but would intersect if either or both objects were extended. In the following example, a rectangle box is drawn from the apparent intersection of lines on the two rectangles as shown in Figure 4–8.

rectangle

Specify first corner point or ⊡: *(invoke the APPint Object Snap, and select line 1 and line 4, as shown in Figure 4–8)*

Specify other corner point or ⊡: *(invoke the Appint Object Snap mode, and select line 2 and line 3, as shown in Figure 4–8).*

 Figure 4–8

Identifying the lines to draw a rectangle

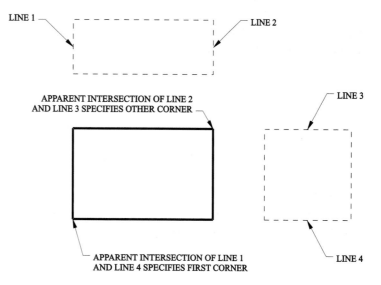

The Extension mode causes a temporary extension line to be displayed when you pass the cursor over the endpoint of objects, so that you can draw objects to and from points on the extension line. As shown in Figure 4–9, LINE A is drawn using extensions of LINE B and ARC C.

▶ **Figure 4–9**

LINE A is drawn using extensions of LINE B and ARC C

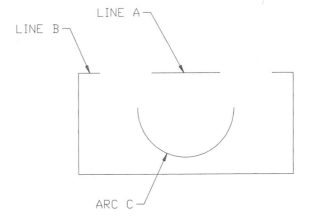

The Parallel mode allows you to draw a line that is parallel to another object. Once the first point of the line has been specified (and the Parallel Object Snap mode is selected), move the cursor over the object to which you wish to make the new line parallel. Then move the cursor near a line from the first point that is parallel to the object selected, and a construction line will appear. While the construction line is visible, specify a point, and the new line will be parallel to the selected object.

Node, Insert, Nearest, From, and None

The Node mode allows you to snap to a point object.

The Insert mode allows you to snap to the insertion point of a block, text string, attribute, or shape.

The Nearest mode lets you select any object (except text and shape) in response to a prompt for a point, and AutoCAD snaps to the point on that object nearest the cursor.

The From mode locates a point offset from a reference point within a command. At an AutoCAD prompt for locating a point, select From Object snap mode, and then specify a temporary reference or base point from which you can specify an offset to locate the next point. Enter the offset location from this base point as a relative coordinate, or use direct distance entry.

The None mode temporarily overrides any Running Object Snaps that may be in effect.

Snapping Between Specified Points

AutoCAD 2006 includes a new Object Snap mode for snapping to a point that is midway between two specified points. You can enter mtp or m2p when prompted to specify a point. For example, if you want to start a line at the midpoint between the centers of two circles, you can invoke the Midway Between Two Points Object Snap mode as follows:

line (ENTER)

Specify first point: mtp (ENTER)

First point of mid: cen

First point of mid: (specify the center of the first circle)

Second point of mid: cen

Second point of mid: (specify the center of the second circle)

Specify next point or ⬇ (specify the end point of the line)

Specify next point or ⬇ (ENTER)

---

note The Midway Between Two Points Object Snap mode is not included on the **Object Snap** tab of the Drafting Settings dialog box, nor on the Object Snap toolbar in the initial release of AutoCAD 2006.

Drawing Objects using Tracking

Tracking, or moving through nonselected point(s) to a selected point, could be called a command "enhancer." To invoke Tracking, whenever AutoCAD prompts you to specify a point, enter **tracking**, **track**, or **tk**. If the desired point can best be specified relative to some known point(s), you can "make tracks" to the desired point by invoking the Tracking option and then specifying one or more points relative to previous point(s) "on the way to" the actual point that the command is prompting for. These intermediate tracking points are not necessarily associated with the object being created or modified by the command. The primary significance of tracking points is that they are used to establish a path to the point you wish to specify as the response to the command prompt. In the following command sequence, a line is drawn from TK1/SP1 (tracking point 1 and starting point 1) to EP1 (end point 1) as shown in Figure 4–10 and line from SP2 (start point 2) to TK2/EP2 (track point 2 and ending point 2) by using the tracking feature:

line (ENTER)

Specify first point: **0,12'** (ENTER) *(specify point TK1/SP1)*

Specify next point or ⬇: **tk** (ENTER) *(invoke the Tracking feature)*

First tracking point: 0,12' (specify point SP1/TK1 again as the first tracking point and also make sure ORTHO is set to off)

Next point (Press enter to end tracking): **@13'8,0** (ENTER) *(locates the second tracking point, TK2, as shown in Figure 4–10)*

Next point (Press enter to end tracking): **@2'2<180** (ENTER) *(locates the third tracking point, TK3, as shown in Figure 4–10)*

Next point (Press enter to end tracking): **@1'6<180** (ENTER) *(locates the fourth tracking point, EP1, as shown in Figure 4–10)*

Next point (Press enter to end tracking): *(press ENTER to exit Tracking; by this you are designating the point to which you have "made tracks" as the response to the prompt that was in effect when you entered Tracking)*

line (ENTER) *(invoke the LINE command)*

Specify first point: **tk** (ENTER) *(invoke the Tracking option)*

First tracking point: 0,12' (ENTER) (specify point TK1/SP1 as the first tracking point and also make sure ORTHO is set to off)

Next point (Press enter to end tracking): **@13'8,0** (ENTER) *(locates the second tracking point, TK2, as shown in Figure 4–10)*

Next point (Press enter to end tracking): **@2'2<180** (ENTER) *(locates the third tracking point, TK3, as shown in Figure 4–10)*

Next point (Press enter to end tracking): **@1'6<0** (ENTER) *(locates the fourth track-ing point, SP2, as shown in Figure 4–10)*

Next point (Press enter to end tracking): *(press* ENTER *to exit Tracking; AutoCAD establishes point SP2)*

Specify next point or ⬇: **tk** (ENTER) *(invoke the Tracking option again)*

First tracking point: 0,12' (specify point TK1/SP1 as the first tracking point and also make sure ORTHO is set to off)

Next point (Press enter to end tracking): **@13'8,0** (ENTER) *(locates the second track-ing point, TK2/EP2, as shown in Figure 4–10)*

Next point (Press enter to end tracking): *(press* ENTER *to exit Tracking, AutoCAD establishes point EP2)*

Specify next point or ⬇: *(press* ENTER *to terminate the* LINE *command)*

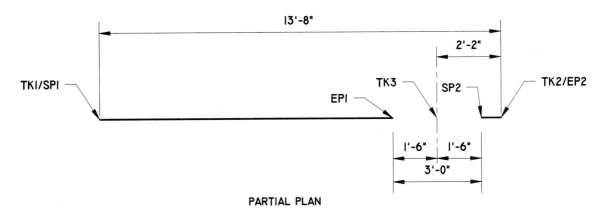

PARTIAL PLAN

▲ Figure 4–10

Example of a partial plan to demonstrate the Tracking option

This example shows an application of the Tracking option in which the points were established with reference to some known points.

---

note If you knew the coordinates of one of the intermediate tracking points, then it probably should be the initial tracking point. The idea behind tracking is to establish a point by means of a path from and through other points. Thus, the shortest path is the best. If the coordinates of TK2 were known, or if you could specify it by some other method, it could become the initial tracking point. Keep this in mind as you learn to use Object Snap. You don't necessarily need to know the coordinates if you can use Object Snap to select a point from which a tracking path could be specified.

Drawing Objects using Direct Distance

The Direct Distance entry feature allows you to specify a point relative to the last point you entered. To invoke Direct Distance, whenever AutoCAD prompts you to specify a point, first move the cursor to specify the direction, and then enter a numeric distance. This feature is usually used with Ortho, Snap, or Polar mode turned on. In the following example, the second point for the line will be located 8 units toward the direction of the cursor. The direct distance that you enter is measured along the path from the last point to the current location of the cursor:

line (ENTER)

Specify first point: *(specify a point)*

Specify next point or ⊡: *(move the cursor in the desired direction) 8* (ENTER)

Polar Tracking

 Figure 4–11

The Drafting Settings dialog box with the Polar Tracking tab displayed

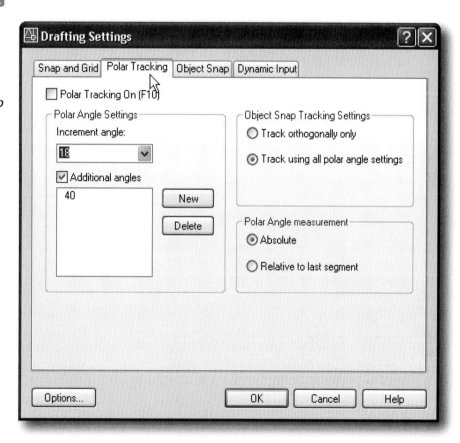

The Polar Tracking feature lets you draw lines and specify point displacements in directions that are multiples of a specified increment angle. You can specify whether the increment angles are measured from the current coordinate system or from the angle of a previous object. You can also add up to ten additional angles to which a line or displacement can be diverted from a base direction. The Object Snap Tracking feature lets you track your cursor from a strategic (Object Snap) point on an object in a specified direction, either orthogonal or preset polar angles.

Setting Polar Tracking Increment and Additional Angles

To set the Polar Tracking increment angle and add additional angles, right-click on the POLAR button located on the Status bar and choose Settings. AutoCAD displays the Drafting Settings dialog box with the **Polar Tracking** tab selected, as shown in Figure 4–11.

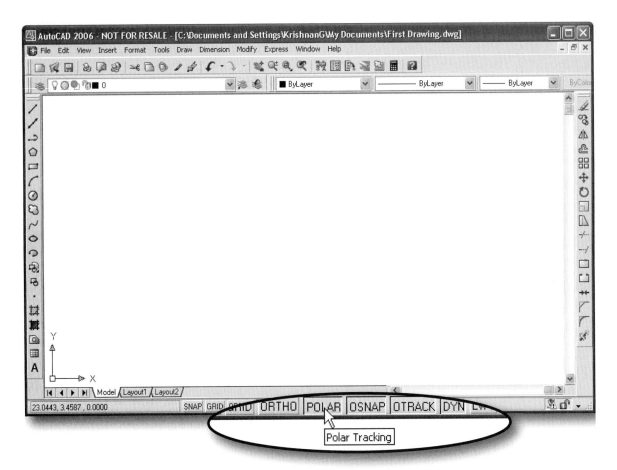

Figure 4–12

Toggling Polar on the Status bar

In the **Polar Angle Settings** section, enter the desired increment angle in the **Increment angle** box. This is the base increment angle whose multiples are used by Polar Tracking. Up to ten additional angles can be added by choosing **New** and then entering the desired additional angle(s). To make the additional angle(s) available when the Polar Tracking mode is set to ON, set the **Additional angles** check box to ON. You can toggle Polar Tracking ON and OFF by choosing POLAR on the Status bar (see Figure 4–12).

note The Polar Tracking and the Ortho modes cannot both be set to ON simultaneously. They can both be set to OFF, or either one can be set to ON.

The **Polar Angle measurement** section sets how polar tracking alignment angles are measured. The **Absolute** selection bases polar tracking angles on the current user coordinate system (UCS) and the **Relative to last segment** selection bases polar tracking angles on the last segment drawn.

With the 18° increment angle and the Polar Angle measurement set to Absolute (see Figure 4–11), you can draw lines and specify displacements at 18°, 36°, 54°, 72°, 90°, 108° ... 324°, 342°, 360° around the compass by just snapping to the construction lines when they appear. The accompanying AutoTrack tooltip will appear when you have snapped to a multiple of the increment angle. The tooltip displays the cursor's distance and direction from the first specified point. You can, with the cursor snapped to one of the Polar Tracking angles, use the Direct Distance option and type a distance from the keyboard and press ENTER. AutoCAD will draw the line or apply the displacement in accordance with the distance entered and the direction set by the cursor.

For example, in Figure 4–13, the line from P1 to P2 was drawn 2 units long at 0°. This could have been done with Direct Distance and either ORTHO set to ON or with Polar Tracking. But with P2 to P3 being at 18° and from P3 to P4 being at 36°, it is easier to use Polar Tracking.

 Figure 4–13

Drawing line segments at multiples of the 18° increment angle

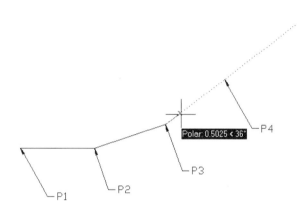

After completing the three segments drawn at multiples of 18°, change the increment angle to 5° and choose **Relative to last segment** in the **Polar Angle measurement** section. Continuing with three more segments, each 1 unit long, the first increment angle will be measured from the last segment, which is at 36°. The tooltip displays "Relative Polar (distance) < 5°" but the resulting angle is 41°. Figure 4–14 shows the three additional line segments drawn using Polar Tracking with the increment angle set to 5° and the Polar Angle measurement set to **Relative to last segment**. The last two segments will actually be at 46° and 51°.

 Figure 4–14

Three line segments drawn using Polar Tracking with the increment angle set to 5° and the Polar Angle measurement set to Relative to last segment

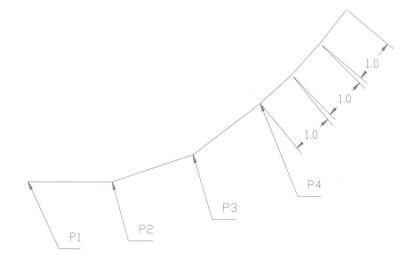

Object Snap Tracking

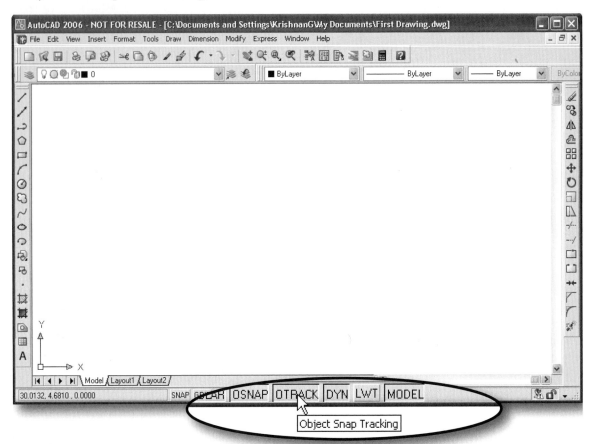

▲ **Figure 4–15**

Toggling the Object Tracking feature on the Status bar

Object Snap Tracking (OTRACK) helps you draw objects at specific angles or in specific relationships to other objects. When you set OTRACK to ON, temporary alignment paths help you create objects at precise positions and angles. Object snap tracking works in conjunction with object snaps and polar angle settings. You must set an object snap before you can track from an object's snap point. You can toggle OTRACK on and off with the OTRACK on the status bar (see Figure 4-15) or the function key F11.

OTRACK includes two tracking options: object snap tracking and polar tracking. Use object snap tracking to track along alignment paths that are based on object snap points. Acquired points display a small plus sign (+), and you can acquire up to seven tracking points at a time. After you acquire a point, horizontal, vertical, or polar alignment paths relative to the point are displayed as you move the cursor over their drawing paths. For example, you can select a point along a path based on an object endpoint or an intersection between objects.

The **Object Snap Tracking Settings** section (see Figure 4–12) lets you select options for Object Snap tracking. Choose **Track orthogonally only** to display orthogonal (horizontal/vertical) Object Snap tracking paths for acquired Object Snap points when Object Snap Tracking is set to ON. Choose **Track using all polar angle settings** to permit the cursor to also track along any polar angle tracking path for acquired Osnap points when Object Snap tracking is ON while specifying points.

Dynamic Input, On-Screen Prompts, and Geometric Values Display

Dynamic Input provides a command interface near the cursor to help you keep your focus in the drafting area. When Dynamic Input is set to ON, tooltips display information near the cursor and are dynamically updated as the cursor moves. When a command is active, a tooltip provides the field for user entry. The actions required to complete a command are the same as when you work from the command line. The difference is that your attention can stay near the cursor.

Dynamic Input can be toggled ON/OFF by choosing **DYN** on the status bar. Dynamic Input has three components: pointer input, dimensional input, and dynamic prompts. To change the format and visibility of Pointer Input, Dimension Input, and Tooltip appearance (Drafting Prompts), right-click **DYN** on the status bar and choose Settings from the shortcut menu. AutoCAD will display the Drafting Settings dialog box with the **Dynamic Input** tab selected (see Figure 4–16).

Figure 4–16

*Drafting Settings dialog box displaying the **Dynamic Input** tab*

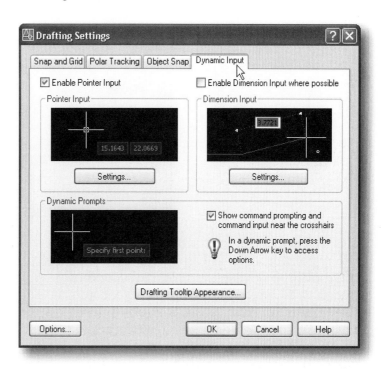

Choosing **Enable Pointer Input** causes the cursor to be displayed with coordinate values in a tooltip. When a command prompts for a point, you can enter coordinate values in the tooltip instead of on the command line. The Preview Area shows an example of pointer input. Choosing **Enable Dimensional Input where possible** turns on dimensional input. When **Enable Pointer Input** and **Enable Dimensional Input where possible** are both checked, dimensional input supersedes pointer input when it is available.

The **Dynamic Prompts** section allows you to set how AutoCAD displays prompts in a tooltip near the cursor when necessary in order to complete the command. You can enter values in the tooltip instead of on the command line. Selecting **Show command prompting and command input near the crosshairs** causes prompts to be displayed in a tooltip near the cursor. The Preview Area shows an example of dynamic prompts.

Pointer Input

The **Pointer Input** section controls how AutoCAD displays coordinate values in a tooltip near the cursor.

Choose **Settings** in the **Pointer Input** section to display the Point Input Settings dialog box (see Figure 4–17), which allows you to control the coordinate format in the tooltips that are displayed and the visibility of the pointer when pointer input is set to ON.

 Figure 4-17

Pointer Input Settings dialog box

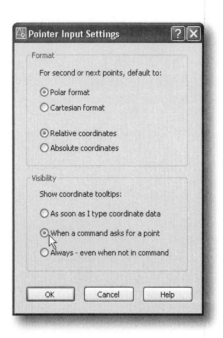

The **Format** section controls coordinate format in the tooltips that are displayed when pointer input is turned on. Selecting **Polar format** (default selection) causes the tooltip to be displayed for the second or next point in polar coordinate format. To switch to a Cartesian format, specify x coordinate and then enter a comma (,). Selecting **Cartesian format** causes the tooltip to be displayed for the second or next point in Cartesian coordinate format. To switch to a polar format, specify distance and enter an angle symbol (<).

Selecting **Relative coordinates** (default selection) causes the tooltip to be displayed for the second or next point in relative coordinate format. To switch to a absolute format, enter a pound sign (#). Selecting **Absolute coordinates** causes the tooltip to be displayed for the second or next point in absolute coordinate format. To switch to a relative format, enter an at sign (@).

---

note   You can use the direct distance method when pointer input is set to absolute coordinates.

The **Visibility** section controls when pointer input is displayed. Selecting **As soon as I type coordinate data** causes tooltips to be displayed only when you start to enter coordinate data (when pointer input is turned on). Selecting **When a command asks for a point** (default selection) causes tooltips to be displayed whenever a command prompts for a point (when pointer input is turned on). Selecting **Always–even when**

not in a command causes tooltips to always be displayed (when pointer input is turned on). Choose **OK** to exit the Pointer Input settings dialog.

 The prompts and responses shown in examples throughout this book and drawing exercises assume that you are using the new dynamic input introduced in AutoCAD 2006. Where they call for you to enter coordinates for second or next points without the @ prefix, they are based on the input being absolute coordinates. Because the default Pointer Input setting is set to Relative coordinates, this causes the @ prefix to be added (without your seeing it on the screen). You should change the setting of the Pointer Input to Absolute coordinates if you wish to try out the examples on your system. Or, if you do not wish to change the setting, you can use an override by prefixing the coordinates with the # symbol. This forces the input to be absolute for second or next points and only works in dynamic input, not at the Command line in the Command window.

Dimension Input

The **Dimension Input** section allows you to set how AutoCAD displays a dimension with tooltips for distance value and angle value when a command prompts for a second point or a distance. The values in the dimension tooltips change as you move the cursor. You can enter values in the tooltip instead of on the Command line.

Choose **Settings** from the **Dimension Input** section to display the Dimension Input Settings dialog box (see Figure 4–18), which controls which tooltips are displayed during grip stretching when dimensional input is turned on.

Figure 4–18

Dimension Input Settings dialog box

Choose one of the three options in the **Visibility** section, to control which tooltips are displayed during grip stretching, when dimensional input is turned on.

Selecting **Show only 1 dimension input field at a time** causes only the Length Change dimensional input tooltip to be displayed, when you are using grip editing to stretch an object.

Selecting **Show 2 dimension input fields at a time** causes the Length Change and Resulting Dimension input tooltips to be displayed when you are using grip editing to stretch an object.

Selecting **Show the following dimension input fields simultaneously** causes the selected dimensional input tooltips to be displayed when you are using grip editing to stretch an object.

The input fields that can be displayed include: Resulting Dimension, Length Change, Absolute Angle, Angle Change, and Arc Radius.

Selecting **Resulting Dimension** causes a length dimension to be displayed and updated as you move the grip.

Selecting **Length Change** causes the change in length to be displayed as you move the grip.

Selecting **Absolute Angle** causes an angle dimension to be displayed and updated as you move the grip.

Selecting **Angle Change** causes the change in the angle to be displayed as you move the grip.

Selecting **Arc Radius** causes the radius of an arc to be displayed and updated as you move the grip.

Drafting Tooltip Appearance

Choose **Drafting Tooltip Appearance** to open the Tooltip Appearance dialog box (see Figure 4–19), which can be used to control the appearance of tooltips.

 Figure 4–19

Tooltip Appearance dialog box

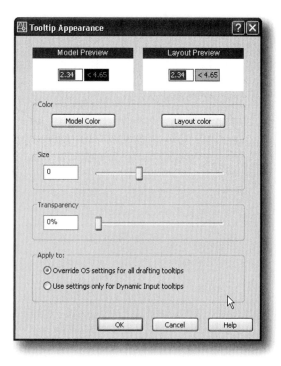

The **Model Preview** and **Layout Preview** show examples of the current tooltip appearance settings in Model space and Layouts, respectively.

Choosing **Model Color** and **Layout Color** causes the Select Color dialog box to be displayed, where you can specify a color for tooltips in model space and in layouts, respectively.

The **Size** slider bar and text box allow you to specify the size for tooltips. The default size is 0. The slider bar is used to make tooltips larger or smaller.

The **Transparency** slider bar and text box allow you to control the transparency of tooltips. The lower the setting, the less transparent the tooltip. A value of 0 sets the tooltip to opaque. The slider bar is used to make tooltips more or less transparent.

The **Apply to** section allows you to specify whether the settings apply to all drafting tooltips or only to Dynamic Input tooltips.

After making changes, choose **OK** to close the Tooltip Appearance dialog box. Choose **OK**, again to close the Drafting Settings dialog box.

CONTROLLING WHAT YOU SEE

There are many ways to view a drawing in AutoCAD. These viewing options vary from on-screen viewing to hard-copy plots. (The hard-copy options are discussed in Chapter 16.) Using the display commands, you can select the portion of the drawing to be displayed, establish 3D perspective views, and much more. By allowing you to view the drawing in different ways, AutoCAD provides a way to draw faster. The commands that are explained in this section are utility commands. They make your job easier and help you to draw more accurately.

ZOOM In and Out

The ZOOM command is like a zoom lens on a camera. You can increase or decrease the viewing area, although the actual size of objects remains constant. As you increase the visible size of objects, you view a smaller area of the drawing in greater detail, as though you were closer. As you decrease the visible size of objects, you view a larger area, as though you were farther away. This ability provides a close-up view for better accuracy and detail or a distant view to get the whole picture.

AutoCAD provides smooth transition (rather than instantaneous) zooms and pans so that you can watch as the view changes from the current view to the selected. This allows you to better keep track of where you are going in relation to where you were. You can change to instantaneous transitions by changing the setting of the VTENABLE system variable.

Zoom Realtime

 Figure 4-20

Invoking Zoom Realtime from the Standard toolbar

The Zoom Realtime selection lets you zoom interactively to a specified extent. After you choose Zoom Realtime from the Standard toolbar, the cursor changes to a magnifying glass with a "±" symbol. To zoom in closer, hold the pick button and move the cursor vertically toward the top of the window. To zoom out farther, hold the pick button and move the cursor vertically toward the bottom of the window. To discontinue the zooming, release the pick button. To exit the Zoom Realtime option, press ENTER or ESC, or choose Enter from the Shortcut menu.

The current drawing window is used to determine the zooming factor. If the cursor is moved by holding the pick button from the bottom of the window to the top of the window vertically, the zoom-in factor would be 50%. Conversely, when holding the pick button from the top of the window and moving vertically to the bottom of the window, the zoom-out factor would be 200%.

When you reach the zoom-out limit, the "-" symbol on the cursor disappears while attempting to zoom out, indicating that you can no longer zoom out. Similarly, when you reach the zoom-in limit, the "+" symbol on the cursor disappears while attempting to zoom in, indicating that you can no longer zoom in.

In addition, you can perform other operations related to zoom and pan by choosing them from the Shortcut menu.

Zoom Window

 Figure 4–21

Invoking Zoom Window from the Zoom toolbar

The Zoom Window selection (see Figure 4–21) displays an area specified by two opposite corners of a rectangular window. The rectangular window can be specified by entering coordinates or by your pointing device. The following command sequence shows an example of using the Zoom Window selection (see Figure 4–22).

Specify first corner: *(specify a point to define the first corner of the window as shown in Figure 4–22)*

Specify opposite corner: *(specify a point to define the diagonally opposite corner of the window as shown in Figure 4–22)*

AutoCAD displays the selected area in the display window as shown in Figure 4–22.

 Figure 4–22

Display of the drawing before and after Zoom Window selection

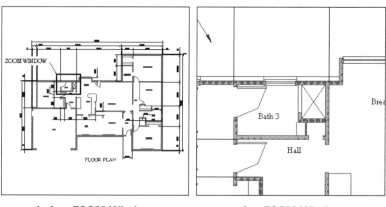

before ZOOM Window   after ZOOM Window

Zoom All

 Figure 4-23

Invoking Zoom All from the Zoom toolbar

The Zoom All selection (see Figure 4–23) zooms to display the entire drawing in the current display window. AutoCAD zooms to the limits or current extents, whichever is greater. The display shows (see Figure 4–24) all objects if the drawing extends outside the limits.

Figure 4-24

Display before and after Zoom All selection

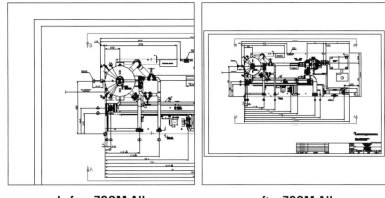

before ZOOM All　　　　　**after ZOOM All**

Zoom Extents and Zoom Object

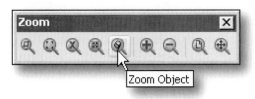　　　

Figure 4-25

Invoking Zoom Objects and Zoom Extents from the Zoom toolbar

The Zoom Extents selection zooms to display the drawing extents and results in the largest possible display of all the objects. The Zoom Object selection zooms the selected object to the largest possible display.

Figure 4–26 illustrates the difference between the Zoom All selection and Zoom Extents selection.

Figure 4-26

Display of the drawing after Zoom All selection and Zoom Extents selection

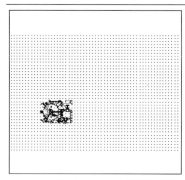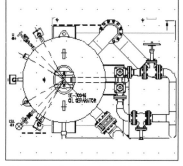

objects displayed with ZOOM All   **objects displayed with ZOOM Extents**

Zoom Previous

The Zoom Previous selection zooms to display the previous view. You can restore up to ten previous views (see Figure 4-27).

 Figure 4-27

Display of the drawing before and after Zoom Previous selection

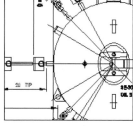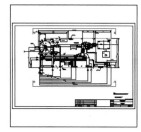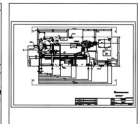

orginal view   **current view**   **after ZOOM Previous**

Zoom Scale

 Figure 4-28

Invoking Zoom Scale from the Zoom toolbar

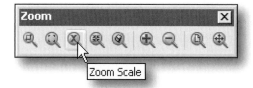

The Zoom Scale selection (see Figure 4-28) displays the drawing at a specified scale factor. The scale factor, when entered as a number (it must be a numerical value and must not be expressed in units of measure), is applied to the area covered by the drawing limits. For example, if you enter a scale value of 3, each object appears three times as large as in the Zoom All view. A scale factor of 1 displays the entire drawing (the full view), which is defined by the established limits. If you enter a value less than 1, AutoCAD decreases the magnification of the full view. For example, if you enter a scale of 0.5, each object appears half its size in the full view, while the viewing area is twice the size in horizontal and vertical dimensions. When you use this option, the object in the center of the screen remains centered. Figure 4-29 shows the difference between a full view and a 0.5 zoom. You can specify a value followed by X to specify the scale relative to the current view. For example, entering **.5x** causes each object to be displayed at half its current size on the screen.

Figure 4-29

Drawing display between a full view and a 0.5 zoom display

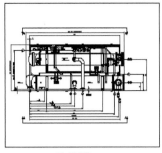 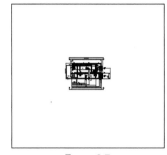

Full View　　　　　**Zoom 0.5**

Zoom Center

 Figure 4-30

Invoking Zoom Center from the Zoom toolbar

The Zoom Center selection (see Figure 4–30) lets you select a new view by specifying its center point and the magnification value or height of the view in current units. A smaller value for the height increases the magnification; a larger value decreases the magnification. The following command sequence shows an example of using the Zoom Center selection (see Figure 4–31).

Specify center point: **8,6** (ENTER)

Enter magnification or height <current height>: **4** (ENTER)

 Figure 4-31

Display before and after Zoom Center selection

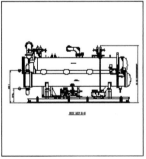 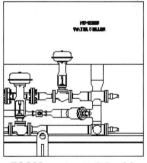

before ZOOM Center　　　　**ZOOM center to 8,6 with hieght of 4"**

In addition to providing coordinates for center point, you can also specify the center point by placing a point on the view window. The height can also be specified in terms of the current view height by specifying the magnification value followed by an X. A response of 3X will make the new view height three times as large as the current height.

Zoom Dynamic

 Figure 4–32

Invoking Zoom Dynamic from the Zoom toolbar

The AutoCAD Zoom Dynamic selection (see Figure 4–32) provides a quick and easy method to move to another view of the drawing. With Zoom Dynamic, you can see the entire drawing and then simply select the location and size of the next view by means of cursor manipulations. Using Zoom Dynamic is one means by which you can visually select a new display area that is not entirely within the current display.

PANNING Around

AutoCAD lets you view a different portion of the drawing in the current view without changing the magnification. The Panning feature allows you to move your viewing area to see details that are currently off screen. You can slide the drawing left, right, up, and down without moving the window.

Pan Realtime

 Figure 4–33

Invoking the Pan Realtime command from the Standard toolbar

The Pan Realtime selection (see Figure 4–33) lets you pan interactively to the logical extent (edge of the drawing space). After you choose Pan Realtime from the Standard toolbar, the cursor changes to a hand cursor. To pan, hold the pick button on the pointing device to lock the cursor to its current location relative to the viewport coordinate system, and move the cursor in any direction. Graphics within the window are moved in the same direction as the cursor. To discontinue the panning, release the pick button.

When you reach a logical extent (edge of the drawing space), a line-bar is displayed on the hand cursor on the side. The line-bar is displayed at the top, bottom, left, or right side of the drawing, depending upon whether the logical extent is at the top, bottom, or side of the drawing.

To exit Pan Realtime, press ESC or ENTER. You can also exit by choosing Exit from the Shortcut menu that is displayed when you right-click with your pointing device. In addition, you can perform other operations related to ZOOM and PAN by choosing the appropriate commands from the Shortcut menu.

Controlling Display with IntelliMouse

AutoCAD also allows you to control the display of the drawing with the small wheel provided with the IntelliMouse (two-button mouse). The tool is readily available and can be used any time while you are in the drawing and editing mode. By rotating the wheel forward you can zoom in, and rotating backward you can zoom out. When you double-click the wheel button, AutoCAD displays the drawing to the extent of the view window. To pan the display of the drawing, press the wheel button and drag the mouse. By default, each increment in the wheel rotation changes the zoom level by 10 percent. The ZOOMFACTOR system variable controls the incremental change, whether forward or backward. The higher the setting, the smaller the change.

Setting Multiple Viewports

The ability to divide the display into two or more separate viewports is one of the most useful features of AutoCAD. Multiple viewports divide your drawing screen into rectangles, permitting several different areas for drawing instead of just one. It is like having multiple zoom lens cameras, with each camera being used to look at a different portion of the drawing. You retain your menus and Command:" prompt area.

Each viewport maintains a display of the current drawing independent of the display shown by other viewports. You can simultaneously display a viewport showing the entire drawing, and another viewport showing a close-up of part of the drawing in greater detail. You can begin drawing (or modifying) an object in one viewport and complete it in another viewport. For example, three viewports could be used in a 2D drawing, two of them to zoom in on two separate parts of the drawing, showing two widely separated features in great detail on the screen simultaneously, and the third to show the entire drawing, as shown in Figure 4–34. In a 3D drawing, four viewports could be used to display simultaneously four views of a wireframe model: top, front, right side, and isometric, as shown in Figure 4–35.

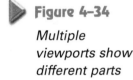

Figure 4–34

Multiple viewports show different parts of the same 2D drawing

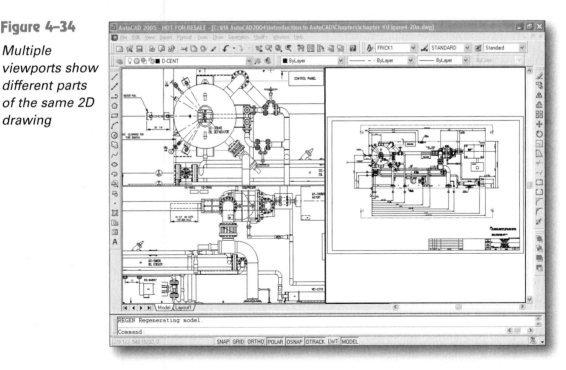

Figure 4–35

*Using viewports
to show
four views
simultaneously
for a 3D
wireframe model*

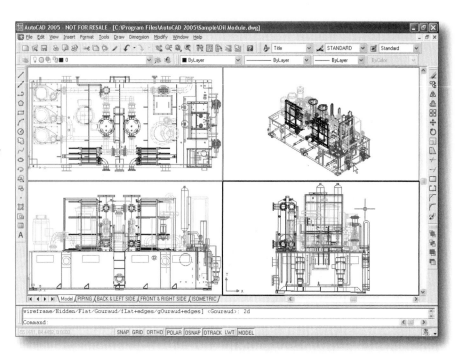

AutoCAD allows you to divide the graphics area of your display screen into multiple, non-overlapping (tiled) viewports, as in Figures 4–34 and 4–35, only when you are in model space (system variable TILEMODE set to 1). The maximum number of active tiled viewports that you can have is set by the system variable MAXACTVP, and the default is 64. In addition, you can create multiple overlapping (floating) viewports when the system variable TILEMODE is set to 0. For a detailed explanation on how to create floating viewports, refer to Chapter 15.

You can work in only one viewport at a time. It is considered the current viewport. A viewport is set to be current one by moving the cursor into it with your pointing device and then pressing the pick button. You can even switch viewports in mid command (except during some of the display commands). For example, to draw a line using two viewports, you must start the line in the current viewport, make another viewport current by clicking in it, and then specify the endpoint of the line in the second viewport. When a viewport is current, its border will be thicker than the other viewport borders. The cursor is active for specifying points or selecting objects only in the current viewport; when you move your pointing device outside the current viewport, the cursor appears as an arrow pointer.

Display commands like ZOOM and PAN and drawing tools like GRID, SNAP, or ORTHO, and UCS icon modes are set independently in each viewport. The most important thing to remember is that the images shown in multiple viewports are all of the same drawing. An object added to or modified in one viewport will affect its image in the other viewports. You are not making copies of your drawing, just viewing its image in different viewports.

When you are working in tiled viewports, visibility of the layers is controlled globally in all the viewports. If you turn off a layer, AutoCAD turns it off in all viewports. For a detailed explanation on using layers, refer to Chapter 6.

Creating Tiled Viewports

AutoCAD allows you to display tiled viewports in various configurations. Display of the viewports depends on the number and size of the views you need to see. By default, whenever you start a new drawing, AutoCAD displays a single viewport that fills the entire drawing area.

 Figure 4–36

Opening the Viewports dialog box from the Layouts toolbar

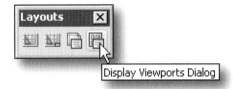

From the Layouts toolbar, choose Display Viewports Dialog (see Figure 4–36). AutoCAD displays a Viewports dialog box with the New Viewport tab selected, similar to Figure 4–37.

 Figure 4–37

Viewports dialog box

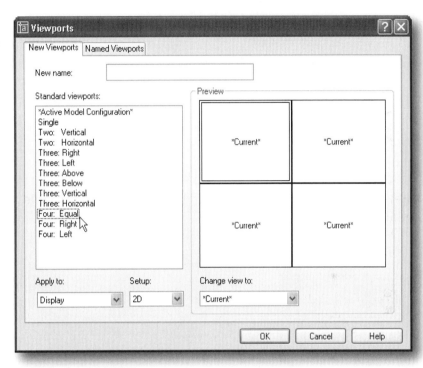

Choose the name of the configuration you want to use from the **Standard viewports** list. AutoCAD displays the corresponding configuration in the preview window. If necessary, you can save the selected configuration by providing a name in the **New name** box.

Choose whether to apply the saved configuration to the entire display or to the current viewport of the selected configuration from the **Apply to** menu. Choose from the **Setup** menu either a 2D or a 3D setup. When you choose 2D, the new viewport configuration is initially created with the current view in all of the viewports. When you choose 3D, a set of standard orthogonal 3D views is applied to the viewports in the configuration. To replace the view in the selected viewport, choose available named views from the **Change view to** list. You can choose a named view or, if you have chosen 3D setup, you can select from the list of standard views.

Choose **OK** to create viewports and close the Viewports dialog box.

The **Named Viewports** tab lists all saved viewport configurations. At any time you can restore one of the saved viewport configurations.

REFRESHING THE SCREEN VIEWING AREA

 Figure 4–38

Invoking the REDRAW command from the View menu

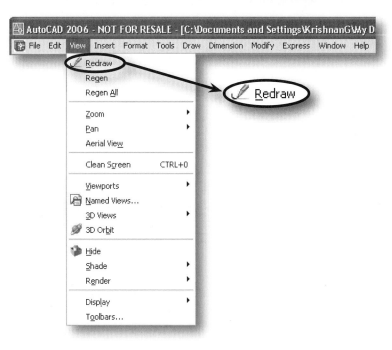

The REDRAW command (see Figure 4–38) is used to refresh the on-screen image. You can use this command whenever you see an incomplete image of your drawing. If you draw two lines in the same place and erase one of the lines, it appears as if both the lines are erased. When the REDRAW command is invoked, the second line will reappear. Also, you can use the REDRAW command to remove the blip marks on the screen. A redraw is considered a screen refresh, as opposed to database regeneration.

 Figure 4–39

Invoking the REGEN command from the View menu

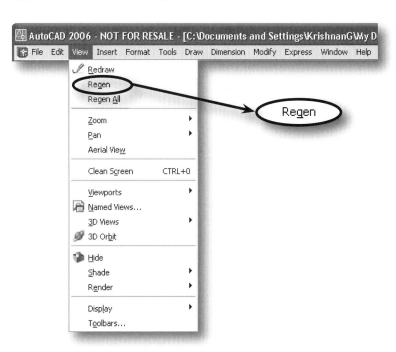

The REGEN command (see Figure 4–39) is used to regenerate the drawing's data on the screen. In general, you should use the REGEN command if the image presented by REDRAW does not correctly reflect your drawing. REGEN goes through the drawing's entire database and projects the most up-to-date information on the screen; this command will give you the most accurate image possible. Because of the manner in which it functions, a REGEN takes significantly longer than a REDRAW.

There are certain AutoCAD commands for which REGEN takes place automatically, unless the REGENAUTO system variable is set to OFF.

BEING NOSY (USING YOUR PROACTIVE CURIOSITY)

AutoCAD provides several commands for displaying useful information about the objects in the drawing. These commands do not create anything, nor do they modify or have any effect on the drawing or objects therein. The only effect on the AutoCAD editor is that on single-screen systems, the screen switches to the AutoCAD Text window (not to be confused with the TEXT command), and the information requested by the particular inquiry command is then displayed on the screen. The FLIP SCREEN feature returns you to the graphics screen so you can continue with your drawing. On most systems this is accomplished with the F2 function key. You can also change back and forth between graphic and text screens with the GRAPHSCR and TEXTSCR commands, respectively, entered at the Command: prompt. The inquiry commands include LIST, ID, and DIST.

Information about Individual Objects

  Figure 4–40

Invoking the LIST command from the Inquiry toolbar

The LIST command (see Figure 4–40) displays information about individual objects stored by AutoCAD in the drawing database. The information includes the following:

- The location, layer, object type, and space (model or paper) of the selected object as well as the color, lineweight, and linetype, if not set to BYLAYER or BYBLOCK
- The length of a line or an arc
- The distance in the main axes between the endpoints of a line, that is, the delta-X, delta-Y, and delta-Z
- The area and circumference of a circle or the area of a closed polyline
- The insertion point, height, angle of rotation, style, font, obliquing angle, width factor, and actual character string of a selected text object
- The object handle, reported in hexadecimal

AutoCAD prompts:

Select objects: *(select the objects and press* ENTER *to terminate object selection)*

AutoCAD lists the information about the selected objects.

Coordinates of a Point

 Figure 4–41

Invoking the Locate Point (ID) command from the Inquiry toolbar

The ID command (see Figure 4–41) is used to obtain the coordinates of a selected point. AutoCAD prompts:

Specify point: *(select a point)*

AutoCAD displays the information about the selected point.

If the BLIPMODE system variable is set to ON, a blip appears on the screen at the specified point, provided it is in the viewing area.

Distance between Two Points

 Figure 4–42

Invoking the DIST command from the Inquiry toolbar

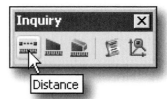

The DIST command (see Figure 4-42) is used to obtain out the distance, in the current units, between two points, either selected on the screen or keyed in from the keyboard. Included in the report are the horizontal and vertical distances (delta-X and delta-Y, respectively) between the points and the angles in and from the XY plane. AutoCAD prompts:

Specify first point: *(specify the first point to measure from)*

Specify second point: *(specify the endpoint to measure to)*

AutoCAD displays the distance between two selected points along with additional data.

SYSTEM VARIABLES

AutoCAD stores the settings (or values) for its operating environment and some of its commands in system variables. Each system variable has an associated type: integer (for switching), integer (for numerical value), real, point, or text string. You can examine and change (unless they are read-only) these variables.

Integers (for Switching) System variables that have limited non-numerical settings can be switched by setting them to the appropriate integer value. Turning the snap ON or OFF is demonstrated in the following example by changing the value of its SNAPMODE system variable.

snapmode (ENTER)
Enter new value for SNAPMODE <0>: **1** (ENTER)

The value is set to 1, which is the same as setting to ON. This sequence may seem unnecessary, because the snap mode is so easily toggled with a press of a function key. Changing the snap with the system variable is inconvenient, but doing so does allow you to view the results immediately.

For any system variable whose status is associated with an integer, the method of changing the status is just like the preceding example. In a similar manner, you can use SNAPISOPAIR to switch one isoplane to another by setting the system variable to one of three integers: 0 is the left isoplane, 1 is the top, and 2 is the right isoplane.

System variables such as APERTURE and AUPREC are changed by using an integer whose value is applied numerically in some way to the setting, rather than just as a switch. For instance, the size of the aperture (the target box that appears for selecting Osnap points) is set in pixels (picture elements) according to the integer value. For example, setting the value of APERTURE to 9 should render a target box that is three times larger than setting it to 3.

System variables that have a real number for a setting, such as VIEWSIZE, are called real.

LIMMIN, LIMMAX, and VIEWCTR are examples of system variables whose settings are points in the form of the X coordinate and Y coordinate.

The string variables have names like CLAYER, for the current layer name, and DWGNAME, for the drawing name.

drawing exercises

Open the Exercise Manual PDF file for Chapter 4 on the accompanying CD for discipline-specific exercises.

review questions

1. The following are AutoCAD tools available, except:

   a. *GRID.*   **d.** *TSNAP.*
   b. *SNAP.*   **e.** *OSNAP.*
   c. *ORTHO.*

2. Which command feature allows you to key in a distance where the direction is determined by the cursor location?

   a. *Direct Distance*   **c.** *Distance Direction*
   b. *Direction*   **d.** *LINE*

3. Which command feature allows you to move through nonselected points(s) to establish a selected point?

   a. *Direct distance*   **c.** *Snap from*
   b. *Osnap*   **d.** *Tracking*

4. After having drawn a 3-point circle, you want to begin a line at the exact center of the circle. What tool in AutoCAD would you use?

   a. *Snap*
   b. *Object Snap*
   c. *Entity Snap*
   d. *Geometric Calculator*

5. Which Osnap option allows you to select the closest endpoint of a line, arc, or polyline segment?

   a. *ENDpoint*
   b. *MIDpoint*
   c. *INSert*
   d. *INSertion point*
   e. *PERpendicular*

6. Which Osnap option allows you to select the point in the exact center of a line?

   a. *ENDpoint*
   b. *MIDpoint*
   c. *CENter*
   d. *INSertion point*
   e. *PERpendicular*

7. When drawing a line, which Osnap option allows you to select the point on a line or polyline segment where the angle formed with the line is 90 degrees?

a. *ENDpoint*
b. *MIDpoint*
c. *CENter*
d. *INSertion point*
e. *PERpendicular*

8. Which Osnap option allows you to select the location where two lines, arcs, or polyline segments cross each other?

a. *NODe*
b. *QUADrant*
c. *TANgent*
d. *Apparent Intersection*
e. *INTersection*

9. Which Osnap option allows you to select the location where a "point" has been established?

a. *NODe*   d. *NEArest*
b. *QUADrant*   e. *INTersection*
c. *TANgent*

10. Which Osnap option allows you to select a point on a circle that is 0, 90, 180, or 270 degrees from the circle's center?

a. *NODe*
b. *QUADrant*
c. *TANgent*
d. *NEArest*
e. *INTersection*

11. Which Osnap option allows you to select a point on any object, except text, that is closest to the cursor's position?

a. *NODe*
b. *QUADrant*
c. *TANgent*
d. *NEArest*
e. *INTersection*

12. Which Osnap option allows you to select the location where two lines or arcs may or may not cross each other in 3D space?

a. *QUADrant*
b. *APParent intersection*
c. *NEArest*
d. *INTersection*

13. How many previous zooms are available with the previous option of the ZOOM command?

a. *4*   c. *Unlimited*
b. *6*   d. *10*

14. Which option of the ZOOM command will display the entire drawing?

    a. *Extents*   **c.** *All*
    b. *Limits*   **d.** *Both a and c*

15. To ensure the entire limits of the drawing are visible on the display, you should perform a ZOOM _____.

    a. *All*
    b. *Previous*
    c. *Extents*
    d. *Limits*

16. The ZOOM command changes the size of not only the visual, but the actual objects in the drawing.

    a. *True*
    b. *False*

17. You can zoom in or out in realtime by moving the cursor horizontally.

    a. *True*
    b. *False*

18. The PAN command:

    a. *Always requires a regeneration.*
    b. *Maintains the current display magnification.*
    c. *Cannot be used transparently.*
    d. *None of the above.*

19. To view a different portion of the drawing without changing its magnification use:

    a. *MOVE.*   **c.** *ZOOM.*
    b. *PAN.*   **d.** *VIEW.*

20. In general, a REDRAW is quicker than a REGEN.

    a. *True*
    b. *False*

21. You are limited to 16 active viewports at one time.

    a. *True*
    b. *False*

SECTION

2

World Power

selecting, copying, and changing objects

chapter

introduction

Earlier chapters explained how to create objects in AutoCAD. This chapter explains how to select, modify, and create new objects from existing objects.

After completing this chapter, you will be able to do the following:

- Use various object selection methods

- Use Modify commands – ERASE, MOVE, COPY, ROTATE, and SCALE

OBJECT SELECTION

Many AutoCAD modify and construct commands prompt you to select one or more objects for manipulation. In most modify commands, the prompt allows you to select any number of objects. In some of the modify commands, however, AutoCAD limits your selection to only one object, for instance, the break, divide, and measure commands. In the case of the fillet and chamfer commands, AutoCAD requires you to select two objects. And, whereas in the dist and id commands AutoCAD requires you to select a point or points, in the area command AutoCAD permits selection of either a series of points or object(s). When you select one or more objects, AutoCAD usually highlights them by displaying them with dashed lines. The group of objects selected for the manipulation is called the selection set.

---

note There is a difference between the highlighting AutoCAD applies to a selected object and the highlighting of an object when the cursor passes over it. The passing-over highlighting only indicates that the object is eligible to be selected. The highlighting disappears when the cursor moves away from the object. When it is highlighted in this manner, you must press the pick button on your mouse/puck to select the object. When you do so, the highlighting that is applied to the object indicates that it is now part of the selection set.

There are several different ways of selecting the objects for manipulation. The selection options include Window, Crossing, Window Polygon (WP), Crossing Polygon (CP), Fence, Previous, Last, Single, and All. Modifications to Selection modes include Add, Remove, and Undo.

All modify and construct commands require a selection set, for which AutoCAD prompts:

>   Select objects:

AutoCAD replaces the screen crosshairs with a small box called the Pickbox. With the target cursor, select individual objects for manipulation. Using your pointing device position Pickbox so it touches only the desired object or a visible portion of it. The Pickbox helps you select the object without having to be very precise. When selection of multiple objects is permissible, each time you select an object, the "Select objects:" prompt reappears. To indicate your acceptance of the selection set, press ENTER at the "Select objects:" prompt.

Sometimes, it is difficult to select objects that are close together or lie directly on top of one another. You can use the pick button to cycle through these objects, one after the other, until you reach the one you want.

You can cycle through objects for selection at the "Select Objects:" prompt by holding down the CTRL button. Select a point as near as possible to the object. Press the pick button on your pointing device repeatedly until the object you want is highlighted, and press ENTER to select the object.

AutoCAD has various options to control the appearance of objects during selection preview. You can make the changes to the appearance of the selection preview from the Visual Effect Settings dialog box (see Figure 5–1), which can be opened from the **Selection** tab of the Options dialog box.

Figure 5–1

Visual Effect Settings dialog box

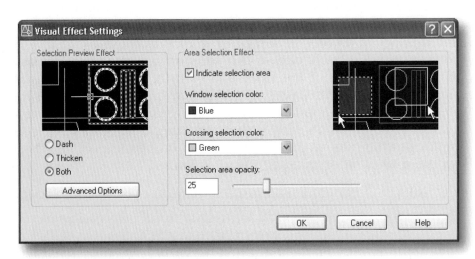

The preview area displays the effects of the current settings.

Selecting **Dash** causes dashed lines to be displayed when the pick box cursor passes over an object. This selection previewing indicates that the object would be selected if you clicked.

Selecting **Thicken** causes thickened lines to be displayed when the pick box cursor passes over an object. This selection previewing indicates that the object would be selected if you clicked.

Selecting **Both** (default setting) causes thickened, dashed lines to be displayed when the pick box cursor passes over an object. This selection previewing indicates that the object would be selected if you clicked.

Choosing **Advanced Options** causes AutoCAD to open the Advanced Preview Options dialog box (see Figure 5–2) from which you can select object types to exclude from selection previewing.

 Figure 5–2

Advanced Preview Options dialog box

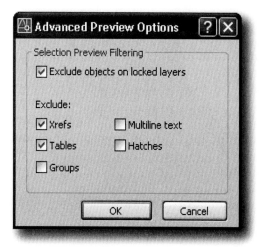

Selecting **Exclude Objects on locked layers** (default set to ON) causes selection previewing to be excluded for objects on locked layers.

Select object types from the **Exclude** section to exclude them from selection previewing. The available object types include Xrefs, Tables, Groups, Multiline Text, and Hatches. By default Xrefs and Tables are set to ON.

The **Area Selection Effect** section of the Visual Effect Settings dialog box controls the appearance of selection areas during selection preview. The preview area displays the effect of the current settings. Select **Indicate selection area** to indicate the selection area with the selected background color. Choose background colors for Window and Crossing selection area from the **Window selection color** and **Crossing selection color** drop-down lists, respectively. The slider bar for **Selection area opacity** controls the degree of transparency background for window selection areas.

Choose **OK** to save the settings and close the Visual Effect Settings dialog box.

Selection by Window

The Window option for selecting objects allows you to select all the objects contained completely within a rectangular window. You define this area by specifying two corner points. At the "Select objects:" prompt, use the pointing device to specify the first corner point. This must be above or below and to the left of the objects you want to select. AutoCAD prompts:

Specify opposite corner: *(specify opposite corner)*

Move your cursor up or down and to the right to create the rectangular area. When the rectangle completely covers the target objects, press the pick button on your pointing device to specify the second corner point. If there is an object that is partially inside the rectangular area, then that object is not included in the selection set. You can select only objects currently visible on the screen. To select a partially visible object, you must include all its visible parts within the window. See Figure 5–3, in which only the lines will be included, not the circles, because a portion of each of the circles is outside the rectangular area.

 Figure 5–3

Selecting objects by means of the Window option

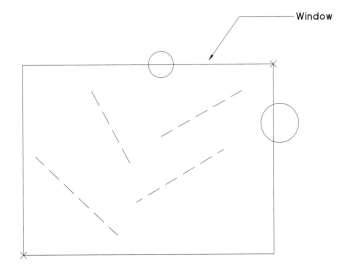

Selection by Crossing

The Crossing option for selecting objects allows you to select all the objects that are crossing the window as well as the objects contained completely in a rectangular area. You define this area by specifying two corner points. At the "Select objects:" prompt, use the pointing device to specify the first corner point. This must be to the right of the objects you want to select. AutoCAD prompts:

Specify opposite corner: *(specify opposite corner)*

Move your cursor to the left to create the rectangular area. When the rectangle covers or crosses a portion of the target objects, press the pick button on your pointing device to specify the second corner point. See Figure 5–4, in which all the lines and circles are included, even though parts of the circles are outside the rectangle.

 Figure 5–4

Selecting objects by means of the Crossing option

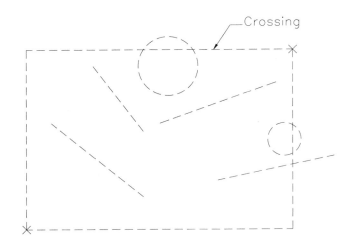

Selection by Window Polygon

The Window Polygon option is similar to the Window option, but it allows you to define a polygon-shaped area rather than a rectangular window. You define the selection area as you specify the points about the objects you want to select. The polygon is formed as you select the points. The polygon can be of any shape but may not intersect itself. When the selected points define the desired polygon, press enter. Only those objects that are totally inside the polygon shape are selected.

To select the Window Polygon option, type **wp** and press ENTER at the "Select objects:" prompt. The Undo option lets you undo the most recent polygon pick point.

Selection by Crossing Polygon

The Crossing Polygon option is similar to the Window Polygon option, but it selects all objects within or crossing the polygon boundary. If there is an object that is partially inside the polygon area, the whole object is included in the selection set.

To select the Crossing Polygon option, type **cp** and press ENTER at the "Select objects:" prompt. The Undo option lets you undo the most recent polygon pick point.

Selection by Fence

The Fence option is similar to the Crossing Polygon option, except that you do not close the last vector of the polygon shape. The selection fence selects only those objects it crosses or intersects. Unlike the Window Polygon and Crossing Polygon methods, the fence can cross over and intersect itself.

To select the Fence option, type **f** and press ENTER at the "Select objects:" prompt.

Previous Selection

The Previous option enables you to perform several operations on the same object or group of objects. AutoCAD remembers the most recent selection set and allows you to reselect it with the Previous option. For example, if you moved several objects and now wish to copy them elsewhere, you can invoke the copy command and respond to the "Select objects:" prompt by entering **p** to select the same objects again. (There is a command called select that does nothing but create a selection set; you can then use the Previous option to select this set at the next prompt to select objects.)

To select the Previous option, type **p** and press ENTER at the "Select objects:" prompt.

Last Selection

The Last option is an easy way to select the most recently created object currently visible. Only one object is designated, no matter how often you use the Last option, when constructing a particular selection set.

To select the Last option, type **l** (the letter, not the number **1**) and press ENTER at the "Select objects:" prompt.

Selection by All

The All option selects all the objects in the drawing, including objects on layers that are turned off. After selecting all the objects, you can use the Remove (**r**) modifier to remove some of the objects from the selection set.

To apply the All modifier, type **all** and press ENTER at the "Select objects:" prompt.

Selection by Multiple Modifier

The Multiple modifier helps you overcome the limitations of selecting objects individually, which is time-consuming in some cases. AutoCAD does a complete scan of the screen each time an object is selected. By using the Multiple modifier, you can pick several objects without delay, and when you press enter, AutoCAD applies all of the points during one scan.

Selecting one or more objects from a crowded group of objects is sometimes difficult. It is often impossible with the Window option. For example, if two objects are very close together and you wish to select them both, AutoCAD normally selects only one no matter how many times you select a point that touches them both. With the Multiple modifier, AutoCAD excludes an object from being selected once it has been included in the selection set. As an alternative, use the Crossing option to cover both objects. If this is not feasible, then the Multiple modifier may be the best choice.

To apply the Multiple modifier, type **m** and press ENTER at the "Select objects:" prompt.

Undoing the Selection

The Undo option allows you to remove the last item(s) selected from the selection set without aborting the "Select objects:" prompt, and then to continue adding to the selection set. It should be noted that if the last option to the selection process includes more than one object, the Undo option selection will remove all the objects from the selection set that were selected by that last option.

To apply the Undo modifier, type **u** and press ENTER at the "Select objects:" prompt.

Adding Objects to Selection

The Add option lets you switch back from the Remove mode in order to continue adding objects to the selection set by however many options you wish to use.

To apply the Add modifier, type **a** and press ENTER at the Select objects: prompt.

Removing Objects from Selection

The Remove option lets you remove objects from the selection set. The Select objects: prompt always starts in the Add mode. The Remove mode is a switch from the Add modifier, not a standard option. Once invoked, the objects selected by whatever and however many options you use will be removed from the selection set. It will be in effect until reversed by the Add modifier.

To apply the Remove modifier, type **r** and press ENTER at the Select objects: prompt.

Selecting a Single Set of Object(s)

The Single option causes the object selection to terminate and the command in progress to proceed after you use only one object selection option. It does not matter if one object is selected or a group is selected with that option. AutoCAD will not abort the command in progress if no object is selected; however, once there is a successful selection, the command proceeds.

To apply the Single modifier, type **s** and press ENTER at the Select objects: prompt.

AutoCAD provides five advanced object selection modes that will enhance object selection: ***Noun/verb selection, Use Shift to add to selection, Press and drag, Implied windowing, and Object grouping***. For detailed explanation, refer to Chapter 10.

MODIFYING OBJECTS

AutoCAD not only allows you to draw objects easily, but it also allows you to modify the objects you have drawn.

Erasing Objects

 Figure 5–5

Invoking the ERASE command from the Modify toolbar

The ERASE command (see Figure 5–5) will probably be the modify command you use most often. Everyone makes mistakes, but in AutoCAD it is easier to erase them. Or, if you are through with an object that you have created as an aid in constructing other objects, you may wish to erase it.

You can use one or more available object selection methods. After selecting the object(s), press ENTER (null response) in response to the next Select objects: prompt to complete the ERASE command. All the objects that were selected will disappear. The following command sequence shows an example of using the ERASE command.

> **erase** (ENTER)
> Select objects: *(select objects with Window selection option as shown in Figure-5–6)*
> Select objects: (ENTER)

AutoCAD deletes the selected object(s) from the drawing.

SELECT OBJECT
BY WINDOW OPTION

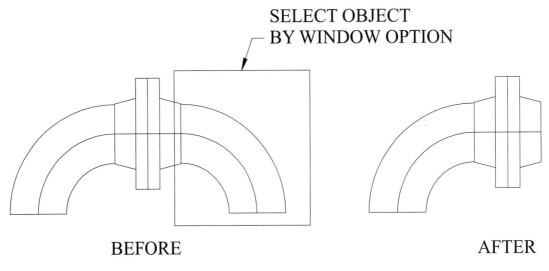

BEFORE   AFTER

 Figure 5–6

Using the ERASE command

Getting Them Back

The oops command restores objects that have been unintentionally erased. Whenever the erase command is used, the last group of objects erased is stored in memory. The oops command will restore the objects; it can be used at any time. It restores only the objects erased by the most recent erase command. See Chapter 3 on the undo command, if you need to step back further than one erase command.

To restore objects erased by the last ERASE command, invoke the OOPS command:

oops (ENTER)

AutoCAD restores the objects erased by the last ERASE command.

Moving Objects

 Figure 5–7

Invoking the MOVE command from the Modify toolbar

The MOVE command (see Figure 5–7) lets you move one or more objects from their present location to a new location without changing orientation or size.

If you specify two points, AutoCAD computes the displacement and moves the selected objects accordingly. If you specify the points on the screen, AutoCAD assists you in visualizing the displacement by drawing a rubber-band line from the first point as you move the crosshairs to the second point. If you provide a null response by pressing ENTER to the second point of displacement, then AutoCAD interprets the point selected as relative to a base point of 0,0,0. The following command sequence shows an example of using the MOVE command (see Figure 5–8).

move (ENTER)

Select objects: *(select objects with Window selection option as shown in Figure 5–8)*

Select objects: (ENTER)

Specify base point or displacement: *(select base point)*

Specify second point of displacement or <use first point as displacement>: *(select displacement determining point)*

AutoCAD moves the selected object(s) from the original location to a new location at the direction and distance determined by the base point/second displacement point vector.

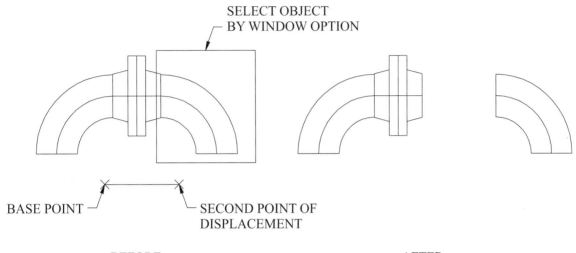

SELECT OBJECT
BY WINDOW OPTION

BASE POINT

SECOND POINT OF
DISPLACEMENT

BEFORE   AFTER

Figure 5-8

Using the MOVE command

Copying Objects

Figure 5-9

Invoking the COPY command from the Modify toolbar

The COPY command (see Figure 5–9) places copies of the selected objects at the specified displacement, leaving the original objects intact. The copies are oriented and scaled the same as the original. If necessary, you can make multiple copies of selected objects. Each resulting copy is completely independent of the original and can be edited and manipulated like any other object. The following command sequence shows an example of using the COPY command (see Figure 5–10).

copy (ENTER)

Select objects: *(select objects with Crossing selection option as shown in Figure 5–10)*

Select objects: (ENTER)

Specify base point or displacement, or ⬇: (*select base point*)

Specify second point of displacement or <use first point as displacement>: (*select displacement determining point*)

Specify second point of displacement: (*specify additional points to copy the selected objects and press* ENTER *to complete the command sequence*)

AutoCAD copies the selected object(s), placing them at a new location displaced from the location of the original objects at a direction and distance determined by the base point/second displacement point vector.

▷ **Figure 5–10**

Using the COPY *command*

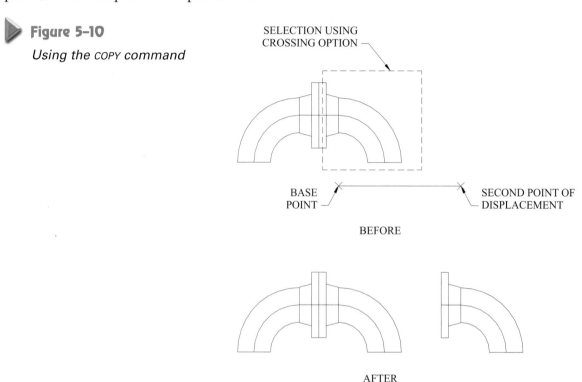

note   The first and second points of displacement do not have to be on or near the object. For example, you can enter 1,1 and 3,4 as the first and second points respectively, causing the objects to be moved or copied 2 units in the X direction and 3 units in the Y direction. To have the same result, you could have entered 0,0 at the first point and 2,3 at the second prompt. Or if, in the examples in Figures 5–6 and 5-8, you were moving or copying the object 24" to the right, you could select a point on the screen for the base point and then enter **@24<0** for the displacement to be relative to the point specified. You can simplify specifying the move or copy displacement vector if you know how far in the X direction and how far in the Y direction you wish to move or copy the selected objects. To do this, enter the coordinates for the second point of displacement at the first prompt (Specify base point of displacement). Then you press enter in response to the second prompt (Specify second point of displacement or <use first point as displacement>). AutoCAD uses the origin (0,0,0) as the first point. This is a stroke-saving procedure. For the example at the beginning of this note, you could have entered **2,3** at the first prompt and pressed enter at the second prompt. It is as if you had entered **0,0,0** at the first prompt and **2,3** at the second prompt. If you wish to move the object in the direction opposite to that of the example, you could enter **-2,-3** at the first prompt and press enter at the second prompt.

Rotating Objects

 Figure 5-11

Invoking the ROTATE command from the Modify toolbar

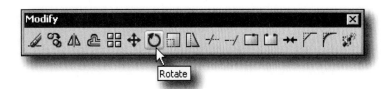

The ROTATE command (see Figure 5–11) changes the orientation of existing objects by rotating them about a specified point, labeled as the base point. Design changes often require that an object, feature, or view be rotated. By default, a positive angle rotates the object in the counterclockwise direction, and a negative angle rotates in the clockwise direction.

The base point can be anywhere in the drawing. The rotation angle determines how far an object rotates around the base point. The following command sequence shows an example of using the ROTATE command (see Figure 5–12).

rotate (ENTER)

Select objects: *(select objects with Window selection option as shown in Figure 5–12)*

Select objects: (ENTER)

Specify base point: *(select base point)*

Specify rotation angle or ⬇: **270**

AutoCAD rotates the selected object(s) from the original location to a new location about the base point for the angle specified.

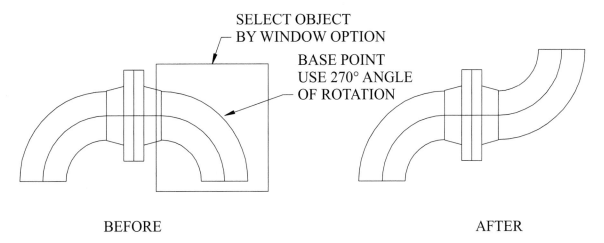

SELECT OBJECT
BY WINDOW OPTION

BASE POINT
USE 270° ANGLE
OF ROTATION

BEFORE   AFTER

 Figure 5-12

Using the ROTATE command

Rotation with Reference Angle

If an object has to be rotated in reference to the current orientation, you can use the Reference option. Specify the current orientation as referenced by the angle, or show AutoCAD the angle by pointing to the two endpoints of an object to be rotated, and specify the desired new rotation. AutoCAD automatically calculates the rotation angle and rotates the object appropriately. This method of rotation is very useful when you want to straighten an object or align it with other features in a drawing.

Scaling Objects

 Figure 5–13

Invoking the SCALE command from the Modify toolbar

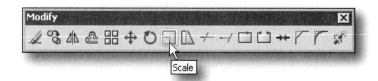

The SCALE command (see Figure 5–13) lets you change the size of selected objects or the complete drawing. Objects are made larger or smaller; the same scale factor is applied to the X, Y, and Z directions. To enlarge an object, specify a scale factor greater than 1. For example, a scale factor of 3 makes the selected objects three times larger. To reduce the size of an object, use a scale factor between 0 and 1. Do not specify a negative scale factor. For example, a scale factor of 0.75 would reduce the selected objects to three-quarters of their current size.

The base point can be anywhere in the drawing. If the base point selected is on the selected object itself, the selected base point becomes an anchor point for scaling. The scale factor multiplies the dimensions of the selected objects by the specified scale. The following command sequence shows an example of using the SCALE command (see Figure 5–14).

> **scale** (ENTER)
>
> Select objects: *(select objects with Window selection option as shown in Figure 5–14)*
>
> Select objects: (ENTER)
>
> Specify base point or displacement: *(select base point)*
>
> Specify scale factor or ⬇: **1.5**

AutoCAD resizes the selected object(s) using the base point and the numeric scale factor specified.

 Figure 5–14

Using the SCALE command

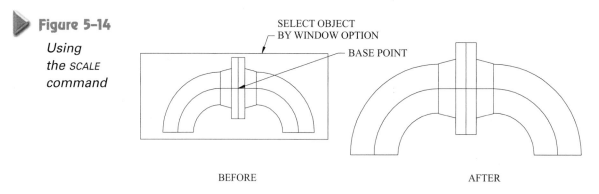

SELECT OBJECT
BY WINDOW OPTION

BASE POINT

BEFORE

AFTER

Scaling to a Reference

The Reference option is used to scale objects relative to the current dimension. Specify the current dimension as a reference length, or select two endpoints of an object to be scaled, and specify the desired new length. AutoCAD will automatically calculate the scale factor and enlarge or shrink the object appropriately.

drawing exercises

Open the Exercise Manual PDF file for Chapter 5 on the accompanying CD for discipline-specific exercises.

review questions

1. To create a six-sided area that would select all the objects completely within it, you should respond to the "Select Objects:" prompt with:

   a. *wp.*
   b. *cp.*
   c. *w.*
   d. *c.*

2. By default, most selection set prompts default to which option:

   a. *All.*
   b. *Add.*
   c. *Window.*
   d. *Window Polygon.*
   e. *None of the Above.*

3. The multiple option selection sets allows you to:

   a. *Select multiple objects that lie on top of each other.*
   b. *Scan the database only once to find multiple objects.*
   c. *Use the window or crossing option.*
   d. *Both a and b.*
   e. *Both b and c.*

4. The Remove option for forming selection sets deletes objects from the drawing in much the same manner as the erase command.

   a. *True*
   b. *False*

5. When erasing objects, if you select a point that is not on any object, AutoCAD will:

   a. *Terminate the ERASE command.*
   b. *Delete the selected objects and continue with the ERASE command.*
   c. *Allow you to drag a window to select multiple objects within the area.*
   d. *Ignore the selection and continue with the ERASE command.*

6. When you use the erase command, AutoCAD deletes each object from the drawing as you select it.

   a. *True*
   b. *False*

7. Once an object is erased from a drawing, which of the following commands could restore it to the drawing?

a. *OOPS*

b. *RESTORE*

c. *REPLACE*

d. *CANCEL*

8. Which command allows you to change the location of the objects and allows a duplicate to remain intact?

a. *CHANGE*   **c.** *COPY*

b. *MOVE*   **d.** *MIRROR*

9. Two lines are drawn. Then the first line is erased and the second line is moved. Executing the *oops* command at the Command: prompt will:

a. *Restore the erased line.*

b. *Execute the line command automatically.*

c. *Replace the second line at its original position.*

d. *Restore both lines to their original positions.*

e. *None of the above.*

10. To efficiently move multiple objects, which option would be more efficient?

a. *Objects*

b. *Last*

c. *Window*

d. *Add*

e. *Undo*

11. The *move* command allows you to:

a. *Move objects to new locations on the screen.*

b. *Dynamically drag objects on the screen.*

c. *Move only the objects that are on the current layer.*

d. *Move an object from one layer to another.*

e. *Both a and b.*

12. Which command allows you to change the size of an object, where the X and Y scale factors are changed equally?

a. *ROTATE*   **d.** *MODIFY*

b. *SCALE*   **e.** *MAGNIFY*

c. *SHRINK*

13. Using the *scale* command, what number would you enter to enlarge an object by 50%?

a. *0.5*

b. *50*

c. *3*

d. *1.5*

14. To avoid changing the location of an object when using the scale command:

   a. *The reference length should be less than the limits.*
   b. *The scale factor should be less than 1.*
   c. *The base point should be on the object.*
   d. *The base point should be at the origin.*

15. The rotate command is used to rotate objects around:

   a. *Any specified point.*
   b. *The point -1,-1 only.*
   c. *The origin.*
   d. *Is only usable in 3D drawings.*
   e. *None of the above.*

AutoCAD offers a means of grouping objects in layers in a manner similar to the manual drafter's separating complex drawings into simpler ones on individual transparent sheets superimposed in a single stack. Under these conditions, the manual drafter would be able to draw on the top sheet only. Likewise, in AutoCAD you can draw only on the current layer. However, AutoCAD permits you to transfer selected objects from one layer to another (neither of which needs to be the current layer). In addition, AutoCAD allows you to assign a color, linetype, and lineweight to each layer.

After completing this chapter, you will be able to do the following:

- Create and assign properties to layers

- Manage layers

- Change properties of the layers

- Set the linetype scale factor

USING LAYERS AND LAYER PROPERTIES

▷ Figure 6–1

Layers toolbar

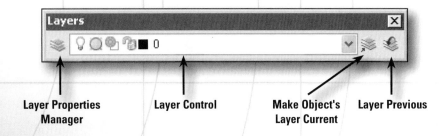

Layer Properties Manager   Layer Control   Make Object's Layer Current   Layer Previous

A common application of the layer feature is to associate similar types of objects by assigning them to the same layer. For example, you can draw text, dimensions, and title blocks on separate layers and then control which are visible and which are not by turning them on and off. Another common application of the layer feature is for drawing construction lines. You can create geometric constructions with objects, such as lines, circles, and arcs. These generate intersections, endpoints, centers, points of tangency, midpoints, and other useful data that might take the manual drafter considerable time to calculate with a calculator or to measure by hand on the board. From these you can create other objects using intersections or other data generated from the layout. Then the layout layer can be turned off (making it no longer visible) or it can be set to not plot. The layer is not lost, but can be recalled (set to ON) for viewing later as required. The same drawing limits, coordinate system, and zoom factors apply to all layers in a drawing.

To draw an object on a particular layer, first make sure that that layer is the "current layer." One, and only one, layer can be current at any time. Whatever you draw will be placed on the current layer. The current layer can be compared to the manual drafter's top sheet on the stack of transparencies. You can always move, copy, or rotate any object, whether it is on the current layer or not. When you copy an object that is not on the current layer, the copy is placed on the layer that the original object is on. This is also true with the mirror or an array of an object or group of objects.

A layer can be visible (ON) or invisible (OFF). Only visible layers are displayed or plotted. If necessary, AutoCAD allows you to set a visible layer not to plot. You can turn layers ON and OFF at any time, in any combination. It is possible to turn OFF the current layer. If this happens and you draw an object, it will not appear on the screen; it will be placed on the current layer and will appear on the screen when that layer is set to ON (provided you are viewing the area in which the object was drawn). This is not a common occurrence, but it can cause concern to both the novice and the more experienced operator who has not faced the problem before. Do not turn the current layer OFF; the results can be very confusing. When the TILEMODE system variable is set to 0, you can make specified layers visible only in selected viewports. Viewports are discussed in Chapter 15.

Each layer in a drawing has an assigned name, color, lineweight, and linetype. The name of a layer can be up to 255 characters long. It can contain letters, digits, and the special characters dollar ($), hyphen (-), underscore (_), and spaces. It is helpful to give a descriptive name appropriate to your application, such as "floor-plan" or "plumbing." The first several characters of the current layer's name are displayed in the layer list box located on the Layers toolbar (see Figure 6–1). You can change the name of a layer any time you wish, and you can delete any unused layer except layer 0.

Object Color

Figure 6–2

Properties toolbar

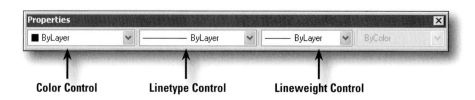

Color Control   Linetype Control   Lineweight Control

You can assign the color of an object either ByLayer, ByBlock or by specifying its color directly, independent of layer. Assigning colors ByLayer makes it easy to identify which objects are on which layer within your drawing. Assigning colors directly to objects provides additional distinctions between objects on the same layer. Color is also used as a way to indicate lineweight for color-dependent plotting. By default, AutoCAD assigns color ByLayer to the newly created objects, as shown in the Properties toolbar (see Figure 6–2). Whenever a new layer is created, it inherits the properties of the currently selected layer in the layer list (color, on or off state, and so on). If necessary, you can change the assigned color to one of the available 255 AutoCAD Color Index (ACI) colors, true colors, and Color Book from the Select Color dialog box.

note AutoCAD allows you to assign color ByLayer or any one of the available colors from the Color control on the Properties toolbar. You are advised not to mix color assignment on any given layer by assigning color directly to objects and also ByLayer. It may cause confusion, especially in a large drawing containing blocks and nested blocks.

Object Linetype

A linetype is a repeating pattern of dashes, dots, and blank spaces. Assigning different linetypes to objects is a means of conveying visual information. Similar to color assignment, you can assign the same linetype to any number of layers. In some drafting disciplines, conventions have been established giving specific meanings to particular dash-dot patterns. If a line is too short to hold even one dash-dot sequence, AutoCAD draws a continuous line between the endpoints. When you are working on large drawings, you may not see the gap between dash-dot patterns in a linetype, unless the scaling for the linetype is set for a large value. This can be done by means of the LTSCALE command explained later in this chapter. The following are some of the linetypes that are provided in AutoCAD in a library file called *acad.lin* (see Appendix G for a listing of available linetypes):

Border	— — — ·· — —	Dashdot	— · — · — · —	Dot	·················
Center	— — · — — ··	Dashed	— — — — — —	Hidden	------------
Continuous	————————	Divide	— ·· — ·· —	Phantom	— — · — — ·

Similar to color assignment, you can assign the linetype of an object either by layer or by specifying its linetype directly to the object independent of layer. By default, AutoCAD assigns linetypes ByLayer to the newly created objects, as shown in the Properties toolbar (see Figure 6–2). Whenever a new layer is created, it inherits the linetype of the currently selected layer in the layer list.

note AutoCAD allows you to assign linetype ByLayer, ByBlock or assign any one of the available linetypes from the Linetype control in the Properties toolbar. You are advised not to mix linetype assignment on any given layer by assigning linetype directly to object and also ByLayer. It may cause confusion, especially in a large drawing containing blocks and nested blocks.

Object Lineweight

Similar to assigning color and linetype, you can also assign a specific lineweight to a layer. Lineweights add width to your objects, both on screen and on paper. Using lineweights, you can create heavy and thin lines to show varying object thicknesses in details. For example, by assigning varying lineweights to different layers, you can differentiate between new, existing, and demolition construction.

AutoCAD allows you to assign a lineweight to a specific layer or an object in either inches or millimeters, with millimeters being the default. If necessary, you can change the default setting by invoking the LINEWEIGHT command.

The initial default lineweight is set 0.01 in (0.25 mm). Appendix G lists all the available lineweights as well as associated industry standards.

Lineweights are displayed differently in model space than in a paper space layout (refer to Chapter 16 for a detailed explanation about lineweights in layouts). In model space, lineweights are displayed in relation to pixels. In a paper space layout, lineweights are displayed in the exact plotting width. You can recognize that an object has a thick or thin lineweight in model space, but the lineweight does not represent an object's real-world width. A lineweight of 0 will always be displayed on screen with the minimum display width of one pixel. All other lineweights are displayed using a pixel width in proportion to its real-world unit value. A lineweight displayed in model space does not change with the zoom factor. For example, a lineweight value that is represented by a width of four pixels is always displayed using four pixels regardless of how close you zoom in to your drawing.

If necessary, you can change the display scale of lineweights of the objects in model space to appear thicker or thinner. Changing the display scale does not affect the lineweight plotting value. However, AutoCAD regeneration time increases with lineweights that are represented by more than one pixel. If you want to optimize AutoCAD performance when working in Model Space, set the lineweight display scale to the minimum value. The LINEWEIGHT command allows you to change the display scale of lineweights.

In model space, AutoCAD allows you to turn OFF the display of lineweights by toggling the LWT toggle button on the Status bar. With the LWT toggle button OFF, AutoCAD displays all the objects with a lineweight of 0 and reduces regeneration time.

When you export drawings to other applications or cutting/copying objects to the Clipboard, objects retain lineweight information.

Similar to color and linetype assignment, you can assign the lineweight of an object either by layer or by specifying lineweight directly, independent of layer. By default, AutoCAD assigns lineweight ByLayer to the newly created objects, as shown in the Properties toolbar (see Figure 6–2). Whenever a new layer is created, it inherits the lineweight of the currently selected layer in the layer list.

___

note   AutoCAD allows you to assign lineweight ByLayer, ByBlock or assign any one of the available lineweights from the Lineweight control in the Properties toolbar. You are advised not to mix lineweight assignment on any given layer by assigning lineweight directly to objects and also ByLayer. It may cause confusion, especially in a large drawing containing blocks and nested blocks.

___

CREATING AND MANAGING LAYERS

▶ **Figure 6-3**

Selecting Layer Properties Manager on the Layers toolbar

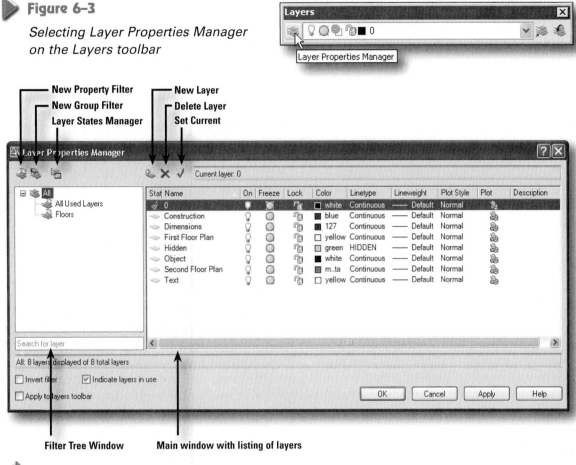

▲ **Figure 6-4**

Layer Properties Manager dialog box with Filter Tree displayed

The Layer Properties Manager dialog box can be used to set up and control layers. The main window lists each layer, along with the status of its associated properties. The Filter Tree window as shown in Figure 6–4 lists the available filters. A layer filter limits the display of layers in the Layer Properties Manager and in the Layer control on the Layers toolbar. In a large drawing, you can use layer filters to display only the layers on which you need to work.

Creating New Layers

By default, AutoCAD provides one layer called 0 (zero) that is set to ON and assigned the color white, the lineweight set to default, and the linetype continuous. To create a new layer, choose **New Layer** from the Layer Properties Manager dialog box or choose New Layer from the Shortcut menu. AutoCAD then creates a new layer by assigning the name "Layer1," and the new layer inherits the properties of the currently selected layer in the layer list (color, on or off state, and so on). When first listed in the layer list box, the name "Layer1" is highlighted and ready to be edited. Just enter the desired name for the new layer. If you accept the name Layer1, you can still change it later. To rename the layer, click anywhere on the line where the layer is listed, and the whole line is highlighted. Then click again on the layer name, and the entire name is high-lighted—you can type a new name and press ENTER. If you click again on the name, a text editing cursor allows you to change only part of the name.

---

note   The layer name cannot contain wild-card characters (such as * and ?). You cannot duplicate existing names.

Making a Layer Current

To make a layer current, first choose the layer name from the layer list box in the Layer Properties Manager dialog box, and then choose **Set Current** located at the top of the dialog box, double-click the icon corresponding to the layer name located under the **Status** column (first column from the left), or choose Set Current from the Shortcut menu.

Visibility of Layers

When you turn a layer OFF, the objects on that layer are not displayed in the drawing area and they are not plotted. The objects are still in the drawing, but they are not visible on the screen. And they are still calculated during regeneration of the drawing, even though they are invisible.

To change the setting for the visibility of selected layer(s), select the icon corresponding to the layer name located under the **On** column (third column from the left) in the Layer Properties Manager dialog box. The icon is a toggle for ON/OFF of layers. It is possible to turn OFF the current layer. If this happens and you draw an object, it will not appear on the screen; it will be placed on the current layer and will appear on the screen when that layer is set to ON (provided you are viewing the area in which the object was drawn).

Freezing and Thawing Layers

In addition to turning the layers OFF, you can freeze layers. The layers that are frozen will not be visible in the view window, nor will they be plotted. In this respect, frozen layers are similar to layers that are OFF. However, layers that are simply turned OFF still go through a screen regeneration each time the system regenerates your drawing, whereas the layers that are frozen are not considered during a screen regeneration. If you want to see the frozen layer later, you simply thaw it, and automatic regeneration of the drawing area takes place.

To change the setting for the visibility of a layer by freezing/thawing, select the icon corresponding to the layer name located under the **Freeze** column (fourth column from the left). The icon is a toggle for the Freeze/Thaw of layers. You cannot freeze the current layer.

Locking and Unlocking Layers

Objects on locked layers are visible in the view window but cannot be modified by means of the modifying commands. However, it is still possible to draw on a locked layer by making it the current layer, and use any of the inquiry commands and Object Snap modes on them.

To Lock or Unlock a layer, select the icon corresponding to the layer name located under the **Lock** column (fifth column from the left). The icon is a toggle for the Lock/Unlock of layers.

Changing the Color of Layers

By default, AutoCAD assigns the color of the currently selected layer in the layer list to the newly created layer. To change the assigned color, choose the icon under the Color column (sixth column from the left) corresponding to the layer name. AutoCAD displays the Select Color dialog box (see Figure 6–5 with the **Index Color** tab displayed), which allows you to change the color of the selected layer(s). You can select one of the 256 colors. Use the cursor to select the color you want, or enter its name or number in the **Color** box. Choose **OK** to accept the color selection. To cancel the selection, choose **Cancel**. AutoCAD 2004 introduced new color capabilities by adding two new tabs in the Select Color dialog box: **True Color** (see Figure 6–6) and **Color Books** (see Figure 6–7). The **True Color** tab specifies color settings using true colors (24-bit color) with either the Hue, Saturation, and Luminance (HSL) color model or the Red, Green, and Blue (RGB) color model. Over 16 million colors are available when using true color functionality. The **Color Books** tab specifies colors using third-party color books (such as PANTONE®) or user-defined color books. Once a color book is selected, the **Color Books** tab displays the name of the selected color book. **True Color** and **Color Books** make it easier to match colors in your drawing with colors of actual materials.

Figure 6–5

Select Color dialog box with Index Color tab displayed

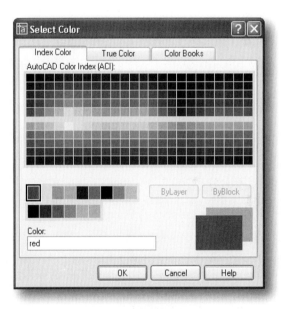

Figure 6–6

Select Color dialog box with True Color tab displayed

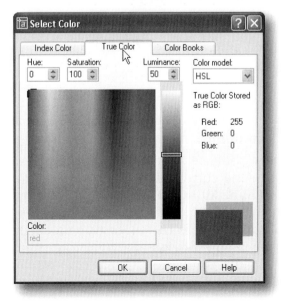

Figure 6–7

Select Color dialog box with Color Books tab displayed

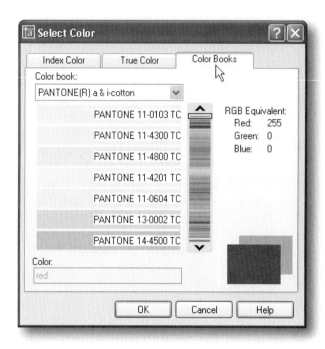

Changing the Linetype of Layers

By default, AutoCAD assigns the linetype of the currently selected layer in the layer list to the newly created layer. To change the assigned linetype, choose the linetype name corresponding to the layer name located under the **Linetype** column (seventh column from the left). AutoCAD displays the Select Linetype dialog box as shown in Figure 6–8, which allows you to change the linetype of the selected layer(s). Select the appropriate linetype from the list box, and choose **OK** to accept the linetype selection.

Figure 6–8

Select Linetype dialog box

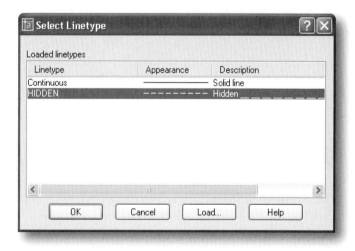

Loading Linetypes

In the Select Linetype dialog box, AutoCAD lists only the linetypes that are loaded in the current drawing. To load additional linetypes in the current drawing, choose **Load**. AutoCAD displays the Load or Reload Linetype dialog box similar to Figure 6–9. AutoCAD lists the available linetypes from *acad.lin,* the default linetype file. Select all the linetypes that need to be loaded, and choose **OK** to load them into the current drawing. If necessary, you can change the default linetype file *acad.lin* to another file by choosing **File** and selecting the desired linetype file.

 **Figure 6-9**

*Load or Reload Linetypes
dialog box*

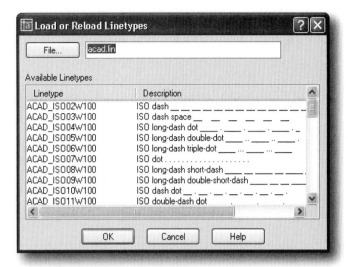

Changing the Lineweight of Layers

By default, AutoCAD assigns the lineweight of the currently selected layer in the layer list to the newly created layer. To change the assigned lineweight, choose the lineweight name corresponding to the layer name located under the **Lineweight** column (eighth column from the left). AutoCAD displays the Lineweight dialog box similar to Figure 6–10, which allows you to change the lineweight of the selected layer(s). Select the appropriate lineweight from the list box, and choose **OK** to accept the lineweight selection.

 Figure 6-10

Lineweight dialog box

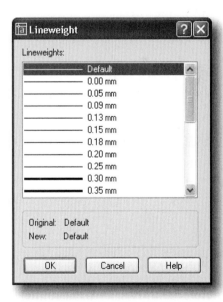

Assigning Plot Styles to Layers

Plot styles are a collection of property settings (such as color, linetype, and lineweight) that can be assigned to a layer or to individual objects. These property settings are contained in a named plot style table. (For a detailed explanation on creating plot styles, refer to Chapter 16.) When applied, the plot style can affect the appearance of the plotted drawing. By default, AutoCAD assigns the plot style Normal to a newly created layer if the Default plot style behavior is set to Named plot styles. To change the assigned plot style, choose the plot style name corresponding to the layer name located under the

Plot Style column (ninth column from the left as shown in Figure 6–11). AutoCAD displays the Select Plot Style dialog box, which allows you to change the Plot Style of the selected layer(s). Select the appropriate plot style from the list box, and choose **OK** to accept the plot style selection. A layer that is assigned a Normal plot style assumes the properties that have already been assigned to that layer. You can create new plot styles, name them, and assign them to individual layers. You cannot change the plot style if the Default plot style behavior is set to Color Dependent plot styles.

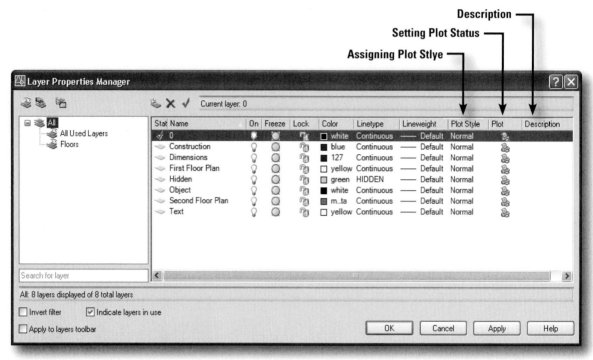

Figure 6–11

Layer Properties Manager dialog box

Setting Plot Status for Layers

AutoCAD allows you to turn plotting ON or OFF for visible layers. For example, if a layer contains construction lines that need not be plotted, you can specify that the layer is not plotted. If you turn OFF plotting for a layer, the layer is displayed but is not plotted. At plot time, you do not have to turn OFF the layer before you plot the drawing.

To change the setting for the plotting of layers, select the icon corresponding to the layer name located under the **Plot** column (tenth column from the left as shown in Figure 6–11). The icon is a toggle for plotting or not plotting layers.

Adding a Description to Layers

In addition to assigning a layer name, AutoCAD allows you to add a description to individual layers. Double-click the appropriate layer name and add the description in the **Description** column (eleventh column from the left as shown in Figure 6–11). To edit a layer's description, select it and choose Change description from the shortcut menu. The description is limited in length to 255 characters.

Saving Layer Properties in a Layer State

At any time during a drawing session, the collective status of all layers properties settings is known as the Layer State. This state can be saved and given a name by which it can be recalled later, thus having every setting of selected properties of every layer revert to what they were when that particular layer state was named and saved. Layer states are saved in files with the extension of *.las*.

To create a new Layer State, in the Layer Properties Manager dialog box choose **Layer States Manager**, which causes the Layer States Manager dialog box to be displayed, as shown in Figure 6–12. From this dialog box, choose **New**, which causes the New Layer State to Save dialog box to be displayed. In the **New Layer State** and **Description** text boxes, enter the name and descriptions respectively and then choose **OK**. This will return you to the Layer States Manager dialog box, and the new Layer State will be saved based on settings of selected properties. Once the desired properties have been selected, choose **Close**. For example, you may wish to save and name a layer state based only on the visibility of the layers. Therefore you would select **On/Off** and when the named layer state is restored, all of the layers will revert to the visibility status at which they were when the layer state was created. Other properties would not be changed.

 Figure 6–12

The Layer States Manager dialog box

A saved Layer State can be restored. To restore a saved layer state, select its name in the Layer States Manager dialog box and then choose **Restore**.

To edit a saved layer state, the procedure is the same as for creating a new Layer State, except that you enter the name of the existing Layer State you wish to be changed in the New Layer State to Save dialog box. AutoCAD will prompt "There is already a layer state named <*layerstatename*> Overwrite existing state?" Choose **Yes**.

To delete a named and saved layer state, in the Layer States Manager dialog box choose **Delete** when the layer state you wish to delete is highlighted in the **Layer states** list.

To import one or more layer states into your drawing, in the Layer States Manager dialog box, choose **Import**. This causes the Import layer state dialog box to be displayed. This dialog box is similar to other Windows file seeking and handling dialog boxes. From this dialog box you can select a saved layer state to import.

To export one or more layer states from your drawing, in the Layer States Manager dialog box choose **Export**. This causes the Export layer state dialog box to be displayed. This dialog box is similar to other Windows file seeking and handling dialog boxes. From this dialog box you can select a saved layer state to export.

Setting Filters for Listing Layers

AutoCAD allows you to create two kinds of filters: Layer Property and Layer Group. *Layer Property filters* are the ones that include layers that have names or other properties in common. For example, you can define a filter that includes all layers that are blue and whose names include the letters "floor." *Layer group filters* are the ones that include the layers that are included into the filter when you define it, regardless of their names or properties.

The tree view in the Layer Properties Manager displays default layer filters and any named filters that you create and save in the current drawing. The icon next to a layer filter indicates the type of filter as shown in Figure 6–13.

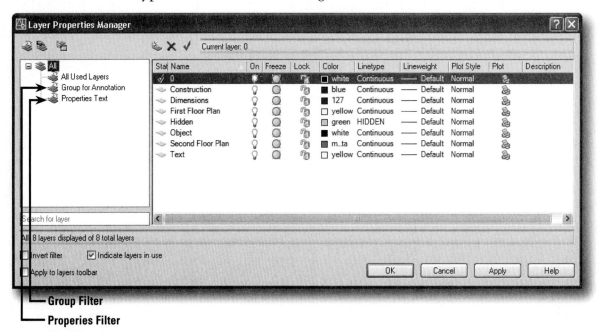

Figure 6–13

Layer Properties Manager with listing of filters

By default, AutoCAD creates three filters in a newly created drawing: The **All** filter selection displays all the layers in the current drawing. The **All Used Layers** filter selection displays all the layers on which objects in the current drawing are drawn. And the **Xref** filter selection displays all the layers being referenced from other drawings if any Xrefs are attached.

Once you have named and defined a layer filter, you can select it in the tree view to display the layers in the list view. You can also apply the filter to the Layers toolbar, so that the Layer control displays only the layers in the current filter.

When you select a filter in the tree view and right-click, options on the Shortcut menu can be used to delete, rename, or modify filters.

Creating a New Layer Property Filter

Choosing **New Property Filter** (or choosing New Properties Filter from the Shortcut menu) causes the Layer Filter Properties dialog box to be displayed, as shown in Figure 6–14.

Figure 6–14

*Layer Filter
Properties dialog box*

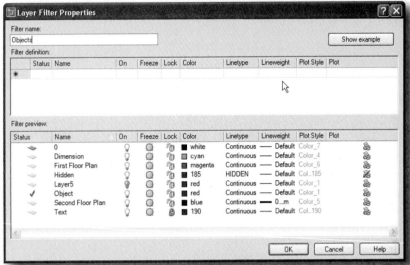

Specify the name of the filter in the **Filter name** box. In the **Filter definition** section, you can use one or more properties to define the filter. For example, you can define a filter that displays all layers that are either green or red and in use. To include more than one color, linetype, or lineweight, duplicate the filter on the next line and select a different setting. Choose **Show example** to display the examples of layer property filter definitions.

The **Status** column can be set to display the In Use icon or the Not In Use icon. In the **Name** column you can use wild-card characters to filter layer names. For example, enter **floor*** to include all layers that start with "floor" in the name. Set any of the one or more corresponding properties to define the filter. AutoCAD Displays the results of the filter as you define in the **Filter preview** section of the dialog box. The filter preview shows which layers will be displayed in the layer list in the Layer Properties Manager when you select this filter. To rename and delete a property filter, first select the property filter in the tree view and then from the Shortcut menu, choose Rename or Delete respectively.

Choose **OK** to save the newly created or modified filter definition or choose **Cancel** to disregard the changes.

Creating a New Layer Group Filter

Choose **New Group Filter** (or choose New Group Filter from the Shortcut menu) to create a new layer group filter. A new layer group filter named GROUP FILTER1 is created in the tree view. Rename the filter to an appropriate name. In the tree view, choose **All** or one of the other layer property filters to display layers in the list view. In the list view, select the layers you want to add to the filter, and drag them to the newly created Group filter name in the tree view. To rename and delete a group filter in the tree view, first select the group filter and then from the Shortcut menu, choose Rename or Delete respectively.

Inverting a Layer Filter

Inverting a filter is another method for listing selected layers. When the **Invert filter** box on the Layer Properties Manager is OFF (not checked), then the layers that are shown in the main window are those that have all the matching characteristic(s) of the specified filter. For example, if you were to create and name a set of filters that specify the color yellow and the linetype dashed, then the list would include only layers whose color is yellow and linetype is dashed. If you set the **Invert filter** box then ON, then all layers would be listed *except* those with both the color yellow and the linetype dashed.

Apply to Layers Toolbar

The **Apply to layers toolbar** option controls the display of layers in the list of layers on the Layers toolbar by applying the current layer filter. When **Apply to layers toolbar** is not checked, then the layers that are filtered (or Invert filtered) are shown in the main window of the Layer Properties Manager dialog box only. When **Apply to layers toolbar** is checked, then the list in the text box for layers in the Layers toolbar will also be filtered (or Invert filtered, depending on the status of the **Invert filter** check box).

Changing the Appearance of the Layer Properties Manager Dialog Box

If necessary, you can drag the widths of the column headings to see additional characters of the layer name, full legend for each symbol and color name, or number in the list box. You can sort the order in which layers are displayed in the list box by choosing the column headings, which causes AutoCAD to list the layers in descending order (Z to A, then numbers). Choosing the column heading again causes AutoCAD to list the layers in ascending order (numbers, A to Z). Choosing the **Status** column headers lists the layers by the property in the list.

Applying and Closing the Layer Properties Manager

After making the necessary changes, choose **Apply** to apply changes that have been made to layers and filters but not close the dialog box. Choose **OK** to accept the changes and close the dialog box. To discard the changes, choose **Cancel**, which closes the dialog box without making any changes to the layer properties.

CHANGING LAYER STATUS FROM THE LAYERS TOOLBAR

You can toggle on/off, freeze/thaw, lock/unlock, or plot/no plot, in addition to making a layer current from the Layer list box provided on the Layers toolbar. Select the appropriate icon next to the layer name you wish to toggle, as shown in Figure 6–15.

Figure 6–15

Layer list box on the Layers toolbar

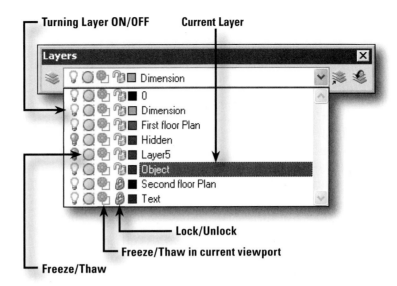

MAKING AN OBJECT'S LAYER CURRENT

Figure 6–16

Choosing the Make Object's Layer Current icon from the Layer toobar

AutoCAD allows you to select an object in the drawing to make its layer the current layer. To do so, choose Make Object's Layer Current (see Figure 6–16) from the Layers toolbar. AutoCAD prompts:

> Select object whose layer will become current: *(select the object in the drawing to make its layer current)*

AutoCAD makes the selected object's layer current.

UNDOING LAYER SETTINGS

Figure 6–17

Invoking the LAYERP command from the Layers toolbar

AutoCAD allows you to undo the last change or set of changes made to layer settings. To do so, choose Layer Previous (see Figure 6–17) from the Layers toolbar.

This command undoes changes you have made to layer settings such as color or linetype. If settings are restored, AutoCAD displays the message "Restored previous layer states." The LAYERP command restores the original properties, but not the original name if the layer is renamed, and it also does not restore if you delete or purge a layer.

SETTING THE LINETYPE SCALE FACTOR

The linetype scale factor allows you to change the lengths of dashes and spaces between dashes and dots linetypes relative to drawing units. The definition of the linetype instructs AutoCAD as to how many units long to make dashes and the spaces between dashes and dots. As long as the linetype scale is set to 1.0, the displayed length of dashes and spaces coincides with the definition of the linetype. The LTSCALE command allows you to set the linetype scale factor. Invoke the LTSCALE command from the Command prompt as follows:

> Command: **ltscale** (ENTER)
>
> Enter new linetype scale factor <1.0000>: *(specify the scale factor)*

Changing the linetype scale affects all linetypes in the drawing. If you want dashes that have been defined as 0.5 units long in the dashed linetype to be displayed as 10 units long, you set the linetype scale factor to 20. This also makes the dashes that were defined as 1.25 units long in the center linetype display as 25 units long, and the short dashes (defined as 0.25 units long) display as 5 units long. Note that the 1.25-unit-long dash in the center linetype is 2.5 times longer than the 0.5-unit-long dash in the dashed linetype. This ratio will always remain the same, no matter what the setting of LTSCALE. So if you wish to have some other ratio of dash and space lengths between different linetypes, you will have to change the definition of one of the linetypes in the *acad.lin* file.

Linetypes are for visual effect. The actual lengths of dashes and spaces are bound more to how they should look on the final plotted sheet than to distances or sizes of any objects on the drawing. An object plotted full size can probably use an LTSCALE setting of 1.0. A 50'-long object plotted on an 18" x 24" sheet might be plotted at a 1/4" = 1'-0" scale factor. This would equate to 1 = 48. An LTSCALE setting of 48 would make dashes and spaces plot to the same lengths as the full-size plot with a setting of 1.0. Changing the linetype scale factor causes the drawing to regenerate.

drawing exercises

Open the Exercise Manual PDF file for Chapter 6 on the accompanying CD for discipline-specific exercises.

review questions

1. All of the following functions can be performed in the Layer Properties Manager, except:

   a. *On.*   **d.** *Lock.*
   b. *Use.*   **e.** *Set.*
   c. *Make.*

2. Which of the following is not a valid option of the Layer Properties Manager dialog box?

   a. *Close*   **d.** *Freeze*
   b. *Lock*   **e.** *Color*
   c. *On*

3. When a layer is ON and THAWed:

   a. *The objects on that layer are visible.*
   b. *The objects on that layer are not visible.*
   c. *The objects on that layer are ignored by a* REGEN.
   d. *The drawing* REDRAW *time is reduced.*
   e. *the objects on that layer cannot be selected*

4. A layer where objects may not be edited or deleted, but are still visible on the screen and may have Object Snap modes used on them, is considered:

   a. *Frozen.*   **d.** *unSet.*
   b. *Locked.*   **e.** *fiXed.*
   c. *On.*

5. How many color choices are available for each layer?

   a. *254*   **c.** *256*
   b. *255*   **d.** *257*

6. In the Layer Properties Manager, which option makes the named layer visible?

   a. *Current*   **c.** *On*
   b. *New*   **d.** *Visible*

7. In the Layer Properties Manager, which option makes the named layer invisible?

   a. *Freeze*   **c.** *Lock*
   b. *Invisible* **d.** *Off*

8. In the Layer Properties Manager, which option makes the named layer visible, but does not allow entities on that layer to be edited with modify commands?

   a. *Freeze*   **c.** *Unlock*
   b. *Lock*   **d.** *Thaw*

9. Which command sets the scale factor to be applied to all linetypes within the drawing?

    a. *DDSCALE*   **c.** *LTSCALE*

    b. *LINETYPE*   **d.** *SCALE*

10. Which is not a valid option of the LAYER command?

    a. *On*   **c.** *Make*

    b. *Use*   **d.** *Lock*

11. When a layer is ON and THAWed:

    a. *The objects on that layer are visible on the monitor.*

    b. *The objects on that layer are not visible on the monitor.*

    c. *The objects on that layer are ignored by a REGEN.*

    d. *The drawing REDRAW time is reduced.*

introduction

You have learned how to draw some of the geometric shapes that make up your design. Now it is time to learn how to annotate your design. When you draw on paper, adding descriptions of the design components and the necessary shop and fabrication notes is a tedious, time-consuming process. AutoCAD provides several text commands and tools (including a spell checker) that greatly reduce the tedium of text placement and the time it takes. Included are features that make it easy to set up tables (rows and columns) for notes on the drawing.

After completing this chapter, you will be able to do the following:

- Create single line text and multiline text using appropriate styles and sizes to annotate drawings

- Edit text objects

- Find and replace text

- Scale and justify text

- Spell check text

- Create and modify tables

DRAWING TEXT

Figure 7-1

Text toolbar

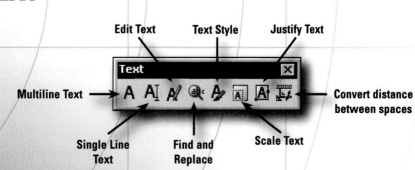

Text is used to label the various components of your drawing and to create the shop or field notes needed for fabricating and constructing your design. AutoCAD includes a large number of text fonts. Text can be stretched, compressed, obliqued, mirrored, or drawn in a vertical column by applying a style. Each text string can be sized, rotated, and justified to meet your drawing needs. You should be aware that AutoCAD considers a text string (all the characters that comprise the line of text) as one object.

AutoCAD has a toolbar just for text-related commands. Included are Multiline Text, Single Line Text, Edit Text, Find and Replace, Text Style, Scale Text, Justify Text, and Convert distance between spaces, as shown in Figure 7–1.

Single Line Text

 Figure 7–2

Invoking the Single Line Text (DTEXT) command from the Text toolbar

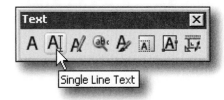

The Single Line Text (DTEXT) command draws text one line at a time, allowing you to specify justification but with no limits on the length of the line of text. To create text one line at a time in the current style, you can invoke the DTEXT command by choosing Single Line Text from the Text toolbar (see Figure 7–2). AutoCAD prompts:

Specify start point of text or ⬇: *(specify start point of text or choose one of the available options from the Shortcut menu)*

Creating Text One Line at a Time

Unless you change the justification from the default, the start point you specify will be the lower left corner of the text. If necessary, you can change the justification point. You can specify the start point in absolute coordinates or by using your pointing device. After you specify the start point, AutoCAD prompts:

Specify height <default>: *(specify the text height)*

This allows you to select the text height. You can accept the default text height by giving a null response, specify the height with two points on the screen by using your pointing device, or you can enter the appropriate text height. Next, AutoCAD prompts:

Specify rotation angle of text <default>: *(specify the rotation angle)*

This allows you to place the text at any angle with reference to 0 degrees (default is three o'clock, or east, measured in the counterclockwise direction). You can accept the default angle by pressing ENTER, or you can specify the angle with two points on the screen by using your pointing device, or you can enter the desired angle. The default value of the rotation angle is 0 degrees the first time, and the text is placed horizontally from the specified start point. If a different angle was specifed in the last use of the TEXT command, the angle defaults to the new angle. The last prompt is:

Enter text: *(type the desired text and press ENTER)*

A cursor appears on the screen at the start point you have selected. After you type the first line of text and press ENTER, you will notice the cursor drop down to the next line, anticipating that you wish to enter more text. If this is the case, enter the next line of text. When you are through typing text strings, press ENTER a second time at the "Enter Text:" prompt to terminate the command sequence. Unlike the Multiline Text, each line becomes a separate object. Pressing ESC exits the command without placing the last line of text entered.

If you are in the DTEXT command and notice a mistake (or simply want to change a character or word), select one side of the character(s) you wish to change with the cursor and use the BACKSPACE and/or DELETE keys as you would in a word processor. You can also add spaces and characters at this point by just typing them in. However, you must remember to press ENTER at the "Enter text:" prompt to accept the text and terminate the command.

For example, the following command sequence shows placement of left-justified text by providing the starting point of the text, as shown in Figure 7–3.

dtext (ENTER)

Specify start point of text or ⊥: *(specify start point of text as shown in Figure 7–3)*

Specify height <default>: **.25**

Specify rotation angle of text: (enter)

Enter text: **Sample Text Left Justified**

Enter text: (enter)

▶ **Figure 7-3**

Using the DTEXT command to place left-justified text by specifying a start point

Sample Text Left Justified

Start Point

Choosing Text Height

How can you determine the size at which text will plot? As mentioned in Chapter 2, it is recommended to draw objects in their actual size, that is, to real-world dimensions. Even in the case of text, you place them at the real-world dimensions. For instance, the architectural elevation of a building 48' x 24' is drawn to actual size and plotted to a scale of 1/2" = 1'-0'. Let's say you wanted your text to plot at 1/4" high. If you were to create your text and annotations at 1/4", they would be so small relative to the elevation drawing itself that you could not read the characters.

Before you begin placing the text, you need to know at what scale you will eventually plot the drawing. In the previous example of an architectural elevation, the plot scale is 1/2" = 1'-0" and you want the text to plot 1/4" high. You need to find a relationship between 1/4" on the paper and the size of the text for the real-world dimensions in the drawing. If 1/2" on the paper equals 12" in the model, then 1/4"-high text on the paper equals 6", so text and annotations should be drawn at 6" high in the drawing to plot at 1/4" high at this scale of 1/2" = 1'-0". Similarly, you can calculate the various text sizes for a given plot scale. Table 7–1 shows the model text size needed to achieve a specific plotted text height at some common scales.

PLOTTED TEXT HEIGHT										
SCALE	FACTOR	1/16"	3/32"	1/8"	3/16"	1/4"	5/16"	3/8"	1/2"	5/8"
1/16" = 1'-0"	192	12"	18"	24"	36"	48"	60"	66"	96"	120"
1/8" = 1'-0"	96	6"	9"	12"	18"	24"	30"	36"	48"	60"
3/16" = 1'-0"	64	4"	6"	8"	12"	16"	20"	24"	32"	40"
1/4" = 1'-0"	48	3"	4.5"	6"	9"	12"	15"	18"	24"	30"
3/8" = 1'-0"	32	2"	3"	4"	6"	8"	10"	12"	16"	20"
1/2" = 1'-0"	24	1.5"	2.25"	3"	4.5"	6"	7.5"	9"	12"	15"
3/4" = 1'-0"	16	1"	1.5"	2"	3"	4"	5"	6"	8"	10"
1" = 1'-0"	12	0.75"	1.13"	1.5"	2.25"	3"	3.75"	4.5"	6"	7.5"
1 1/2" = 1'-0"	8	0.5"	.75"	1"	1.5"	2"	2.5"	3"	4"	5"
3" = 1'-0"	4	0.25"	.375"	0.5"	0.75"	1"	1.25"	1.5"	2"	2.5"
1" = 10'	120	7.5"	11.25"	15"	22.5"	30"	37.5"	45"	60"	75"
1" = 20'	240	15"	22.5"	30"	45"	60"	75"	90"	120"	150"
1" = 30'	360	22.5"	33.75"	45"	67.5"	90"	112.5"	135"	180"	225"
1" = 40'	480	30"	45"	60"	90"	120"	150"	180"	240"	300"
1" = 50'	600	37.5"	56.25"	75"	112.5"	150"	187.5"	225"	300"	375"
1" = 60'	720	45"	67.5"	90"	135"	180"	225"	270"	360"	450"
1" = 70'	840	52.5"	78.75"	105"	157.5"	210"	262.5"	315"	420"	525"
1" = 80'	960	60"	90"	120"	180"	240"	300"	360"	480"	600"
1" = 90'	1080	67.5"	101.25"	135"	202.5"	270"	337.5"	405"	540"	675"
1" = 100'	1200	75"	112.5"	150"	225"	300"	375"	450"	600"	750"

Table 7–1 *Text Height Corresponding to Specific Plotted Text Size at Various Scales*

Choosing Justification

The Justify option allows you to locate text by using one of the 14 available justification modes. When you select this option, AutoCAD prompts:

> Enter an option ⊡: *(select one of the available options or choose one of the available options from the Shortcut menu)*

The Center option allows you to select the center point for the baseline of the text. Baseline refers to the line along which the bases of the capital letters lie. Letters with descenders, such as g, q, and y, dip below the baseline. After providing the center point, enter the text height and rotation angle.

For example, the following command sequence shows placement of center-justified text, by providing the center point of the text, as shown in Figure 7–4.

> **dtext** (ENTER)
>
> Specify start point of text or ⊡: *(choose Justify from the Shortcut menu)*
>
> Enter an option ⊡: *(choose Center from the Shortcut menu)*

Specify height <default>: **.25**

Specify rotation angle of text:<default>:÷ **0**

Enter text: **Sample Text Center Justified**

Sample Text Center Justified
                    └─ Center Point

Sample Text Middle Justified
                    └─ Middle Point

Sample Text Right Justified
              End Point─┘

The Middle option allows you to center the text both horizontally and vertically at a given point as shown in Figure 7–4. After providing the middle point, enter the text height and rotation angle.

The Right option allows you to place the text with reference to its lower right corner (right justified) as shown in Figure 7–4. Here, you provide the point where the text will end. After providing the right point, enter the text height and rotation angle.

The Align option allows you to place the text by designating the endpoints of the baseline. AutoCAD computes the text height and orientation so that the text just fits proportionately between two points. The overall character size adjusts in proportion to the height. The height and width of the character will be the same.

For example, the following command sequence shows placement of text using the Align option as shown in Figure 7–5.

   dtext (ENTER)

   Specify start point of text or ⊡: *(choose Justify from the Shortcut menu)*

   Enter an option ⊡: *(choose Align from the Shortcut menu)*

   Specify first endpoint of text baseline: *(specify the first point)*

   Specify second endpoint of text baseline: *(specify the second point)*

   Enter text: **Sample Text Aligned**

The Fit option is similar to the Align option, but in the case of the Fit option, AutoCAD uses the current text height and adjusts only the text's width, expanding or contracting it to fit between the points you specify.

For example, the following command sequence shows placement of text using the Fit option as shown in Figure 7–5.

> **dtext** (ENTER)
> Specify start point of text or ⊡: *(choose Justify from the Shortcut menu)*
> Enter an option: *(choose Fit from the Shortcut menu)*
> Specify first endpoint of text baseline: *(specify the first point)*
> Specify second endpoint of text baseline: *(specify the second point)*
> Specify height <default>: **0.25**
> Enter text: **Sample Text Fit**

▶ **Figure 7–5**

Using the Align and Fit options of the DTEXT command to place text

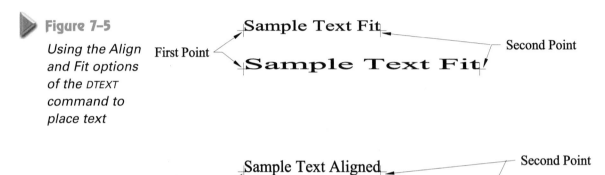

Following are the additional options provided in combination explained earlier:

TL	top left
TC	top center
TR	top right
ML	middle left
MC	middle center
MR	middle right
BL	bottom left
BC	bottom center
BR	bottom right

Choosing a Text Style

All text in an AutoCAD drawing has a text style associated with it. When you enter text, AutoCAD uses the current text style, which sets the font, size, obliquing angle, orientation, and other text characteristics. The Standard style is the default style, and it is assigned with TXT font. With the Style option, you can make another text style current or modify a style. To modify an existing text style or to create a new text style, refer to Chapter 13.

Multiline Text – Creating Text by the Paragraph

 Figure 7-6

Invoking the MTEXT command from the Text toolbar

The MTEXT command draws text by "processing" the words in paragraph form; the width of the paragraph is determined by the user-specified rectangular boundary. It is an easy way to have your text automatically formatted as a multiline group, with left, right, or center justification applied to the group. Features include indents and tabs, making it easier to correctly align your text for tables and numbered lists.

The default text editor is the In-Place Text Editor, but you can elect to use an alternate editor by specifying the editor with the MTEXTED system variable. You can use any text editor, such as Microsoft Notepad, that saves files in ASCII format.

To create paragraph style text in the current style, you can invoke the MTEXT command by choosing Multiline Text from the Text toolbar (see Figure 7–6). AutoCAD prompts:

Specify first corner: *(specify the first corner of the rectangular boundary)*
Specify opposite corner or ⬇: *(specify the opposite corner of the rectangular boundary, or choose one of the available options from the Shortcut menu)*

When you drag the cursor after specifying the first corner of the rectangular boundary, referred to as the In-Place Text Editor, AutoCAD displays an arrow within the rectangle to indicate the direction of the paragraph's text flow. After you specify the opposite corner of the bounding box, AutoCAD displays the Text Formatting toolbar and In-Place Text Editor, shown in Figure 7–7.

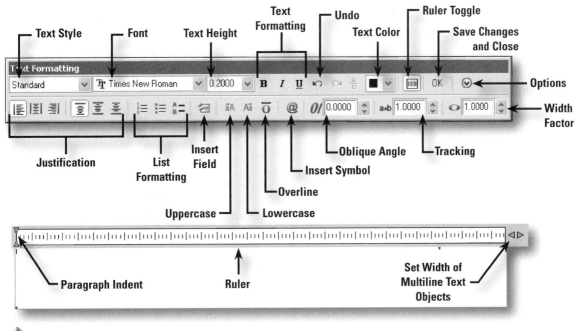

▲ **Figure 7-7**

Text Formatting toolbar, Options toolbar, ruler, and In-Place Text Editor

Before you specify the opposite corner of the In-Place Text Editor, you can choose one of the property settings available from the shortcut menu, which include Height, Justify, Line spacing, Rotation, Style, and Width.

Choosing Height sets the character height in drawing units. Once the text height is specified, AutoCAD returns to the previous prompt to specify the opposite corner of the In-Place Text Editor.

Choosing Justify allows you to place text in one of the nine available justification points, similar to Single Line Text, explained earlier.

Choosing Line spacing sets the spacing between two lines of text. Two options are available: At least and Exactly. The At least option adjusts lines of text automatically based on the height of the largest character in the line. The Exactly option forces the line spacing to be the same for all lines of text in the multiline text object.

Choosing Rotation sets the rotation angle for new or selected text, in the current unit of angle measurement.

Choosing Style allows you to select one of the available styles for the new text.

Choosing Width sets the paragraph width for new text. Once the width is specified, AutoCAD returns to the previous prompt to specify the opposite corner of the In-Place Text Editor.

To edit all or part of the text in the In-Place Text Editor, highlight the text to be edited. The text can be highlighted in the following three ways: holding down the pick button while dragging across the selected text, double-clicking to select an entire word, or triple-clicking to select an entire line of text.

You can create one or more paragraphs of multiline text in the In-Place Text Editor. You can also insert text from a file saved in ASCII or RTF format.

The In-Place Text Editor is transparent so that, as you create text, you can see whether the text overlaps other objects. To turn off transparency while you work, click the bottom edge of the ruler. You can also make the background of the finished multiline text object opaque and set its color.

You can set tabs and indent text to control the appearance of the multiline text object and create lists. Once entered, text can be selected by holding down the pick button while dragging the cursor across the selected text, double-clicking to select the entire word, or triple-clicking to select the entire line of text.

You can also insert special symbols and fields in multiline text. A field is text that is set up to display data that might change. When the field is updated, the latest value of the field is displayed.

The In-Place Text Editor includes a Text Formatting toolbar, Ruler, and a Shortcut menu.

Text Formatting Toolbar

The Text Formatting toolbar at the top of the user-specified rectangle, as shown in Figure 7–7, is a ruler similar to those in word processors. Individual characters, strings of characters, or the entire text in the In-Place Text Editor can be selected for applying formatting styles such as bold, underline, and italics.

The options provided on the Text Formatting toolbar (see Figure 7–7) include character formatting: Style, Font, Text Height, Bold, Italic, Underline, Undo, Redo, Stack, and Text Color.

The **Style** option allows you to apply an existing style to new text or selected text. If you apply a new style to an existing text object, AutoCAD overrides character formatting such as font, height, bold, and italic attributes. Styles that have backwards or upside-down effects are not applied. To modify an existing text style or to create a new text style, refer to Chapter 13.

The **Font** option allows you to specify a font for new text or changes the font of selected text. All of the available TrueType fonts and SHX fonts are listed in the drop-down list box.

The **Text Height** option sets the character height in drawing units. The default value for the height is based on the current style. If the text height of the current style is set to 0, then the initial value of the height is set to the value in the Height box. Each multiline text object can contain a text string of varying text size. When you highlight the text string in the dialog box, AutoCAD displays the selected text height in the list box. If necessary, you can specify a new height in addition to those listed.

The **Bold** option allows you to turn ON and OFF bold formatting for new text or selected text. The Bold option is available only to the characters that belong to a TrueType font.

The **Italic** option allows you to turn ON and OFF italic formatting for new text or selected text. The Italic option is available only to the characters that belong to a TrueType font.

The **Underline** option allows you to turn ON and OFF underlining for new text or selected text.

The **Undo** option undoes the last edit action in the Multiline Text Editor dialog box that includes changes in the content of the text string or formatting.

The **Redo** option cancels the previous Undo.

The **Stack** option is used to place one part of a selected group of text over the remaining part. Before you use the Stack option, the selected text must contain a forward slash (/) to separate the top part (to the left of the /) from the bottom part (to the right of the /). The slash will cause the horizontal bar to be drawn between the upper and lower parts, necessary for fractions with center justification. Instead of a slash (/), you can use the caret (^) symbol. In this case, AutoCAD will not draw a horizontal bar between the upper and lower parts with left justification, which is useful for placing tolerance values.

The **Color** option sets the color for new text or changes it for the selected text. You can assign the color by ByLayer, by ByBlock or one of the available colors.

The **Ruler** option causes the ruler at the top of the bounding box to be displayed or hidden.

Choose **OK** to accept the text and close the In-Place Text Editor. You can also click in the drawing outside the editor to save changes and exit the editor. To close the In-Place Text Editor without saving changes, press ESC.

Choosing the Options icon at the right end of the Text Formatting toolbar causes an abbreviated Text Editor shortcut menu to be displayed. The features listed in the Text Editor shortcut menu are described below under Text Editor Shortcut Menu.

The Options toolbar, attached to the bottom of the Text Formatting toolbar, includes buttons for quick access to some of the options that are also available in the Text Editor shortcut menu. These options include **Left**, **Center**, and **Right** justification, **Top**, **Middle**, and **Bottom** justification, **Numbering** (lists), **Bullets** (lists), and **Uppercase Letters** (lists), **Insert Field**, **Uppercase**, **Lowercase**, **Overline**, **Symbol**, **Oblique Angle**, **Tracking**, and **Width Factor**.

Choosing **Left**, **Center**, or **Right** sets the horizontal justification for new text or selected text.

Choosing **Top**, **Middle**, or **Bottom** sets the vertical justification for new text or selected text.

Choosing **Numbering** causes AutoCAD to establish a numbered list beginning with the number 1 for new text or selected text.

Choosing **Bullets** causes AutoCAD to establish a bulleted list for new text or selected text.

---

note Numbered, lettered, and bulleted lists can be created in the multiline text object. AutoCAD, with the Auto-list option, automatically converts text lines in the specified type of list and increments the numbers and letters appropriately. If one line in the list is removed or added, the numbers or letters of the lines that follow are adjusted accordingly. If you enter a special character, such as an asterisk or hyphen, it will be automatically repeated in subsequent bulleted lines.

You can access the Numbered, Lettered, and Bulleted lists from the Bullets and Lists flyout in the Text Editor shortcut menu, which has the options Restart, Continue, and Auto-list.

Choosing Restart causes the current numbered or lettered line in a list to start over with a "1" or "A" respectively.

Choosing Continue causes the current numbered or lettered line in a list to start with the next number or letter. This is necessary if creation of the lines needs to be continued from where it was interrupted.

Choosing Auto-list when you have started a line with a letter or number followed by a period automatically converts it to a numbered or lettered list designation. For example, if you start a line with "5." or "C." with the Auto-list option on, the next line will begin with "6." or "D." respectively.

Choosing Lettered and then Uppercase from the flyout menu causes AutoCAD to establish an alphabetized list using uppercase letters beginning with the letter A for new or selected text. To use lowercase letters for a list, right-click,choose Bullets and Lists, Lettered and then choose Lowercase from the shortcut menu flyout.

Ruler

At the top of the bounding box is a ruler similar to those in word processors. The ruler allows you to set indents and tabs and adjust the width of the bounding box. To set indents, use the arrows on the left of the ruler. To set the indent of the first line of a paragraph, drag the top arrow to the desired point. To set the indent of the rest of the lines in the paragraph, drag the bottom arrow to the desired point. To set a tab, specify a point on the ruler. To change the width of the bounding box, drag the double arrow on the right end of the ruler to the desired width. You can also right-click in the ruler and select Indents and Tabs or Set Mtext Width from the shortcut menu. These menu items open dialog boxes in which numeric values can be entered for setting indents, tabs, and bounding-box width.

Text Editor Shortcut Menu

In addition to the options available on the toolbars, you can also access additional options from the shortcut menu available within the Text Editor.

Undo, **Redo**, **Cut**, **Copy**, and **Paste** options operate in the same manner as similar commands in other Windows-based word processors and text-handling programs.

The **Learn About Mtext** option causes the New Features Workshop dialog box to be displayed from which you can access help and information about the MTEXT command.

The **Show Toolbar**, **Show Options**, or **Show Ruler** options turn the corresponding toolbars and Ruler ON and OFF.

The **Opaque Background** option allows you to turn the opaque background ON and OFF inside the bounding box.

Inserting Fields

Choosing Insert Field displays the Field dialog box (see Figure 7–8), where you can select a field to insert in the text. When the dialog box closes, the current value of the field is displayed in the MTEXT object. The options available in the Field dialog box change with the field category and field name. For instance, to insert the current system date, first select Date from the **Field Names** list, as shown in Figure 7–8. AutoCAD lists the available options for the selected field in the **Examples** list. The value displayed in the Field dialog box reflects the format that you select. The **Field Expression** displays the expression that underlies the field. The field expression cannot be edited, but you can learn how fields are constructed by reading this area.

Figure 7-8

Field dialog box

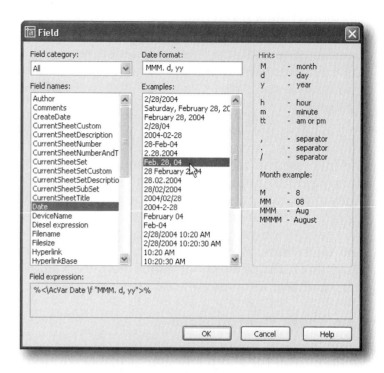

The **Field category** list box allows you to filter the fields that are displayed in the **Field names** list box into the following categories: All, Date & Time, Document, Linked, Objects, Other, Plot, and SheetSet. The **Field names** list box is visible for all categories, but the other text boxes vary with the category. The **Field names** list box allows you to select from the list of available fields according to the filter selected in the **Field category** list box.

Choose All from the **Field category** list box to list all the available fields in the **Field names** list box.

Choose Date & Time from the **Field category** list box to list fields related to the current drawing: CreateDate, Date, PlotDate, and SaveDate. The text boxes that appear when the Date & Time category is selected are **Date format** and **Examples**. The Date format text box displays the format of the example that is selected in the **Examples** text box. For example, if you choose 1/12/05 in the **Examples** text box, then M/d/yy is displayed in the **Date format** text box.

Choose Document from the **Field category** list box to list fields related to working in a document: Author, Comments, Filename, Filesize, HyperlinkBase, Keywords, LastSavedBy, Subject, and Title. The top right text box that appears when the Document category is selected corresponds to the field selected in the **Field names** list box. The items listed in the **Format** list box are (none), Uppercase, Lowercase, First capital, and Title case, except the Filesize item, which has Bytes, Kilobytes, and Megabytes listed.

Choose Linked from the **Field category** list box to list fields related to text display for the hyperlinks. The top right text box that appears when the Linked category is selected is **Text to display**, which allows to you enter a text string that will be inserted in the ultiline text box and a link to the specified hyperlink when selected. Choosing **Hyperlink** causes the Insert Hyperlink dialog box to be displayed. Creation of Hyperlinks is explained in Chapter 21.

Choose Objects from the **Field category** list box to list fields related to named objects: BlockPlaceholder, Formula, NamedObject, and Object.

Choose Other from the **Field category** list box to list fields related to SystemVariable and DieselExpression.

Choose Plot from the **Field category** list box to list fields related to plot-related variables: DeviceName, Login, PageSetupName, PaperSize, PlotDate, PlotOrientation, PlotScale, and PlotStyleTable.

Choose SheetSet from the **Field category** list box to list fields related to sheet set variables.

Select the Field name and corresponding format and choose **OK** to place the field in the In-place text editor.

Symbol

The Symbol option, when selected, causes a shortcut menu to be displayed from which symbols for Degrees, Plus/Minus, and Diameter Degrees can be selected, along with the Non-Breaking Space. Also on the shortcut menu is the Other option, which causes the Character Map dialog box to be displayed. The Character Map dialog box operates in the same manner as the similar dialog box in other Windows-based word processors and text handling programs.

In addition to the options provided in the Multiline Text Editor dialog box for drawing special characters, you can draw them by means of the control characters. The control characters for a symbol begin with a double percent sign (%%). The next character you enter represents the symbol. The control sequences defined by AutoCAD are presented in Table 7–2.

Special Character or Symbol	Control Character Sequence	Example	
		Text String	Control Character Sequence
° (degree symbol)	%%d	104.5°F	104.5%%dF
± (plus/minus tolerance symbol)	%%p	34.5±3	34.5%%p3
⌀ (diameter symbol)	%%c	56.06⌀	56.06%%c
% (single percent sign; necessary only when it must precede another control sequence	%%%	34.67%±1.5	34.67%%%%P1.5
Special coded symbols (where nnn stands for a three-digit code)	%%nnn	@	%%064

Table 7–2 *Control Character Sequences for Drawing Special Characters and Symbols*

Importing Text

The Import Text option, when selected, causes the Select File dialog box to be displayed, from which a file of ASCII or RTF format can be imported. The file imported is limited to 32K.

AutoCAD provides additional options accessible after the first corner of the boundary box has been specified. They include Height, Justify, Line Spacing, Rotation, Style and Width.

The Height, Style, and Justification options have been discussed previously in this section.

The Width option sets the paragraph width for new or selected text. If it is set to the No Wrap option, the resulting multiline text object appears on a single line.

The Rotation option sets the rotation angle for new or selected text, in the current unit of angle measurement.

The Line Spacing option specifies line spacing for the text objects. Two options are available for line spacing: At least and Exactly. The At least selection adjusts lines of text automatically based on the height of the largest character in the line. When At least is selected, lines of text with taller characters have added space between lines. The Exactly selection forces the line spacing to be the same for all lines of text in the multiline text object. Spacing is based on the text height of the object or text style.

 note Exact line spacing is recommended when you use MTEXT to create a table.

Setting Indents and Tabs

The Indents and Tabs option, when selected, causes the Indents and Tabs dialog box to be displayed as shown in Figure 7–9. From this dialog box you can set indentation for paragraphs and the first lines of paragraphs and set tab stops.

 Figure 7-9

The Indents and Tabs dialog box

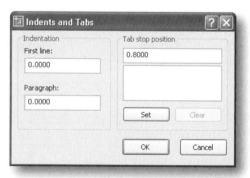

In the **Indentation** section of the Indents and Tabs dialog box, the **First line** text box lets you set the indentation of the first line of the current or selected paragraphs. The **Paragraph** text box lets you set the indentation for the current or selected paragraph. In the **Tab stop position** section, the upper text box lets you set tab positions, which are then recorded in the lower text box. Choosing **Set** causes the values in the upper text box to be copied to and listed in the lower text box. Choosing **Clear** causes the selected tab stop to be removed from the list.

Setting Justification

The Justification option, when selected, causes a shortcut menu to be displayed from which the customary AutoCAD Text justification and alignment for new or selected text options can be selected. Text is centered, left or right justified, with respect to the left and right text boundaries, and aligned from the middle, top, or bottom of the paragraph with respect to the top and bottom text boundaries. If the selected justification is one of the "top" options (Top Left, Top Center, Top Right), excess text will "spill" out of the bottom of the specified boundary box. If the selected justification is one of the "bottom" options (Bottom Left, Bottom Center, Bottom Right), excess text will "spill" out of the top of the specified boundary box.

You can also set the justification by selecting the Justify option when AutoCAD prompts for the opposite corner of the rectangular boundary. The available justification options are the same as in the DTEXT command. Once justification is specified, AutoCAD returns to the previous prompt until the opposite corner of the rectangular boundary is specified.

Find and Replace

The Find and Replace option, when selected, causes the Find and Replace dialog box to be displayed (see Figure 7–10), which includes the **Find what** and **Replace with** boxes, and **Match Case** and **Match whole word only** check boxes. These are used to search for specified text strings and replace them with new text.

Type the text string to be searched for in the **Find what** box, and then choose the **Find Next** to start the search. AutoCAD highlights the appropriate text string in the bounding box. To continue the search, choose **Find Next** again.

 Figure 7–10

Find and Replace dialog box

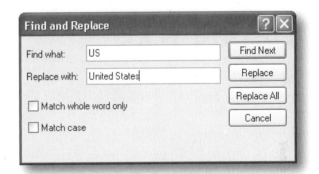

In the **Replace with** box, type the text string that you want as replacement for the text string in the **Find what** box. Then choose **Replace** to replace the highlighted text with the text in the **Replace with** box. If you choose **Replace All**, all instances of the specified text will be replaced.

When the **Match Case** check box is selected, AutoCAD finds text only if the case of all characters in the text object are matched to that of the text characters in the **Find what** box. When it is not selected, AutoCAD finds a match for the specified text string regardless of the case of the characters.

When the **Match whole word only** check box is selected, AutoCAD finds text only if the text string is a single word. If the text is part of another text string, it is ignored. When it is not selected, AutoCAD finds a match for the specified text string whether it is a single word or part of another word.

Select All

The Select All option, when selected, causes all the text in the multiline text object to be selected and highlighted.

Change Case

The Change Case option, when selected, causes a Shortcut menu to be displayed with options for UPPERCASE and lowercase. Selecting UPPERCASE causes any selected and highlighted text to be uppercase. Selecting lowercase causes any selected and highlighted text to be lowercase.

AutoCAPS

The AutoCAPS option, when selected, operates in a manner similar to pressing the CAPS LOCK key, toggling Caps Lock on and off.

Remove Formatting

The Remove Formatting option, when selected, removes any bold, italics, or underlining formatting to selected text.

Combine Paragraphs

The Combine Paragraphs option, when selected, combines selected paragraphs into a single paragraph and replaces each paragraph return with a space.

After making the necessary changes to the available options choose OK to place the text in the drawing.

EDITING TEXT

Figure 7-11

Invoking the DDEDIT command from the Text toolbar

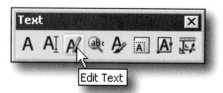

The DDEDIT command allows you to edit text and attributes. An attribute is informational text associated with a block. See Chapter 17 for a detailed discussion of blocks and attributes. To edit text, you can invoke the DDEDIT command from the Text toolbar (see Figure 7-11). AutoCAD prompts:

Select an annotation object or ⬇: *(select the text or attribute definition, or choose one of the available options from the Shortcut menu)*

If you select text created by means of the MTEXT command, AutoCAD displays the Multiline Text Editor with the selected text objects. Make the necessary changes in the text, and choose **OK** to keep the changes.

If you select a text string created by means of a DTEXT command, AutoCAD displays the Edit Text dialog box with the selected text objects. Make the necessary changes in the text string and choose **OK** to apply the changes.

AutoCAD continues to prompt you to select a new text string to edit, or you can select the U option to undo the last change made to the text. To terminate the command sequence, give a null response (press ENTER).

FINDING AND REPLACING TEXT

Figure 7–12

Invoking the FIND command from the Text toolbar

The FIND command (see Figure 7–12) is used to find a string of specified text and replace it with another string of specified text. AutoCAD displays the Find and Replace dialog box, similar to Figure 7–13.

Figure 7–13

Find and Replace dialog box

In the **Find text string** text box, type the string that you wish to find. In the **Replace with** text box, type the new text string. Choose **Find**, and an instance of the string to be replaced is displayed in the **Context** area of the **Search results** section of the dialog box. Select **Replace** to replace this instance with the string in the **Replace with**. To skip over this instance without replacing it, select **Find**. To replace all instances of the string entered in the **Find text string**, select **Replace All**.

In the **Search in** box, you can direct AutoCAD to search either in the Entire drawing or the Current selection, for the string to be replaced. **Select objects** returns you to the Graphics screen to select text objects. AutoCAD then searches for the string to be replaced. Selecting **Options** causes the Find and Replace Options dialog box to appear, similar to Figure 7–14.

Figure 7–14

Find and Replace Options dialog box

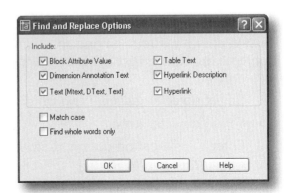

The **Include** section of the Find and Replace Options dialog box lets you filter the type of text to be included in the Find search. Optional categories for text to be included are **Block Attribute Value**, **Dimension Annotation Text**, **Text (Mtext and Dtext)**, **Table Text**, **Hyperlink Description**, and **Hyperlink**. The **Match case** check box, when checked, causes AutoCAD to include text strings whose case matches that of the specified text string to search. The **Find whole words only** check box, when checked, causes AutoCAD to include only whole words that match the specified text string.

SCALING TEXT

 Figure 7-15

Invoking the SCALETEXT command from the Text toolbar

The SCALETEXT command (see Figure 7–15) is used to increase or reduce the size of the selected text without changing their locations.

AutoCAD prompts:

> Select objects: *(select a text object)*
>
> Enter a base point option for scaling <Right>: *(choose a base point for scaling or press* ENTER *to leave base point unchanged or select an option from the shortcut menu)*
>
> Specify new height or ⬇ <3/16">: *(specify new height)*

With the base point prompt, choose one of the optional justification points to serve as base point for scaling, which is used individually for each selected text object. The base point for scaling can be established based on one of several standard justification insertion point locations for text options. Even though the options are the same as when you choose an insertion point, the justification of the text objects is not affected.

JUSTIFYING TEXT

 Figure 7-16

Invoking the JUSTIFYTEXT command from the Text toolbar

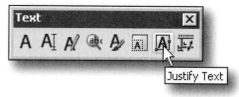

The JUSTIFYTEXT command (see Figure 7–16) lets you change the justification point of a text string without having to change its location. You can choose single line text objects, multiline text objects, leader text objects, and attribute objects. AutoCAD prompts:

> Select objects: *(select a text object)*
>
> Select objects: *(enter)*
>
> Enter a justification option <Left>: *(select one of the available options to change the current justification)*

SPELL CHECKING

The SPELL command is used to correct the spelling of text objects created with the DTEXT or MTEXT command in addition to attribute values in blocks. Invoke the SPELL command from the Tools menu, AutoCAD prompts:

Select objects: *(select one or more text strings, and press* ENTER *to terminate object selection)*

AutoCAD displays the Check Spelling dialog box, similar to Figure 7–17, only when it finds a dubious word in the selected text objects.

Figure 7–17

Check Spelling dialog box

AutoCAD displays the name of the current dictionary in the top of the Check Spelling dialog box. If necessary, you can change to a different dictionary by choosing **Change Dictionaries** and selecting the appropriate dictionary from the Change Dictionaries dialog box.

AutoCAD displays each misspelled word in the **Current Word** section and lists the suggested alternate spellings in the **Suggestions** list box. Choose **Change** to replace the current word with the selected suggested word, or choose **Change All** to replace all instances of the current word. Alternatively, choose **Ignore** to skip the current word, or choose **Ignore All** to ignore all subsequent entries of the current word.

Choose **Add** to include the current word (up to 63 characters) in the current or custom dictionary. Choose **Lookup** to check the spelling of the word in the **Suggestions** box.

After completion of the spelling check, AutoCAD displays a message informing you that the spelling check is complete.

CONTROLLING THE DISPLAY OF TEXT

The QTEXT command is a utility command for DTEXT and MTEXT that is designed to reduce the redraw and regeneration time of a drawing. Regeneration time becomes a significant factor if the drawing contains numerous text objects and attribute information and/or if a fancy text font is used. Using QTEXT, the text is replaced with rectangular boxes of a height corresponding to the text height. These boxes are regenerated in a fraction of the time required for the actual text.

If a drawing contains many text and attribute items, it is advisable to set QTEXT to ON. However, before plotting the final drawing, or to review text details, set the QTEXT command to OFF and invoke the REGEN command. You can invoke the QTEXT command from the On-screen prompt:

> **qtext** (ENTER)
> Enter mode [ON/OFF] *(select one of the available options)*

TABLES

AutoCAD's TABLE command makes it easy to create tables that contain text in row and column format customarily found on drawings for listing revisions, finish schedules, specifications, and other structured textual information. A combination of table characteristics such as row and column sizes, border lineweights, text alignments and associated text styles, and colors can be saved in a Table Style with a specified name to be recalled and applied to tables when required.

Inserting Tables

Figures 7–18 and 7–19 show typical examples of tabular information included in drawings. After you invoke the TABLE command (see Figure 7–20), AutoCAD displays the Insert Table dialog box as shown in Figure 7–21.

ROOM FINISH SCHEDULE				
ROOM NO.	ROOM DESCRIP	WALLS	FLOOR	CEILING
100	ENTRY	PLASTER	CARPET	SUSP. ACOUS.
101	HALL	GYPSUM	CARPET	GYPSUM
102	RECEPTION	PANELLING	VINYL TILE	GYPSUM
103	OFFICE	GYPSUM	CARPET	GYPSUM

 Figure 7–18

Example of Room Finish Schedule drawn as table from top down

3	CLOSET DIM	T.A.S.	01/23/04
2	WDW DETAIL	G.V.K.	01/12/04
1	DOOR 101	A.B.C.	01/08/04
REVISION NO.	DESCRIPTION	BY	DATE
REVISIONS			

 Figure 7–19

Example of Revision History drawn as table from bottom up

 Figure 7–20

Invoking the TABLE command from the Draw toolbar

Figure 7-21

Insert Table dialog box

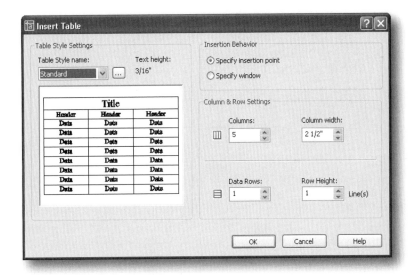

The **Table Style name** text box in the **Table Style Settings** section of the Insert Table dialog box enables you to select the table style to be applied. Refer to Chapter 13 for information about creating and modifying table styles.

The **Insertion Behavior** section of the Insert Table dialog box allows you to specify which characteristics of the table controls insertion and determines how selections can be made in the **Column & Row Settings** section.

Choosing **Specify window** creates a table based on the constraints set in the **Column & Row Settings** section of the table. You can specify the number of columns with the column width automatic, or you can specify the column width with the number of columns automatic. You can specify the number of **Data Rows** with the **Row height** automatic, or you can specify the number of lines for the Row height and the number of **Data Rows** automatic.

Choosing **Specify window** creates a table based on the constraints set in the **Column & Row Settings** section of the table. You can specify the number of columns with the column width automatic, or you can specify the column width with the number of columns automatic. You can specify the number of Data Rows with the Row height automatic, or you can specify the number of lines for the Row height and the number of Data Rows automatic.

After setting up the Table Style and Column/Row configuration, choose **OK** to close the Insert Table dialog box. AutoCAD then prompts you to "Specify first corner:" and shows a phantom image of the potential table attached to the cursor following its movement.

Once you specify the first corner, if you have chosen **Specify insertion point** in the **Insertion Behavior** section, AutoCAD draws the table, shows the text entering cursor in the heading at the justification specified, and displays the Text Formatting toolbar. If you have chosen **Specify window** in the **Insertion Behavior** section, AutoCAD prompts you to "Specify second point:" allowing you to drag the cursor to determine the number of columns and rows desired. Then AutoCAD displays the Text Formatting dialog box.

While the Text Formatting toolbar is being displayed, text can be input into each cell by typing at the keyboard or pasting from the Clipboard. You can move from a cell to the cell below by pressing the ENTER key, to the cell above by pressing SHIFT + ENTER, to the cell to the right by pressing the TAB key, or to the cell to the left by pressing SHIFT + TAB.

Editing Text in a Cell

The TABLEDIT command edits text in a table cell. AutoCAD prompts:

Command: **tabledit**

Pick a table cell: *(Click inside a table cell, and enter text or use the toolbar or the select one of the available options from the Shortcut menu to make changes)*

You can also edit text by selecting the cell in which the text has to be changed and choosing Edit Cell Text from the Shortcut menu. AutoCAD displays the Text Formatting toolbar; make the necessary changes and choose OK to close the text editor.

Additional options are available from the Shortcut menu when a text is selected in a cell. The options include the following: **Cell Alignment** allows you to change the alignment of the selected text in a cell; **Cell Borders** sets the properties of the borders of table cells; **Match Cell** applies the properties of a selected table cell to other table cells; **Insert Block** allows you to insert a block or drawing that is stored locally or in a network; **Insert Columns** allows you to insert a column right or left of the selected cell; **Delete Columns** deletes the selected column; **Insert Rows** allows you to insert a row above or below the selected cell; **Delete Rows** deletes the selected row(s); **Delete Cell Contents** deletes the text objects in the selected cell(s).

Modifying Tables

Similar to modifying the individual cells, AutoCAD allows you to modify the table. First select the table, and then select the available options from the Shortcut menu. The options include the following: **Size Columns Equally** resizes the columns to equal width; **Size Rows Equally** resizes the rows to equal width; **Remove all property overrides** removes any property overrides applied to the selected table; **Export** allows you to export all the text objects in the table to a text file in *.csv* format.

drawing exercises

Open the Exercise Manual PDF file for Chapter 7 on the accompanying CD for discipline-specific exercises.

review questions

1. While you use the DTEXT command, AutoCAD will display the text you are typing:
  a. *In the command prompt area.*
  b. *In the drawing screen area.*
  c. *Both a and b.*
  d. *Neither a nor b.*

2. By default, the Start Point option of DTEXT indicates which of the following?
  a. *The lower left corner of the text*   **c.** *The lower right corner of the text*
  b. *The center of the text*   **d.** *The upper left corner of the text*

3. The default text rotation angle in AutoCAD (measured in a counterclockwise direction) is which of the following?
  a. *0 degrees*   **d.** *The Last value used*
  b. *Three o'clock*   **e.** *All of the above*
  c. *East*

4. Which of the following options allows you to place text by designating the endpoints of the baseline?
  a. *Align*   **c.** *Both a and b*
  b. *Fit*   **d.** *None of the above*

5. Which of the following options uses the current text height, but adjusts the text width?
  a. *Align*   **c.** *Half width*
  b. *Fit*   **d.** *Width*

6. Additional lines of text placed while still in the DTEXT command will appear:
  a. *Directly below the previous line of text.*
  b. *At a new point specified by moving the crosshairs cursor to a new point.*
  c. *Either a or b.*
  d. *None of the above.*

7. Which of the following commands "processes" text in paragraph form?
  a. *TEXT*   **c.** *MTEXT*
  b. *DTEXT*   **d.** *All of the above*

8. Text can be highlighted by doing which of the following?
  a. *Holding down the pick button while dragging across the selected text*
  b. *Double-clicking to select an entire word*
  c. *Triple-clicking to select a paragraph*
  d. *All of the above*

9. The Import Text option of the Multiline Text Editor can import:

    a. *ASCII formats.*   **c.** *TIF formats.*

    b. *RTF formats.*   **d.** *Both a and b.*

10. The QTEXT utility command reduces regeneration time when:

    a. *The drawing contains a great amount of text.*

    b. *A great amount of attribute information is included in the drawing.*

    c. *Fancy text fonts are used.*

    d. *All of the above.*

11. The degree symbol, the plus/minus tolerance symbol and the diameter symbol can be placed with text using which of the following?

    a. *Unicode characters*

    b. *Control codes*

    c. *The Symbol option of the Multiline Text Editor*

    d. *All of the above*

12. AutoCAD considers a text string (all the characters that comprise a line of text) as a single object.

    a. *True*   **b.** *False*

13. Text imported into the AutoCAD Multiline Text Editor is limited to a 16KB file size.

    a. *True*   **b.** *False*

14. The justification options for MTEXT are the same as those for DTEXT commands.

    a. *True*   **b.** *False*

15. An arrow in the Text Boundary Window indicates the direction of flow in DTEXT.

    a. *True*   **b.** *False*

the size of it all

introduction

Earlier chapters explained how to draw objects in AutoCAD, communicating their shape with combinations of lines, circles, arcs, and other drawing elements. Dimensions communicate the size of objects and the location of features on an object.

After completing this chapter, you will be able to do the following:

- Draw linear dimensioning

- Draw aligned dimensioning

- Draw angular dimensioning

- Draw diameter and radius dimensioning

- Draw arc length dimensioning

- Draw ordinate dimensioning

- Find area of a closed object

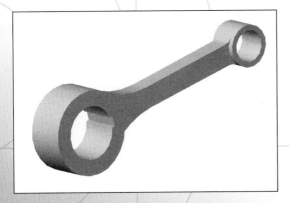

 Figure 8-1

Pictorial view of lever arm

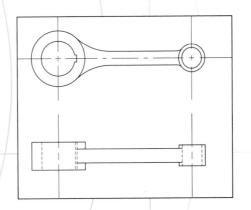

 Figure 8-2

Orthogonal views of lever arm

Figure 8–1 is a three-dimensional drawing of a lever arm. This "pictorial" rendering gives a viewer a good idea of what the object looks like, especially if the viewer has trouble understanding engineering drawings. Figure 8–2 is a typical engineering drawing of the same object. Because of the symmetry of the object, two views are sufficient. But in order to communicate size and location information necessary to make the object, the drawing in Figure 8–3 has the vital data that the manufacturer needs. Although other information such as material and finish specifications and tolerances will be needed, this section covers how to use AutoCAD to produce dimensions and notes for this type of drawing, and also drawings for other disciplines such as architecture, civil engineering, and surveying.

AutoCAD provides commands to draw the full range of dimension types: Linear, Angular, Diameter/Radius, and Ordinate. Each type includes primary and secondary commands. For example, Linear dimensioning includes Horizontal, Vertical, Aligned, and Rotated. Figure 8–3 has both horizontal and vertical formats of the Linear dimension type.

---

note   Horizontal and Vertical formats of Linear dimensions refer to their direction in the drawing, not their orientation on the object or in real space. Horizontal dimensions are aligned with the X axis (orthogonal left to right) in the drawing coordinate system and Vertical dimensions are aligned with the Y axis (orthogonal up and down).

▶ **Figure 8-3**

Dimensioned drawing of lever arm

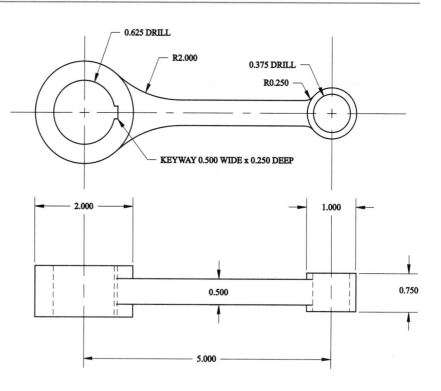

DIMENSION TERMINOLOGY

Figure 8–4 shows the terms for the different components of a typical dimension in AutoCAD.

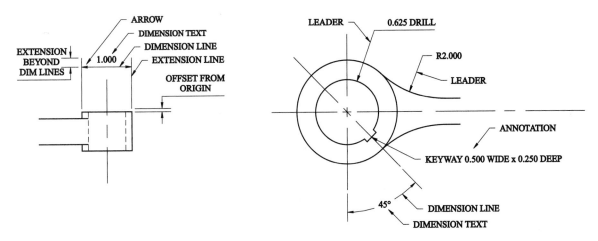

 Figure 8–4

Components of a typical dimension

LINEAR DIMENSIONS

 Figure 8–5

Invoking the Linear Dimension command from the Dimension toolbar

Linear dimensions include the Horizontal, Vertical, Aligned, and Rotated formats. The Command that draws Horizontal and Vertical formats of the Linear dimensions can be invoked from the Dimension toolbar shown in Figure 8–5. Examples of these formats of dimensioning are shown in Figure 8–6.

Figure 8–6

Horizontal and Vertical Linear Dimensions

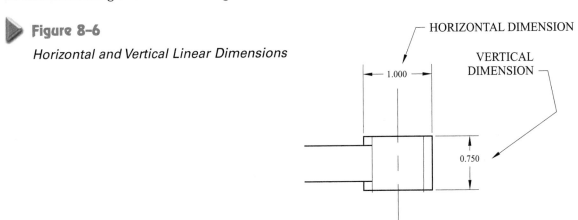

Horizontal and Vertical Dimensions by Selecting Critical Points

After invoking the Linear Dimension command, select the three critical points on the drawing (see Figure 8–7) to cause the proper dimension to be drawn: "first extension line origin," "second extension line origin," and "dimension line location" as prompted for in the following command sequence:

Specify first extension line origin or <select object>: *(specify point A)*

Specify second extension line origin: *(specify point B)*

Specify dimension line location or ⏬: *(specify point C)*

▶ **Figure 8–7**

Selecting the three critical points for the horizontal dimension

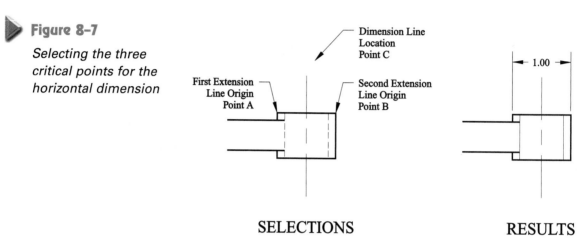

SELECTIONS   RESULTS

Points A and B indicate the location for the first extension line origin and second extension line origin respectively, and while you are moving the cursor to specify the point C, AutoCAD shows a dynamic preview (in dashed lines that follow the cursor movement) of what the dimensions will look like. Once the third point is specified, the dimension is drawn.

---

note   The points specified for the first extension line origin and second extension line origin must be selected with an appropriate Object Snap mode in effect, such as Endpoint or Intersection. This assures that the dimension will be accurate (not just close). This also assures that if the dimension is Associative (explained later in the chapter), changes to the object will be accurately reflected in changes to the dimension. For the point specified for the dimension line location, only the Y component (in the case of horizontal dimensions) of the coordinates is used to locate the dimension line. That is, moving the cursor perpendicular to a line through points A and B (vertically) determines how far away the dimension line is located. Moving the cursor parallel to a line through points A and B (horizontally) has no effect on the location of the dimension line.

The vertical dimension as shown in Figure 8–8 can be drawn in a manner similar to the horizontal dimension, using points A, B, and C in response to the prompts for the three critical points.

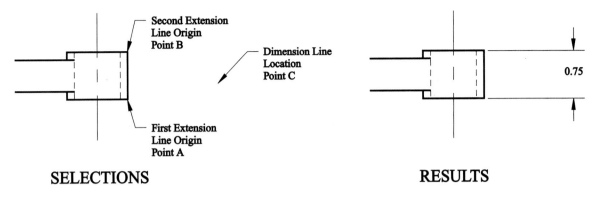

SELECTIONS   RESULTS

 Figure 8-8

Selecting the three critical points for the vertical dimension

Horizontal and Vertical Dimensions by Selecting an Object

Instead of specifying two end points, you can select an object, and AutoCAD automatically determines the origin points of the first and second extension lines. The following command sequence shows an example of drawing a linear dimension by selecting an object (see Figure 8–9).

Specify first extension line origin or <select object>: (ENTER)

Select object to dimension: *(select Line A)*

Specify dimension line location or ⊡: *(select point for dimension line location)*

Figure 8-9

Selecting the Object for the horizontal dimension

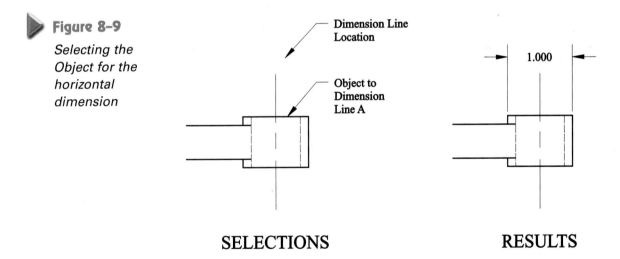

SELECTIONS   RESULTS

The vertical dimension can be drawn in a manner similar to the horizontal dimension by selecting a vertical line as the object to be dimensioned.

Dynamic Horizontal/Vertical Dimensioning

Dynamic horizontal/vertical dimensioning is an option available after specifying two points that are not on the same horizontal or vertical line. That is, they can be considered diagonally opposite corners of an imaginary rectangle with both width and height. After the two points are specified, you are prompted to specify the location of the dimension line. You will also be shown a preview image of where the dimension will be drawn from the cursor location relative to the imaginary rectangle formed by the two points. If the cursor is above the top line or below the bottom line of the rectangle, then the dimension will be horizontal. If the cursor is to the right of the right side or to the left of the left side of the rectangle, then the dimension will be vertical. If the cursor is dragged to one of the outside quadrants or inside the rectangle, it will maintain the type of dimension in effect before the cursor was moved.

Changing Dimension Text

The Mtext and Text options allow you to change the measured dimension text. The Angle option allows you to change the rotation angle of the dimension text. After responding appropriately for Text or Angle, AutoCAD repeats the prompts for the dimension line location.

Forcing a Horizontal or Vertical Dimension

The Horizontal option forces a horizontal dimension to be drawn (even when the dynamic drag switching calls for a vertical dimension). Likewise, the Vertical option forces a vertical dimension to be drawn (even when the dynamic drag switching calls for a horizontal dimension).

Rotated Dimensions

The Rotated option allows you to draw the dimension at a specified angle that is neither horizontal nor vertical; nor is the desired angle at the angle determined by the first two points specified, as it is in the case of aligned dimensioning (see the "Aligned Dimensions" section in this chapter). Figure 8–10 shows a situation where Rotated Dimension is applicable. Here is a case where the desired dimensions are from point A to point C and from point A to point D. However, the desired angle of dimensioning is the angle created by a line from point A to point B. The dimension from point A to point C can be started by choosing the Rotated option, after specifying points A and C. Then points A and B are specified to determine the angle.

Choose Rotated from the Shortcut menu, and then AutoCAD prompts:

Specify angle of dimension line <0>: *(Specify the dimension line angle or specify points A and B to determine the angle)*

The points specified to determine the angle are not parallel to the direction of the dimensioned distance.

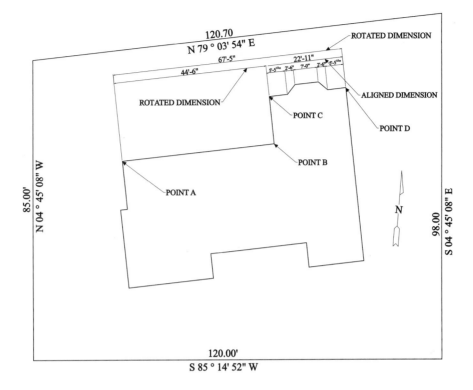

▷ **Figure 8-10**

*Example of
using rotated
dimensioning*

ASSOCIATIVE DIMENSIONS

Dimensions in AutoCAD can be drawn associative, non-associative, or exploded, depending on the setting of the DIMASSOC dimensioning system variable. The setting for associative dimensions is 2 (default), non-associative is 1, and exploded dimensions is 0.

You associate a dimension (DIMASSOC dimensioning system variable set to 2) with an object by selecting points on the object (using an Object Snap mode) when prompted to do so during a dimensioning command. If the object is subsequently modified in a manner that changes the location(s) of one or both of the selected points, the associated dimension is automatically updated to correctly indicate the new distance or angle. Associative dimensioning does not support multilines. The association between a dimension and a block is lost when the block is redefined.

The elements of an associative dimension are drawn as a single object. Therefore, if any one of its members is selected for modifying, all members are highlighted and subject to being modified. This is similar to the manner in which member objects of a block reference are treated. (See Chapter 17 for information about Blocks.) In addition to the customary visible parts, AutoCAD draws point objects at the extension line origins, where the measurement actually occurs on the object. If you have dimensioned the width of a rectangle with an associative dimension and then specified one end of the rectangle to stretch, the dimension will also be stretched, and the Dimension Text will be changed to correspond to the new measurement.

The non-associative dimensions are drawn while the DIMASSOC dimensioning system variable is set to 1. If the object that was dimensioned is selected for modification without the dimension being selected, the dimension itself will remain unchanged. But if both dimension and the object are selected for modification, then dimension text will reflect the new measurement. The elements of a non-associative dimension are drawn as a single object, similar to an associative dimension.

The DIMDISASSOCIATE command converts selected dimensions that are associated with geometric objects to non-associative dimensions.

The DIMREASSOCIATE command allows you to change a non-associative dimension to associate to geometric objects. You can also change the existing associations in an associative dimension.

Exploded dimensions are drawn while the DIMASSOC dimensioning system variable is set to 0, and the members are drawn as separate objects. If one of the components of the dimension is selected for modifying, that component will be the only one modified.

note An associative dimension can be converted to an exploded dimension with the EXPLODE command, and the DIMDISASSOCIATE command can convert an associate dimension to a non-associative dimension. Once the dimension is exploded, you cannot recombine the separate parts back into the associative dimension from which they were exploded (except by means of the UNDO command if feasible), but you can convert a non-associate dimension to an associate dimension with the help of the DIMREASSOCIATE command. Note that when you explode an associative dimension, the measurement-determining points (nodes) remain in the drawing as point objects.

Figure 8–11 shows an example of a dimensioned object that has been revised. If the second extension line origin of the dimension on the right is moved horizontally with the STRETCH command, the associative dimension and the dimension text will reflect the new location. But the change involved relocation in the X and Y direction as shown in Figure 8–11 (revised object). Instead of having to erase and then redraw the original dimension, the dimension can be disassociated and then reassociated.

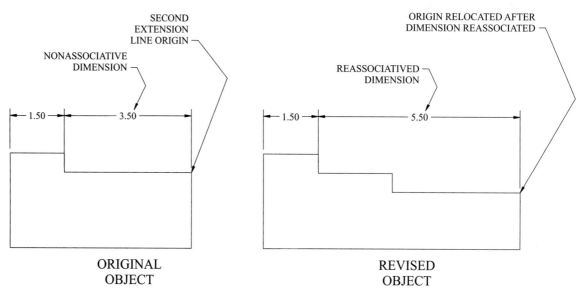

Figure 8–11

Non-associative dimension that has been reassociated

ALIGNED DIMENSIONS

Invoke the Aligned Dimension command from the Dimension toolbar (see Figure 8–12) to draw aligned dimension. When dimensioning at a non-orthogonal angle, it may be necessary to align the dimension with an object line or a line determined by two specified points. The Aligned Dimension command creates an aligned linear dimension by selecting three critical points on the drawing (see Figure 8–13) to cause the proper dimension to be drawn: "first extension line origin", "second extension line origin", and "dimension line location" as prompted in the following command sequence:

Specify first extension line origin or <select object>: *(specify point A)*

Specify second extension line origin: *(specify point B)*

Specify dimension line location or [Mtext/Text/Angle]: *(specify point C)*

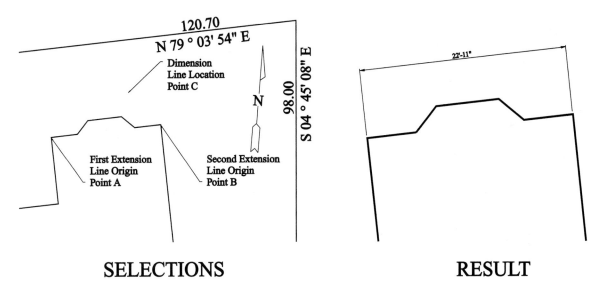

SELECTIONS   RESULT

▲ **Figure 8–13**

Drawing aligned dimensioning for a line drawn at an angle

ORDINATE DIMENSIONING

 Figure 8–14

Invoking the Ordinate Dimension command from the Dimension toolbar

Invoke the Ordinate Dimension command from the Dimension toolbar (see Figure 8–14) to draw ordinate dimensions. Ordinate dimensions measure the perpendicular distance from an origin point called the datum to a dimensioned feature, such as a hole in a part. AutoCAD uses the mutually perpendicular X and Y axes of the World Coordinate System or current User Coordinate System as the reference lines from which to base the X or Y coordinate displayed in an ordinate dimension (sometimes referred to as a datum dimension). In the following examples, Figure 8–15 is valid when the base of the rectangle lies on the X axis, giving it a Y value of 0.0000, and Figure 8–16 is valid when the left side of the rectangle lies on the Y axis, giving it an X value of 0.0000. If this were not the case, where the objects are located in the drawing and you still wished to have their values to be 0.0000, a new coordinate system would have to be created (even if temporarily for drawing the dimensions).

Specify feature location: *(specify a point)*

Although the default prompt is "Specify feature location:", AutoCAD is actually looking for a point that is significant in locating a feature point on an object, such as the endpoint/intersection where planes meet or the center of a circle representing a hole or shaft. Therefore, an Object Snap mode, such as Endpoint, Intersection, Quadrant, or Center, will normally need to be invoked when responding to the "Specify feature location:" prompt. Specifying a point determines the origin of a single orthogonal leader that will point to the feature when the dimension is drawn. AutoCAD prompts:

Specify leader endpoint or [Xdatum/Ydatum/Mtext/Text/Angle]: *(specify a point or select one of the available options from the Shortcut menu)*

If the Ortho mode is set to ON, the leader will be a single horizontal line for a Ydatum ordinate dimension, as shown in Figure 8–15, or a single vertical line for an Xdatum ordinate dimension, as shown in Figure 8–16.

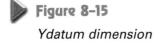 Figure 8–15

Ydatum dimension

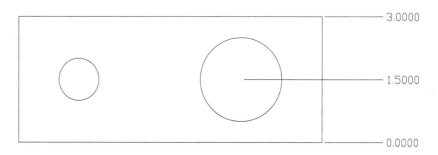

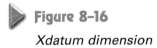

Figure 8–16

Xdatum dimension

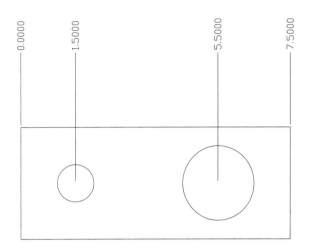

If the Ortho mode is set to OFF, the leader will be a three-part line consisting of orthogonal lines on each end joined by a diagonal line in the middle. It may be necessary to use the non-orthogonal leader if the text has to be offset to keep from interfering with other objects in the drawing. The type of dimension drawn (Ydatum or Xdatum) depends on which is greater of the horizontal and vertical distances between the specified feature location point and the leader endpoint point. A preview image of the dimension is displayed during specification of the leader endpoint.

Choose Xdatum or Ydatum from the Shortcut menu, and AutoCAD then draws an Xdatum dimension or Ydatum dimension, regardless of the location of the leader endpoint relative to the feature location point.

RADIUS DIMENSIONING

Figure 8–17

Invoking the Radius Dimension command from the Dimension toolbar

Invoke the Radius Dimension command from the Dimension toolbar (see Figure 8–17) to draw radius dimenions. The Radius dimensioning feature provides commands to create radius dimensions, as shown in Figure 8–18, for a circle and an arc. The type of dimensions that AutoCAD utilizes depends on the Dimensioning System Variable settings (refer to Chapter 13 for how to change the Dimensioning System Variable settings to draw appropriate radius dimensioning).

Figure 8–18

Radius dimensioning of a circle and an arc

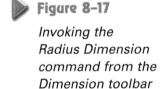

After invoking the Radius Dimension command, select the circle as shown in Figure 8–19 and then a point to locate the text and the end of the leader as prompted for in the following command sequence to cause the proper dimension to be drawn:

Select arc or circle: *(select the circle object)*
Specify dimension line location or ⬇: *(specify a point to draw the dimension leader line or select one of the available options from the Shortcut menu)*

▶ **Figure 8–19**

Radius dimensioning of a circle

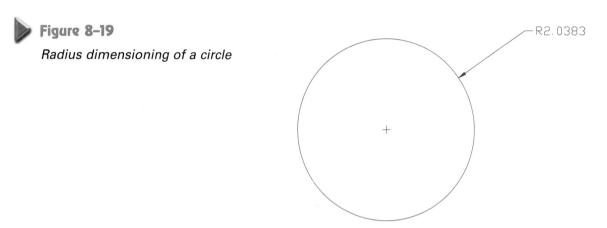

DIAMETER DIMENSIONING

▶ **Figure 8–20**

Invoking the Diameter Dimension command from the Dimension toolbar

Invoke the Diameter Dimension command from the Dimension toolbar (see Figure 8–20) to draw diameter dimensions. The Diameter dimensioning feature provides commands to create diameter dimensions, as shown in Figure 8–21 for a circle and an arc. The type of dimensions that AutoCAD utilizes depends on the Dimensioning System Variable settings (refer to Chapter 13 for how to change the Dimensioning System Variable settings to draw an appropriate diameter dimensioning).

▶ **Figure 8–21**

Diameter dimensioning of a circle and an arc

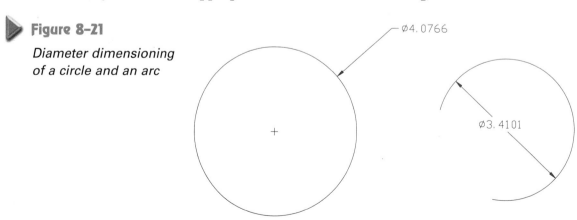

Specifying a point determines the location of the diameter dimension. The dimension for an arc of less than 180 degrees cannot be forced to where neither end of the dimension is on the arc.

After invoking the Diameter Dimension command, select the circle as shown in Figure 8–22 and then a point to locate the text and the end of the leader as prompted for in the following command sequence to cause the proper dimension to be drawn:

Select arc or circle: *(select an arc or a circle to dimension)*

Specify dimension line location or ⬇: *(specify a point to draw the dimension or select one of the available options from the Shortcut menu)*

▶ **Figure 8–22**

Diameter dimensioning of a circle

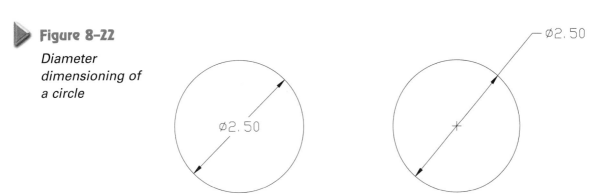

ARC LENGTH DIMENSIONING

▶ **Figure 8–23**

Invoking the Arc Length from the Dimension toolbar

Invoke the Arc Length from the Dimension toolbar (see Figure 8–23) to draw arc length dimensions (see Figure 8-24).

▶ **Figure 8–24**

Arc Length Dimensioning

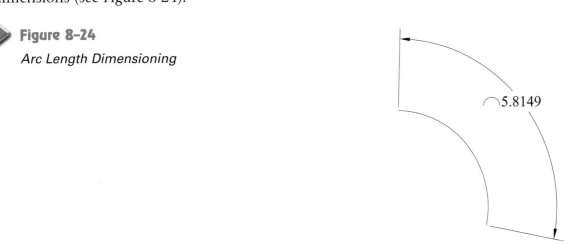

The Mtext, Text, and Angle options are the same as those in linear dimensioning, explained earlier in this section.

Selecting Partial allows you to specify a point other than the second endpoint of the arc, and AutoCAD will dimension the length of the portion of the arc between the first endpoint of the arc and the point specified (see Figure 8–25).

Figure 8-25

Partial Arc Length dimensioning of an arc

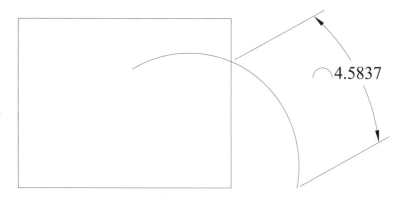

⌒4.5837

ANGULAR DIMENSIONING

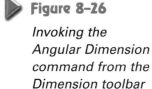

Figure 8-26

Invoking the Angular Dimension command from the Dimension toolbar

Invoke the Angular Dimension command from the Dimension toolbar (see Figure 8–26) to draw angular dimensions. The Angular dimensioning feature allows you to draw angular dimensions using three points (vertex, point, point), between two nonparallel lines, on an arc (between the two endpoints of the arc, with the center as the vertex), and on a circle (between two points on the circle, with the center as the vertex).

The default method of angular dimensioning is to select an object.

If the object selected is an arc, as shown in Figure 8–27, AutoCAD automatically uses the arc's center as the vertex and its endpoints for the first angle endpoint and second angle endpoint respectively to determine the three points of a Vertex/Endpoint/Endpoint angular dimension. AutoCAD prompts:

Specify dimension arc line location or ⊡: *(specify the location for the dimension arc line or select one of the available options from the Shortcut menu)*

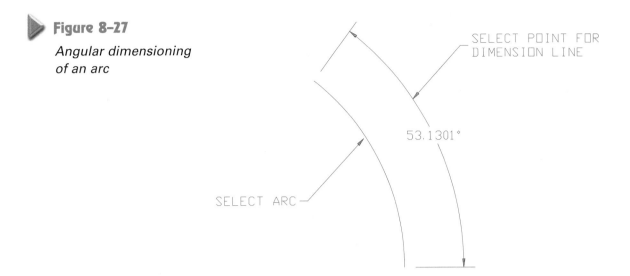

Figure 8-27

Angular dimensioning of an arc

If the object selected is a line, as shown in Figure 8–28, then AutoCAD prompts:

Select second line: *(select a line object)*

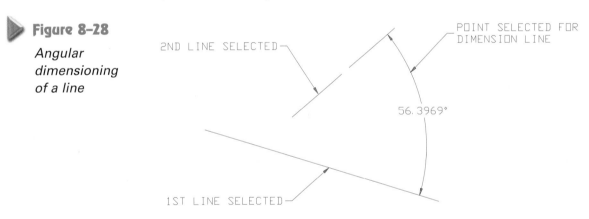

Figure 8-28

Angular dimensioning of a line

AutoCAD uses the actual or apparent intersection of the two lines as the vertex for drawing a Vertex/Vector/Vector angular dimension. You are then prompted to specify the location of the dimension arc, which will always be less than 180 degrees.

If the dimension arc is beyond the end of either line, AutoCAD adds the necessary radial extension line(s). Then AutoCAD prompts:

Specify dimension arc line location or ⊡: *(specify the location for the dimension arc line or select one of the available options from the Shortcut menu)*

If you specify a point for the location of the dimension arc, AutoCAD automatically draws extension lines and draws the dimension text.

If you provide a null response instead of selecting an arc, a circle, or two lines for angular dimensioning, AutoCAD allows you to do three-point angular dimensioning. The following command sequence shows an example of drawing angular dimensioning by providing three data points, as shown in Figure 8–29.

Select arc, circle, line, or <specify vertex>: (ENTER)

Specify angle vertex: *(Point 1)*

Specify first angle endpoint: *(Point 2)*

Specify second angle endpoint: *(Point 3)*

Specify dimension arc line location or ⬇: *(specify the dimension arc line location or select one of the available options from the Shortcut menu)*

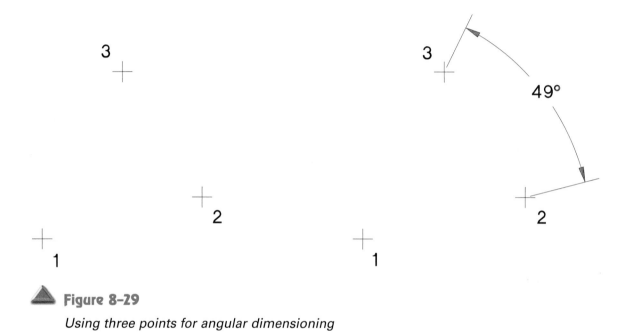

▲ Figure 8–29

Using three points for angular dimensioning

SETTING DIMENSION SCALE

Dimensioning should achieve three major objectives: first, dimensions should convey accurate sizes and locations with respect to objects and their features; second, all of the components of the dimensions and notes should comply with conventions applicable to the type of drawing; third, the dimensions and their components should be clear, readable, and the proper size on the plotted sheet.

AutoCAD's default dimensioning components are proportioned to be the most acceptable sizes when plotted to full scale, that is, a rectangle that is drawn 2" wide by 4" high will be plotted at its true size, 2" wide by 4" high. And if the dimensions are drawn without making any changes to the Dimensioning variables from the way they came "out-of-the-box" with AutoCAD, they will closely comply with the conventions for mechanical drawings. The drawing in Figure 8–30 shows the sizes of the default dimensioning components when plotted at full scale.

Figure 8–30

Sizes of AutoCAD's default dimensioning components plotted to full scale

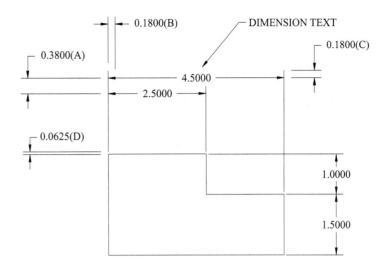

With reference to Figure 8–30, components and spacing in the default dimensioning are as follows:

- Spacing between dimension lines in Ordinate dimensioning (A) = 0.3800".
- Arrow length (B) = 0.1800".
- Dimension lines end in closed filled arrows.
- Extension line past dimension line (C) = 0.1800".
- Extension line offset from origin (D) = 0.0625".
- Dimension text height = 0.1800".
- Dimension text is placed centered between extension lines.
- Dimension line is broken for text to be in line.
- Dimension precision is four decimal places.
- Dimension text is horizontal in vertical dimensions.

When plotted to full scale, the size of the components of the default dimensions are clear and readable. However, if you wish to draw an object that is 24" wide by 36" high on a letter size (8 1/2" x 11") sheet, it will have to be plotted at a reduced scale. If the plotted object is reduced to one-fourth of its true size, it will be plotted 6" wide and 9" high. This will fit on the 8 1/2" x 11" sheet. This reduction can be done during plotting as an option in the PLOT command. However, the dimension components will be reduced proportionately, making them practically unreadable. Instead of the text height being 3/16" high, they will plot to less than 1/16" high.

In order to assure that dimensions will be drawn in a reduced scale drawing with all of the components plotted at the same clear and readable size as they were at full scale, you can change the size of all the components by changing the DIMSCALE system variable. In the example above, because the drawing will be reduced by a factor of 4, the value of the DIMSCALE system variable should be set to 4, as shown in the following command sequence:

dimscale (ENTER)
Enter new value for DIMSCALE <1.0000>: **4**

In a similar manner, if the drawing is an architectural plan or elevation that needs to be plotted at a scale of 1/4" = 1'- 0", which is a ratio of 1:48, then the value of the DIMSCALE system variable must be set at 48. This means that if the Dimension Text Height in the current Dimension Style is set at 3/16" (the size it will plot at full scale with the DIMSCALE set to 1.0), then it will plot at 3/16" when plotted at a reduced scale of 1:48 with the DIMSCALE set to 48.

---

 note   Changing the DIMSCALE value affects the size of the dimensioning components only. It does not affect the value of the dimensions. If an object is 33" long, the dimension text indicating its length will read "33.00" (when drawn and dimensioned at full scale) regardless of the setting of the DIMSCALE system variable. With the Dimension Text Height set to its default of 3/16" and the DIMSCALE set to 48, the dimension text will be drawn to scale at a height of 9" (3/16" times 48) but will be plotted at a height of 3/16" when plotted to a scale of 1/4"=1'- 0".

The dimensioning components can be changed individually by entering the associated dimension variable name at the Command: prompt or on the appropriate tab of the Modify Dimension Style dialog box, accessible through the Dimension Style Manager dialog box. Detailed explanation is provided in Chapter 13.

FINDING THE AREA

▷ **Figure 8-31**

Invoking the Area command from the Inquiry toolbar

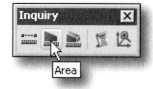

The AREA command is used to report the area (in square units) and perimeter of a selected closed geometric figure, such as a circle, polygon, or closed polyline. You may also specify a series of points that AutoCAD considers a closed polygon, compute, and report the area.

Invoke the AREA command from the Inquiry toolbar, AutoCAD prompts:

Specify first corner point or ⬇: *(specify a point or select one of the available options from the Shortcut menu)*

The default option calculates the area when you select the vertices of the objects. If you want to know the area of a specific object such as a circle, polygon, or closed polyline, choose the Object option.

The following command sequence is an example of finding the area of a polygon using the Object option, as shown in Figure 8-32.

Specify first corner point or or ⬇: *(choose Object from the shortcut menu)*
Select objects: *(select an object, as shown in Figure 8-32)*

Figure 8-32

Finding the area of a polygon using the AREA command Object option

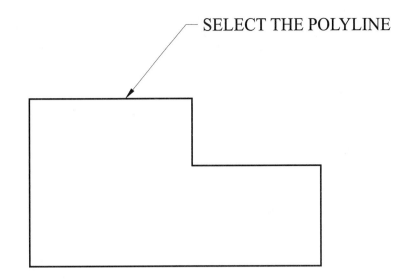

SELECT THE POLYLINE

The Add option allows you to add selected objects to form a total area; then you can use the Subtract option to remove selected objects from the running total.

The following example demonstrates the application of the Add and Subtract options. In this example, the area is determined for the closed shape after subtracting the area of the four circles, as shown in Figure 8-33.

area (ENTER)

Specify first corner point or ⊡: *(choose Add from the shortcut menu)*

Specify first corner point or ⊡: *(choose Object from the shortcut menu)*(ADD mode) Select objects: *(select the polyline, as shown in Figure 8-33)*

Area = 9.2500, Perimeter = 14.0000

Total area = 9.2500

(ADD mode) Select objects:

Specify first corner point or ⊡: *(choose Subtract from the shortcut menu)*

Specify first corner point or ⊡: *(choose Object from the shortcut menu)*

(SUBTRACT mode) Select objects: *(select circle A, as shown in Figure 8-33)*

Area = 0.1556, Circumference = 1.3984

Total area = 9.0944

(SUBTRACT mode) Select objects: *(select circle B, as shown in Figure 8-33)*

Area = 0.1556, Circumference = 1.3984

Total area = 8.9388

(SUBTRACT mode) Select objects: *(select circle C, as shown in Figure 8-33)*

Area = 0.2766, Circumference = 1.8645

Total area = 8.6621

(SUBTRACT mode) Select objects: *(select circle D, as shown in Figure 8-33)*

Area = 0.2766, Circumference = 1.8645

Total area = 8.3855

(SUBTRACT mode) Select objects: (ENTER)
Specify first corner point or ⊡: (ENTER)

Figure 8-33

Using the Add and Subtract options of the AREA command

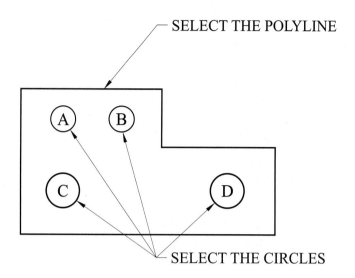

SELECT THE POLYLINE

SELECT THE CIRCLES

drawing exercises

Open the Exercise Manual PDF file for Chapter 8 on the accompanying CD for discipline-specific exercises.

review questions

1. Dimension types available in AutoCAD include:
   a. *Linear.*   **d.** *Radius.*
   b. *Angular.*   **e.** *All of the above.*
   c. *Diameter.*

2. The associative dimension drawn with the DIMASSOC variable set to 1 has all of its separate parts drawn as separate objects.
   a. *True*   **b.** *False*

3. The Linear Dimensioning command allows you to draw horizontal, vertical, and aligned dimensions.
   a. *True*   **b.** *False*

4. To draw a linear dimension you can (1) specify the first extension line origin, (2) locate the dimension line, and then (3) specify the second extension line.
   a. *True*   **b.** *False*

5. The Angular Dimension command allows you to draw angular dimensions between two parallel lines.
   a. *True*   **b.** *False*

6. By default, the dimension text for a radius dimension is preceded by:
   a. *Radius.*
   b. *Rad.*
   c. *R.*

7. The angular dimensioning command allows you to draw angular dimensions by:
   a. *Selecting an arc, circle, or line.*   **c.** *Selecting three points.*
   b. *Selecting an angle vertex.*   **d.** *Both a and c.*

8. When dimensioning a circle or an arc, AutoCAD allows you to use a:
   a. *Radius dimension only.*   **c.** *Either a radius or a diameter dimension.*
   b. *Diameter dimension only.*   **d.** *None of the above.*

9. You can enter the dimensioning commands:
   a. *From the keyboard.*   **c.** *Both a and b.*
   b. *From the Dimensioning toolbar.*   **d.** *None of the above.*

10. The linear dimension command allows you to dimension:
    a. *Horizontally.*   **c.** *Aligned with the object.*
    b. *Vertically.*   **d.** *All of the above.*

11. The DIMALIGNED command is used to dimension non-orthogonal objects.

   a. *True* **b.** *False*

12. The radius dimensioning icon will always be a leader extending from the center of the circle past the circle's edge.

   a. *True* **b.** *False*

13. By default, the dimension text for a radius is preceded by an "R".

   a. *True* **b.** *False*

14. Extension lines will always be perpendicular to the object that is dimensioned.

   a. *True* **b.** *False*

15. An arrowhead is used some times to indicate the end of a dimension line.

   a. *True* **b.** *False*

16. Dimensioning text can include numbers, words, characters, and symbols.

   a. *True* **b.** *False*

17. Dimensions in AutoCAD can be associative or non-associative.

   a. *True* **b.** *False*

18. While using the ordinate dimensions, if Ortho is turned on, then the leader will be a single horizontal line.

   a. *True* **b.** *False*

19. While in the ordinate dimensioning mode, if Ortho is turned off, then the leader will be a straight line perpendicular from the object.

   a. *True* **b.** *False*

A Bigger World

creating complex objects

introduction

This chapter introduces some of the basic commands and concepts in AutoCAD that can be used to create more complex geometric objects, such as construction lines, closed shapes, and multiple and segmented objects.

After completing this chapter, you will be able to do the following:

- Create construction lines using the XLINE and RAY commands

- Construct closed shapes using the ELLIPSE, DONUT, SOLID, REVCLOUD, and WIPEOUT commands

- Construct multiple and segmented objects using PLINE, MLINE, and SPLINE commands

DRAWING CONSTRUCTION LINES

This section covers straight-line objects that can be used primarily as guidelines from which to create other objects.

Construction Line Drawn by Xline

 Figure 9-1

Invoking the XLINE command from the Draw toolbar

The xline command allows you to draw lines that extend infinitely in both directions from the point selected when being created. After you invoke the xline command from the Draw toolbar (see Figure 9–1), AutoCAD prompts:

Specify a point or ⬇: *(specify a point or choose one of the available options from the Shortcut menu)*

When you specify a point to define the root of the construction line, this point becomes the conceptual midpoint of the construction line. AutoCAD prompts:

> Specify through point: *(specify a point through which the construction line should pass)*

AutoCAD draws a line that passes through two points and extends infinitely in both directions. AutoCAD continues to prompt for additional through points to draw construction lines. To terminate the command sequence, press enter or the spacebar.

The **Horizontal** option allows you to draw a construction line through a point that you specify and parallel to the X axis of the current UCS.

The **Vertical** option allows you to draw a construction line through a point that you specify and parallel to the Y axis of the current UCS.

The **Angle** option allows you to draw a construction line at a specified angle. AutoCAD prompts:

> Enter angle of xline or ⬇: *(specify an angle at which to place the construction line)*
> Specify through point: *(specify a point through which the construction line should pass)*

AutoCAD draws the construction line through the specified point, using the specified angle.

The **Reference** option allows you to draw a construction line at a specific angle for a selected reference line. The angle is measured counterclockwise from the reference line.

The **Bisect** option allows you to draw a construction line through the first point bisecting the angle determined by the second and third points, with the first point being the vertex. AutoCAD prompts:

> Specify angle vertex point: *(specify a point for the vertex of an angle to be bisected and through which the construction line will be drawn)*
> Specify angle start point: *(specify a point to determine one line of an angle to be bisected)*
> Specify angle endpoint: *(specify a point to determine second line of angle to be bisected)*

The construction line lies in the plane determined by the three points.

The **Offset** option allows you to draw a construction line parallel to and at the specified distance from the line object selected and on the side selected. AutoCAD prompts:

> Specify offset distance or ⬇: *(specify an offset distance or choose one of the available options from the Shortcut menu)*
> Select a line object: *(select a line, pline, ray, or xline)*
> Specify side to offset? *(specify a point to draw a construction line parallel to the selected object)*

The **Through** option allows you to specify a point through which a construction line is drawn to the line object selected.

Construction Line Drawn by Ray

The ray command allows you to draw lines that extend infinitely in one direction from the point selected when the line is being created. Invoke the ray command from Draw menu, AutoCAD prompts:

Specify start point: *(specify the start point to draw the ray)*

Specify through point: *(specify a point through which you want the ray to pass)*

Specify through point: *(specify a point to draw additional rays or press* ENTER *to terminate the command sequence)*

The ray is drawn starting at the first point and extending infinitely in one direction through the second point. AutoCAD continues to prompt for through points until you provide a null response to terminate the command sequence.

CLOSED SHAPES

In addition to the rectangle and circle, AutoCAD lets you draw more advanced closed shapes with the ellipse, donut, revcloud, and wipeout commands. These commands simplify the creation of ellipses, filled circles, revision clouds, and blank areas respectively.

Drawing Ellipses

 Figure 9-2

Invoking the ELLIPSE command from the Draw toolbar

AutoCAD allows you to draw an ellipse or an elliptical arc with the ellipse command by various methods. After you invoke the ellipse command (see Figure 9–2), AutoCAD prompts:

Specify axis endpoint of ellipse or ⬇: *(specify axis endpoint of the ellipse to be drawn or choose one of the available options from the Shortcut menu)*

Drawing an Ellipse by Defining Axis Endpoints

The Defining Axis Endpoints option allows you to draw an ellipse by specifying the endpoints of the axes. AutoCAD prompts for two endpoints of the first axis. The first axis can define either the major or the minor axis of the ellipse. Then AutoCAD prompts for an endpoint of the second axis as the distance from the midpoint of the first axis to the specified point. An ellipse is drawn with the command sequence as shown in the following example. AutoCAD prompts:

Specify axis endpoint of ellipse or : **1,1**

Specify other endpoint of axis: **5,1**

Specify distance to other axis or : **3,2**

AutoCAD draws an ellipse whose major axis is 4.0 units long in a horizontal direction and whose minor axis is 2.0 units long in a vertical direction, as shown in Figure 9–3.

▶ **Figure 9–3**

An ellipse drawn by specifying the major and minor axes

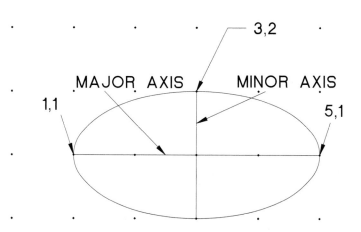

Drawing an Ellipse by Defining the Center of the Ellipse

The Center option allows you to draw an ellipse by defining the center point and axis endpoints. First, AutoCAD prompts for the ellipse center point. Then AutoCAD prompts for an endpoint of an axis as the distance from the center of the ellipse to the specified point. The first axis can define either the major or the minor axis of the ellipse. An ellipse is drawn with the command sequence as shown in the following example. AutoCAD prompts:

Specify axis endpoint of ellipse or : *(choose Center from the Shortcut menu)*

Specify center of ellipse: **3,1**

Specify endpoint of axis: **1,1**

Specify distance to other axis or : **3,2**

AutoCAD draws an ellipse similar to the previous example, with a major axis 4.0 units long in a horizontal direction and a minor axis 2.0 units long in a vertical direction.

Drawing an Ellipse by Specifying the Rotation Angle

AutoCAD allows you to draw an ellipse by specifying a rotation angle after defining two endpoints of one of the two axes. The rotation angle defines the major-axis-to-minor-axis ratio of the ellipse by rotating a circle about the first axis. AutoCAD draws a circle if you set the rotation angle to 0 degrees. An ellipse is drawn with the command sequence as shown in the following example. AutoCAD prompts:

Specify axis endpoint of ellipse or ⊻: **3,–1**

Specify other endpoint of axis: **3,3**

Specify distance to other axis or ⊻: *(choose Rotation from the Shortcut menu)*

Specify rotation around major axis: **30**

See Figure 9–4 for examples of ellipses with various rotation angles.

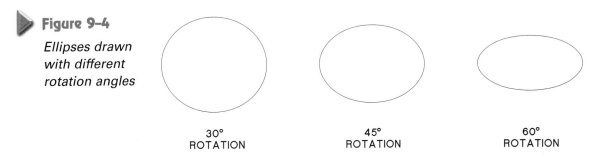

▶ **Figure 9–4**

Ellipses drawn with different rotation angles

30°
ROTATION

45°
ROTATION

60°
ROTATION

Drawing an Elliptical Arc

The Arc option allows you to draw an elliptical arc. After you specify the major and minor axis endpoints, AutoCAD prompts for the start and end angle points for the elliptical arc to be drawn. Instead of specifying the start angle or the end angle, you can toggle to the Parameter option, which prompts for the Start parameter and End parameter point locations. AutoCAD creates the elliptical arc using the following parametric vector equation:

$$p(u) = c + a^X + \cos(u) + b^X \sin(u)$$

Here c is the center of the ellipse, and a and b are its major and minor axes, respectively. Instead of specifying the end angle, you can also specify the included angle of the elliptical arc to be drawn.

The ellipse command, with the Arc option, is accessible directly from the Draw toolbar, or you can select the Arc option while in the ellipse command. An elliptical arc is drawn with the command sequence as shown in the following example. AutoCAD prompts:

Specify axis endpoint of ellipse or ⊻: *(choose Arc from the Shortcut menu)*

Specify axis endpoint of elliptical arc or ⊻: **1,1**

Specify other endpoint of axis: **5,1**

Specify distance to other axis or ⊻: **3,2**

Specify start angle or ⊻: **3,2**

Specify end angle or ⊻: **1,1**

AutoCAD draws an elliptical arc with the start angle at 3,2 and the ending angle at 1,1.

Drawing Isometric Circles (or Isocircles)

By definition, Isometric Planes (*iso* meaning "same" and *metric* meaning "measure") are all being viewed at the same angle of rotation (see Figure 9–4). The angle is approximately set to 54.7356 degrees. AutoCAD uses this angle of rotation automatically when you wish to represent circles in one of the isoplanes by drawing ellipses with the Isocircle option.

Normally, a circle 1 unit in diameter being viewed in one of the isoplanes will project a short axis dimension of 0.577350 units. One of its diameters parallel to an isoaxis will project a dimension of 0.816497 units. A line drawn in isometric that is parallel to one of the three main axes will also project a dimension of 0.816497 units. We would like these lines and circle diameters to project a dimension of exactly 1.0 unit. Therefore, you automatically increase the entire projection by a fudge factor of 1.22474 (the reciprocal of 0.816497) in order to be able to use true dimensioning parallel to one of the isometric axes.

This means that circles 1 unit in diameter will be measured along one of their isometric diameters rather than along their long axis. This facilitates using true lengths as the lengths of distances projected from lines parallel to one of the isometric axes. So an isocircle 1 unit in diameter will project a long axis that is 1.224744871 units and a short axis that is 0.707107 (0.577350 x 1.22474) units. These "fudge" factors are built into AutoCAD isocircles.

The Isometric Circle method is available as one of the options of the ellipse command when you are in the isometric Snap mode (see Figure 9–5).

> Specify axis endpoint of ellipse or ⊡: *(choose Isocircle from the Shortcut menu)*
>
> Specify center of isocircle: *(select the center of the isometric circle)*
>
> Specify radius of isocircle or ⊡: *(specify the radius or choose one of the available options from the Shortcut menu)*

 Figure 9–5

Ellipses drawn using the Isocircle option of the
ELLIPSE *command*

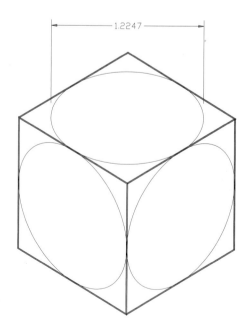

If you override the last prompt default by choosing Diameter from the Shortcut menu, the following prompt will appear:

Circle diameter: *(specify the desired diameter)*

 The Iso and Diameter options will work only when you are in the Isometric Snap mode.

Drawing Solid-Filled Circles

▶ **Figure 9–6**

Invoking the DONUT command from the Draw menu

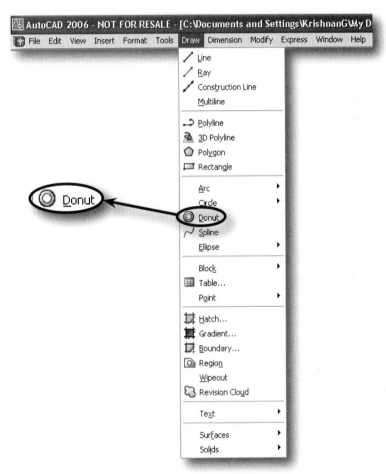

The doughnut (or donut) command (see Figure 9–6) lets you draw solid-filled circles and rings by specifying outer and inner diameters of the filled area. The fill display depends on the setting of the fillmode system variable. A doughnut-shaped object is drawn with the command sequence in the following example. AutoCAD prompts:

Specify inside diameter of donut <current>: *(specify a distance, or press enter to accept the current setting)*

Specify outside diameter of donut <current>: *(specify a distance, or press enter to accept the current setting)*

Specify center of donut or <exit>: *(specify a point to draw the donut)*

You can specify the inside and outside diameters of the donut to be drawn by specifying two points at the appropriate distance apart on the screen for either or both diameters, and AutoCAD will use the measured distance for the diameter(s). Or you can enter the distances from the keyboard.

You can select the center point by entering its coordinates or by selecting it with your pointing device. After you specify the center point, AutoCAD prompts for the center of the next donut and continues prompting for subsequent center points. To terminate the command, enter a null response.

note   Be sure the FILLMODE system variable is set to ON (a value of 1). Check at the Command: prompt by typing **fill** and pressing ENTER. If FILLMODE is set to OFF (value of 0), then the PLINE, TRACE, DONUT, and SOLID commands display only the outline of the shapes. With FILLMODE set to ON, the shapes you create with these commands appear solid. If FILLMODE is reset to ON after it has been set to OFF, you must use the REGEN command in order for the screen to display as filled any unfilled shapes created by these commands. Switching between ON and OFF affects only the appearance of shapes created with the PLINE, TRACE, DONUT, and SOLID commands. Solids can be selected or identified by specifying the outlines only. The interior of a solid area is not recognized as an object when specified with the pointing device.

The following command sequence shows placement of a solid-filled circle, as shown in Figure 9–7 (figure on the left), by use of the donut command.

Specify inside diameter of donut <0.5000>: **0**

Specify outside diameter of donut <1.0000>:**1**

Specify center of donut or <exit>: **3,2**

Specify center of donut or <exit>: (ENTER)

▷ **Figure 9–7**

Using the DONUT command to place a solid-filled circle and a filled circular shape

INSIDE DIA. 0   INSIDE DIA..5

The following command sequence shows placement of a filled circular shape, as shown in Figure 9–7 (figure on the right), by use of the donut command.

Specify inside diameter of donut <0.0000>: **0.5**

Specify outside diameter of donut <1.0000>: **1**

Specify center of donut or <exit>: **6,2**

Specify center of donut or <exit>: (ENTER)

Drawing Solid-Filled Polygons

 Figure 9–8

*Invoking the 2D
Solid command
from the Surfaces
flyout menu of
the Draw menu*

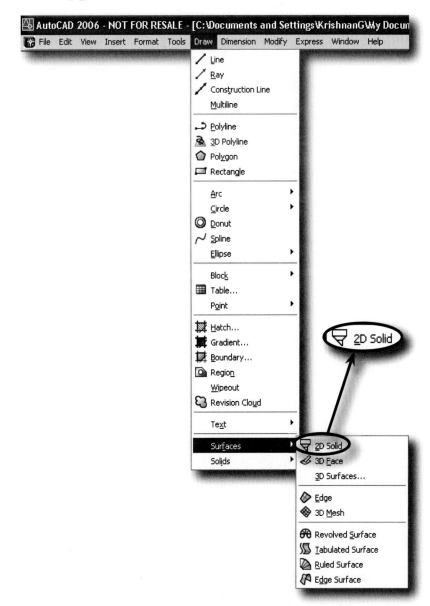

The solid command (see Figure 9–8) creates a solid-filled straight-sided area whose outline is determined by points you specify on the screen. Two important factors should be kept in mind when using the solid command: first, the points must be selected in a specified order or else the four corners generate a bowtie instead of a rectangle; second, the polygon generated has straight sides. (Closer study reveals that even filled donuts and pline-generated curved areas are actually straight-sided, just as arcs and circles generate as straight-line segments of small enough length to appear smooth.) A solid-filled polygonal object is drawn with the command sequence in the following example. AutoCAD prompts:

Specify first point: *(specify a first point)*

Specify second point: *(specify a second point)*

Specify third point: *(specify a third point diagonally opposite the second point)*

Specify fourth point or <exit>: *(specify a fourth point, or press* ENTER *to exit)*

When you specify a fourth point, AutoCAD draws a quadrilateral area. If instead you press enter, AutoCAD creates a filled triangle.

For example, the following command sequence shows how to draw a quadrilateral area, as shown in Figure 9–9.

Specify first point: *(pick PT. 1)*
Specify second point: *(pick PT. 2)*
Specify third point: *(pick PT. 3)*
Specify fourth point or <exit>: *(pick PT. 4)*
Specify third point: (ENTER)

 Figure 9–9

The pick order for creating a quadrilateral area with the SOLID command

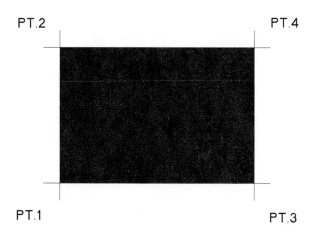

To create the solid shape, the odd-numbered picks must be on one side and the even-numbered picks on the other side. If not, you get an effect such as shown in Figure 9–10.

Figure 9–10

Results of using the SOLID command when odd/even points are not specified correctly

You can use the solid command to create an arrowhead or triangle, such as the one shown in Figure 9–11. Polygon shapes can be created with the solid command by keeping the odd picks along one side and the even picks along the other side of the object, as shown in Figure 9–12.

Figure 9–11

Using the SOLID command to create a solid triangular shape

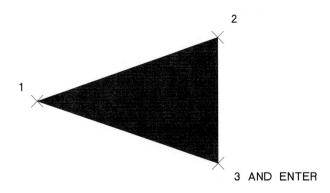

Figure 9–12

Using the SOLID command to create a polygonal shape

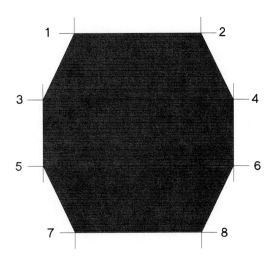

Revision Clouds

Figure 9–13

Invoking the REVCLOUD command from the Draw toolbar

The revcloud command (see Figure 9–13) allows you to draw a connected series of arcs encircling objects in a drawing to signify an area on the drawing that has been revised, as shown in Figure 9–14. A revision cloud is drawn with the command sequence in the following example. AutoCAD prompts:

Specify start point or ⊡:*(specify the start point of the revision cloud)*
Guide crosshairs along cloud path... *(move the cursor crosshairs along the path of the desired revision cloud)*

When the revision cloud is almost complete and the cursor crosshairs approach the starting point, the cloud is automatically closed without requiring any additional action or input. AutoCAD prompts:

Revision cloud finished.

Figure 9–14

An example of using the REVCLOUD command

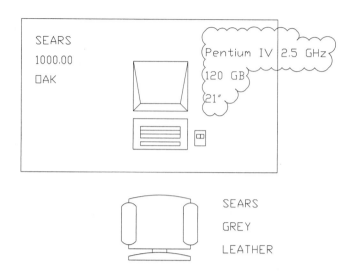

The size range of the arcs can be set for the desired appearance when prompted for minimum and maximum arc lengths. This value is then multiplied by the value of the dimension variable dimscale to compensate for drawings with different scale factors. The Object option allows you to select a closed shape (polyline, rectangle, circle, etc.) from which AutoCAD creates a revision cloud. You are then prompted:

Reverse direction [Yes/No] <No>: *(choose Yes or No)*

Selecting Yes causes the arcs to be redrawn on the opposite side of the line/arc defining the cloud. Selecting No causes the revision cloud to remain as drawn.

Wipeout

Figure 9–15

Invoking the WIPEOUT command from the Draw menu

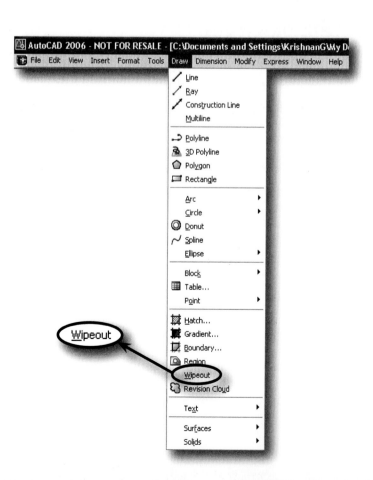

The wipeout command (see Figure 9–15) allows you to create an area on the screen that obscures previously drawn objects within its boundary. These areas can be displayed with or without a visible boundary (called a frame). A wipeout area is drawn with the command sequence in the following example. AutoCAD prompts:

> Specify first point or ⤓:(*specify the start point of the wipeout area*)
>
> Specify next point: (*specify the second point*)
>
> Specify next point or ⤓: (*specify next point or choose one of the available options from the Shortcut menu*)

Having drawn a connected series of lines, you can give a null reply (press enter) to terminate the wipeout command. The series of responses determines the polygonal boundary of the wipeout object from a series of points.

Setting the Frame Display

The Frames option determines whether the edges of all wipeout objects are displayed or hidden.

AutoCAD prompts

> Enter mode ON/OFF <existing setting>: (*Enter **on** or **off***)

Choosing ON displays all wipeout frames, and choosing OFF suppresses the display of all wipeout frames.

Converting a Polyline to Wipeout Boundary

The Polyline option allows you to select a polyline, which determines the polygonal boundary of the wipeout area.

AutoCAD prompts:

> Select a closed polyline: (*Use an object selection method to select a closed polyline*)
>
> Erase polyline? [Yes/No]<No>: (*choose Yes or No*)

Choose Yes to erase the polyline that was used to create the wipeout object. Choose No to retain the polyline.

---

note The area and its frame created by the WIPEOUT command will cover existing objects. Objects drawn after and covering the area will not be hidden by the area. If one or more of the objects being covered is modified, by the MOVE command for example, then it will no longer be covered. Likewise, if the wipeout area is modified, it will then cover all objects that overlap it.

DYNAMIC MULTIPLE AND COMBINED SEGMENTS

AutoCAD provides commands for creating three very powerful types of complex objects: polylines, multilines, and splines. Because the objects are complex, the commands that are used to create them are naturally more complicated. But once the commands are mastered, you will be able to create these objects so much more quickly and accurately than trying to generate them by drawing simpler objects and then trying to combine those simpler objects. Also, these complex objects have characteristics not possible to replicate without using the command that created them. And, perhaps most important, these complex objects come with editing commands that are especially time saving and distinct.

Drawing Polylines

 Figure 9–16

Invoking the PLINE command from the Draw toolbar

The "poly" in polyline refers to a single object with multiple connected straight-line and/or arc segments. The polyline is drawn by invoking the pline command (see Figure 9–16) and then selecting a series of points. In this respect, pline functions much like the line command. However, when completed, the segments act like a single object when operated on by modify commands. You can specify the endpoints using only 2D (X,Y) coordinates.

The versatile pline command also draws lines and arcs of different widths, linetypes, tapered lines, and a filled circle. The area and perimeter of a 2D polyline can be calculated.

By default, polylines are drawn as optimized polylines. The optimized polyline provides most of the functionality of 2D polylines, but with much improved performance and reduced drawing file size. The vertices are stored as an array of information on one object. When you use the pedit command to edit the polyline to spline fitting or curve fitting, the polyline loses its optimization feature, and vertices are stored as separate entities, but it still behaves as a single object when operated on by modify commands. A polyline as shown in Figure 9–17 is drawn with the command sequence in the following example. AutoCAD prompts:

Specify start point: **2,2**

Current line-width is 0.0000

Specify next point or ⬇: **4,2**

Specify next point or ⬇: **5,1**

Specify next point or ⬇: **7,1**

Specify next point or ⬇: **8,2**

Specify next point or ⬇: **10,2**

Specify next point or ⬇: **10,4**

Specify next point or ⬇: **9,5**

Specify next point or ⬇: **8,5**

Specify next point or ⬇: **7,4**

Specify next point or ⬇: **5,4**

Specify next point or ⬇: **4,5**

Specify next point or ⬇: **3,5**

Specify next point or ⬇: **2,4**

Specify next point or ⬇: *(choose Close from the Shortcut menu)*

The Close and Undo options work similarly to the corresponding options in the line command.

 Figure 9–17

Example of connected line segments drawn using the PLINE command

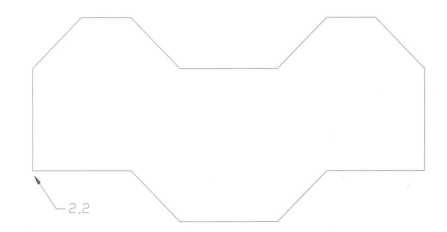

Setting Width of a Segment

The Width option specifies the width of the next line segment. After selecting a starting point, you can enter **w** or choose the Width option from the Shortcut menu to specify a starting and an ending width for a wide segment. When you select this option, AutoCAD prompts:

Specify starting width <default>: *(specify starting width)*

Specify ending width <default>: *(specify ending width)*

You can specify a width by entering a value at the prompt or by selecting a width-determining point on the screen. When you specify points on the screen, AutoCAD uses the distance from the starting point of the polyline to the point selected as the starting width. The starting width you enter becomes the default for the ending width. If necessary, you can change the ending width to another width, which results in a tapered segment or an arrow. The ending width, in turn, becomes the uniform width for all subsequent segments until you change the width again. The starting and ending points of wide line segments are at the center of the line.

The following command sequence presents an example of connected lines with tapered width, as shown in Figure 9–18, drawn by means of the pline command.

Specify start point: **2,2**

Current line-width is 0.0000

Specify next point or ⊠: *(choose Width from the Shortcut menu)*

Specify starting width <default>: **0**

Specify ending width <default>: **.25**

Specify next point or ⊠: **2,2.5**

Specify next point or ⊠: **2,3**

Specify next point or ⊠: *(choose Width from the Shortcut menu)*

Specify starting width <0.2500>: (ENTER)

Specify ending width <0.2500>: **0**

Specify next point or ⊠: **2,3.5**

Specify next point or ⊠: *(choose Width from the Shortcut menu)*

Specify starting width <0.0000>: (ENTER)

Specify ending width <0.0000>: **.25**

Specify next point or ⊠: **2.5,3.5**

Specify next point or ⊠: **3,3.5**

Specify next point or ⊠: *(choose Width from the Shortcut menu)*

Specify starting width <0.2500>: (ENTER)

Specify ending width <0.2500>: **0**

Specify next point or ⊠: **3.5,3.5**

Specify next point or ⊠: *(choose Width from the Shortcut menu)*

Specify starting width <0.0000>: (ENTER)

Specify ending width <0.0000>: **.25**

Specify next point or ⊠: **3.5,3**

Specify next point or ⊠: **3.5,2.5**

Specify next point or ⊠: *(choose Width from the Shortcut menu)*

Specify starting width <0.2500>: (ENTER)

Specify ending width <0.2500>: **0**

Specify next point or ⊠: **3.5,2**

Specify next point or ⊠: *(choose Width from the Shortcut menu)*

Specify starting width <0.0000>: (ENTER)

Specify ending width <0.0000>: **.25**

Specify next point or ⊠: **3,2**

Specify next point or ⬇: **2.5,2**

Specify next point or ⬇: *(choose Width from the Shortcut menu)*

Specify starting width <0.2500>: (ENTER)

Specify ending width <0.2500>: **0**

Specify next point or ⬇: *(choose Close from the Shortcut menu)*

 Figure 9–18

Example of connected line segments with tapered width drawn using the PLINE command

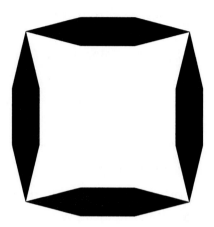

Setting Halfwidth of a Segment

The Halfwidth option is similar to the Width option, including the prompts, except that it lets you specify the width from the center of a wide polyline to one of its edges. In other words, you specify half of the total width. For example, it is easier to input 1.021756 as the halfwidth than to figure out the total width by doubling. You can specify a halfwidth by selecting points on the screen in the same manner as used to specify the full width.

Adding Arc Segments

The Arc option allows you to draw arc segments to the polyline. By default, AutoCAD prompts for the endpoint of the arc as shown in the following prompt:

Specify endpoint of arc or ⬇: *(specify end point of arc or choose one of the available options from the Shortcut menu)*

If you respond with a point, it is interpreted as the endpoint of the arc. The endpoint of the previous segment is the starting point of the arc, and the starting direction of the new arc will be the ending direction of the previous segment (whether the previous segment is a line or an arc). This resembles the arc command's Start, End, Direction (S,E,D) option, but requires only the endpoints to be specified or selected on the screen.

The Close, Width, Halfwidth, and Undo options are similar to the corresponding options for the straight-line segments described earlier.

The Angle option lets you specify the included angle by prompting:

> Specify included angle: *(specify an angle)*

The arc is drawn counterclockwise if the value is positive, clockwise if it is negative. After the angle is specified, AutoCAD prompts for the endpoint of the arc.

The Center option lets you override with the location of the center of the arc, and AutoCAD prompts:

> Specify center point: *(specify center point)*

When you provide the center point of the arc, AutoCAD prompts for additional information:

> Specify endpoint of arc or [Angle/Length]: *(specify end point of arc or choose one of the available options from the Shortcut menu)*

If you respond with a point, it is interpreted as the endpoint of the arc. Selecting Angle or Length allows you to specify the arc's included angle or chord length.

The Direction option lets you override the direction of the last segment, and AutoCAD prompts:

> Specify the tangent direction for the start point of arc: *(specify the direction)*

If you respond with a point, it is interpreted as the starting point of the direction, and AutoCAD prompts for the endpoint for the arc.

The Line option reverts to drawing straight-line segments.

The Radius option allows you to specify the radius by prompting:

> Specify radius of arc: *(specify the radius of the arc)*

After the radius is specified, you are prompted for the endpoint of the arc.

The Second point option causes AutoCAD to use the three-point method of drawing an arc by prompting:

> Specify second point on arc: *(specify second point)*

If you respond with a point, it is interpreted as the second point, and then you are prompted for the endpoint of the arc. This resembles the arc command's Three-point option.

The Length option continues the polyline in the same direction as the last segment for a specified distance.

Drawing Multiple Parallel Lines

 **Figure 9-19**

Invoking the MLINE command from the Draw menu

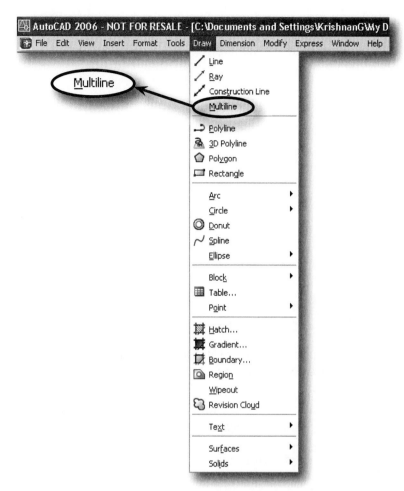

The mline command (see Figure 9-19) allows you to draw multiple parallel line segments, similar to polyline segments that have been offset one or more times. In addition, you can also modify the intersections of two or more multilines or create gaps with the mledit command.

The properties of each element of a multiline are determined by the multiline style that is current when the multiline is drawn. The properties of the multiline that can be determined by the style include whether to display the line at the joints (miters) and ends and whether to close the ends with a variety of half circles, connecting inner and/or outer elements. In addition, the style controls the element properties, such as color, linetype, and offset distance between two parallel lines, and you can assign as many as 16 parallel lines. The mlstyle command allows you to create and edit a multiline style. The mledit and mlstyle commands are discussed in Chapter 11 and Chapter 13 respectively.

Examples of applying the mline command are shown in Figure 9–20.

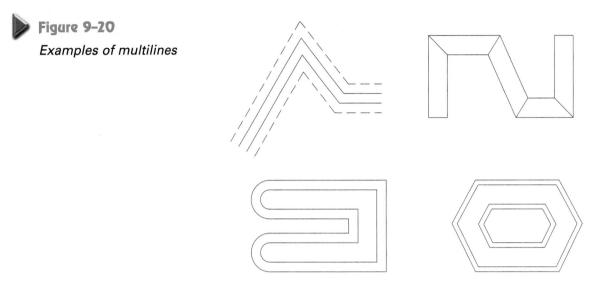

Multiline segments are drawn with the command sequence as shown in the following example. AutoCAD prompts:

Specify start point or ⊡: *(specify start point or select one of the available options from the Shortcut menu)*

The Start point option (default) lets you specify the starting point of the multiline, known as its origin. Once you specify the starting point for a multiline, AutoCAD prompts:

Specify next point:

When you respond by selecting a point, the first multiline segment is drawn according to the current style. You are then prompted:

Specify next point or ⊡:

If you specify a point, the next segment is drawn, along with segments of all other elements specified by the current style. After two segments have been drawn, the prompt will include the Close option:

Specify next point or ⊡:

Choosing the Close option causes the next segment to join the starting point of the multiline, fillets all elements, and exits the command.

Selecting Undo after any segment is drawn (and the mline command has not been terminated) causes the last segment to be erased, and you are prompted again for a point.

Setting Justification

The Justification option determines the relationship between the elements of the multiline and the line you specify by way of the placement of the points. The justification is set by selecting one of the three available options: Top, Zero, and Bottom.

The Top option causes the element with the greatest offset value to be drawn on the line of selected points. All other elements will be to the right of the line of points as viewed from the starting point to the ending point of each segment. In other words, if the line is drawn left to right, the line of points (and the element with the greatest offset value) will be at the top of (above) all other elements.

The Zero option causes the baseline to coincide with the line of selected points. Elements with positive offsets will be to the right of, and those with negative offsets to the left of, the line of selected points, as viewed from the starting point to the ending point of the each segment.

The Bottom option causes the element with the least offset value to be drawn on the line of selected points. All other elements will be to the left of the line of points as viewed from the starting point to the ending point of each segment. In other words, if the line is drawn left to right, the line of points (and the element with the least offset value) will be at the bottom of (below) all other elements.

Figure 9–21 shows the location for various justifications.

Figure 9-21

Location of various justifications

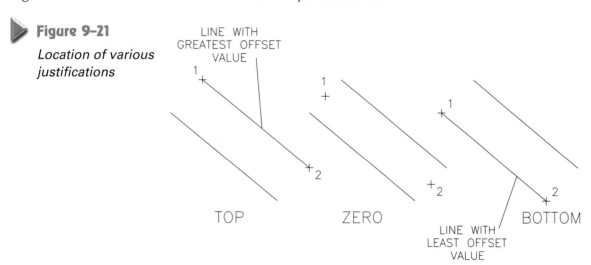

Setting Multiline Spacing

The Scale option determines the value used for offsetting elements when drawing them relative to the values assigned to them in the style. For instance, if the scale is changed to 3.0, elements that are assigned 0.5 and –1.5 will be drawn with offsets of 1.5 and –4.5, respectively. If a negative value is given for the scale, then the signs of the values assigned to them in the style will be changed (positive to negative and negative to positive). The value can be entered in decimal form or as a fraction. A 0 (zero) scale value produces a single line.

Setting Current Multiline Style

The Style option sets the current multiline style from the available styles.

Drawing Spline Curves

Figure 9-22

Invoking the SPLINE command from the Draw menu

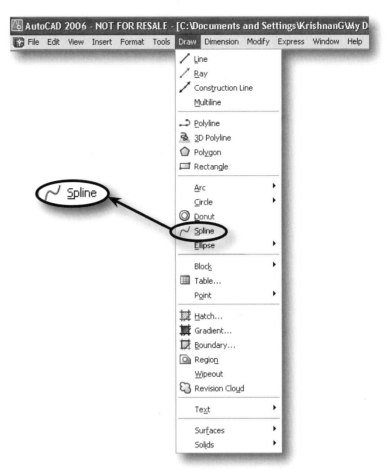

The spline command (see Figure 9–22) is used to draw a curve through or near points in a series. The type of curve is a nonuniform rational B-spline (NURBS). This type is used for drawing curves with irregularly varying radii, such as topographical contour lines.

The spline curve is drawn through a series of two or more points, with options either to specify end tangents or to use Close to join the last segment to the first. Another option lets you specify a tolerance, which determines how close to the selected points the curve is drawn. A spline curve is drawn with the command sequence as shown in the following example. AutoCAD prompts:

Specify first point or 💽: *(specify a point or type **o** for the Object option)*

The default option lets you specify the point from which the spline starts and to which it can be closed. After you specify the first point, a preview line appears. You will then be prompted:

Specify next point:

When you respond by specifying a point, the spline segments are displayed as a preview spline, curving from the first point, through the second point, and ending at the cursor. You are then prompted:

Specify next point or 💽 <start tangent>:

If you specify a point, the next segment is added to the spline. This will occur with each additional point specified until you use the Close option or enter a null response by pressing enter.

Choosing a null response terminates the specifying of segment determining points. You are then prompted for a start tangent determining point as follows:

Specify start tangent:

If you specify a point (for tangency), its direction from the start point determines the start tangent. If you press enter, the direction from the first point to the second point determines the tangency. After the start tangency is established, you are prompted:

Specify end tangent:

If you specify a point (for tangency), its direction from the endpoint determines the end tangent. If you press enter, the direction from the last point to the previous point determines the tangency.

The Close option uses the original starting point of the first spline segment as the endpoint of the last segment and terminates segment placing. You are then prompted:

Specify tangent:

You can specify a point to determine the tangency at the connection of the first and last segments. If you press enter, AutoCAD calculates the tangency and draws the spline accordingly. You can also use the Perp or Tan option to cause the tangency of the spline to be perpendicular to or tangent to a selected object.

The Fit tolerance option lets you vary how the spline is drawn relative to the selected points. You are then prompted:

Specify fit tolerance <current>:

Entering 0 (zero) causes the spline to pass through the specified points. A positive value causes the spline to pass within the specified value of the points.

Converting a Polyline to Spline Curve

The Object option is used to change spline-fit polylines to splines. This can be used for 2D or 3D polylines, which will be deleted depending on the setting of the delobj system variable.

drawing exercises

Open the Exercise Manual PDF file for Chapter 9 on the accompanying CD for discipline-specific exercises.

review questions

1. In order to draw two rays with different starting points, you must use the RAY command twice.

   a. *True*   **b.** *False*

2. When you place xlines on a drawing, they:

   a. *May affect the limits of the drawing.*
   b. *May affect the extents of the drawing.*
   c. *Always appear as construction lines on layer 0.*
   d. *Can be constructed as offsets to an existing line.*
   e. *None of the above.*

3. Ellipses are drawn by specifying:

   a. *The major and minor axes.*
   b. *The major axis and a rotation angle.*
   c. *Any three points on the ellipse.*
   d. *Any of the above.*
   e. *Both a and b.*

4. A polyline:

   a. *Can have width.*
   b. *Can be exploded.*
   c. *Is one object.*
   d. *All of the above.*

5. Polylines are:

   a. *Made up of line and arc segments, each of which is treated as an individual object.*
   b. *Are connected sequences of lines and arcs.*
   c. *Both a and b.*
   d. *None of the above.*

6. The following are all options of the PLINE command except:

   a. *Undo.*   **d.** *Ltype.*
   b. *Halfwidth.*   **e.** *Width.*
   c. *Arc.*

7. In regard to using the SOLID command, which of the following statements is true?

   a. *The order of point selection is unimportant*
   b. *Fill must be turned ON in order to use the SOLID command*
   c. *The points must be selected on existing objects*
   d. *The points must be selected in a clockwise order*
   e. *None of the above*

8. What command is commonly used to create a filled rectangle?
   a. *FILL-ON*
   b. *PLINE*
   c. *RECTANGLE*
   d. *LINE*
   e. *SOLID*

9. The following are all options of the ELLIPSE command except:
   a. *Radius.*   **c.** *Center.*
   b. *Arc.*   **d.** *Axis Endpoint.*

10. A polyline:
   a. *Can be exploded.*   **c.** *Can have width.*
   b. *Is one object.*   **d.** *All of the above.*

11. XLINE and RAY commands allow you to draw lines that extend infinitely in one or both directions.
   a. *True*   **b.** *False*

12. Xlines can be moved, copied, and rotated like any other objects.
   a. *True*   **b.** *False*

13. The RAY command allows you to draw lines that extend infinitely in one direction from the point selected when the line is being created.
   a. *True*   **b.** *False*

14. The DONUT command lets you draw solid-filled circles and rings by specifying outer and inner diameters of the filled area.
   a. *True*   **b.** *False*

15. In order for the DONUT command to work, the FILLMODE system variable must be set to ON.
   a. *True*   **b.** *False*

16. If FILLMODE is set to OFF, then the PLINE, TRACE, DONUT, and SOLID commands display the outline of the shapes.
   a. *True*   **b.** *False*

17. When creating a solid-filled object with Solid command, the points can be selected in any order.
   a. *True*   **b.** *False*

18. A solid-filled polygon is generated with straight sides.
   a. *True*   **b.** *False*

duplication and alteration

chapter

introduction

This chapter introduces some of the basic commands and concepts in AutoCAD that can be used to create objects from existing objects, modify objects with advanced commands, and modify complex objects.

After completing this chapter, you will be able to do the following:

- Create new objects from existing ones using the ARRAY, OFFSET, MIRROR, FILLET, and CHAMFER commands

- Modify objects using the LENGTHEN, STRETCH, TRIM, BREAK, and EXTEND commands

- Modify complex objects such as multilines, splines and polylines

- Change and match properties of objects

CREATING OBJECTS FROM EXISTING OBJECTS

AutoCAD not only allows you to draw objects easily, but also allows you to create additional objects from existing objects. Chapter 5 discussed the simplest of these, the COPY command. This section discusses five additional commands that will make your job easier: ARRAY, OFFSET, MIRROR, FILLET, and CHAMFER.

Array – Creating a Pattern of Objects as They Are Copied

 Figure 10–1

Invoking the ARRAY command from the Modify toolbar

The ARRAY command (see Figure 10–1) is used to make multiple copies of selected objects in either rectangular or polar arrays (patterns). In the rectangular array, you can specify the number of rows, the number of columns, and the spacing between rows and columns (row and column spacing may differ). The whole rectangular array can be rotated at a selected angle. In the polar array, you can specify the angular intervals, the number of copies, the angle that the group covers, and whether or not the objects maintain their orientation as they are arrayed. Invoke the ARRAY command from Modify toolbar (see Figure 10–1) to display the Array dialog box (see Figure 10–2).

 Figure 10-2

Array dialog box with Rectangular Array selected

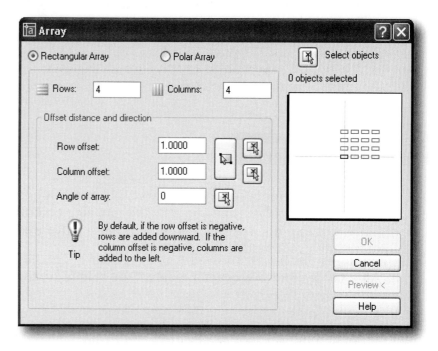

In the Array dialog box, you can select the type of array (rectangular or polar). For the rectangular type of array, you can select the number and spacing of rows and columns of the array and the angle of the array. For the polar type of array, you can select the center of the array, the angle through which objects are arrayed, the number of objects in the array, and whether the arrayed objects are rotated as they are arrayed or they keep the orientation of the original object when arrayed.

Rectangular Array

In the Array dialog box, select **Rectangular Array** as shown in Figure 10–2. Specify the number for rows and columns in their respective boxes. At least one of them must be equal to two or greater.

note   The number for rows and columns you enter include the original object. A positive number for the column and row offset causes the elements to array toward the positive x and y directions, respectively. Negative numbers for the column and row spacing cause the elements to array toward the negative x and y directions, respectively.

In the **Offset distance and direction** section of the Array dialog box, specify the value for the spacing between objects in rows in the **Row offset** box. Or you can select the button to the far right of the **Row offset** box (with a single arrow and crossmark icon). This will allow you to specify the row offset (spacing) by selecting two points on the screen with the cursor. The two points can be in any direction. AutoCAD will use the distance between them for the row offset.

Specify the value for the spacing between objects in columns in the **Column offset** box. Or you can select the button to the far right of the **Column offset** box (with a single arrow and crossmark icon). This will allow you to specify the column offset (spacing) by selecting two points on the screen with the cursor. The two points can be in any direction. AutoCAD will use the distance between them for the row column.

---

note   The offset you specify determines the distance between corresponding points of adjacent objects and not the space between adjacent objects. For example, if you arrayed 3" diameter circles with 2" offsets, the adjacent circles would overlap 1".

You can also specify the row and column offsets in one maneuver by first selecting the large button just to the right of the **Row offset** and **Column offset** boxes. Then select two points on the screen with the cursor that specify the opposite corners of a rectangle called a unit cell. AutoCAD uses the width of the unit cell as the horizontal distance(s) between columns and the height as the vertical distance(s) between rows.

The array can be rotated by entering the angle in the **Angle of array** box. Or you can select the button to the right of the **Angle of array** box (with a single arrow and crossmark icon). This will allow you to specify the rotation angle by selecting two points on the screen with the cursor. The two points can be anywhere on the screen. AutoCAD will use the direction between them (measured from the first point selected to the second) for the angle that the array will be rotated.

The array preview window on the right of the dialog box shows the rows and columns selected. Choose **Select objects** to return to the drawing screen in order to select the objects to be arrayed. Once one or more objects are selected, the number of objects is shown above the array sample window, and you can preview the resulting array on the screen by choosing **Preview**.

Once the settings have been satisfactorily specified, choose **OK**.

Polar Array

In the Array dialog box, first select **Polar Array** as shown in Figure 10–3. Specify X and Y coordinate for the center of the polar array in their respective boxes. Or you can choose the button with the arrow and crossmark, which will return you to the drawing screen to specify the center point with your cursor.

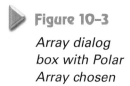

Figure 10-3

Array dialog box with Polar Array chosen

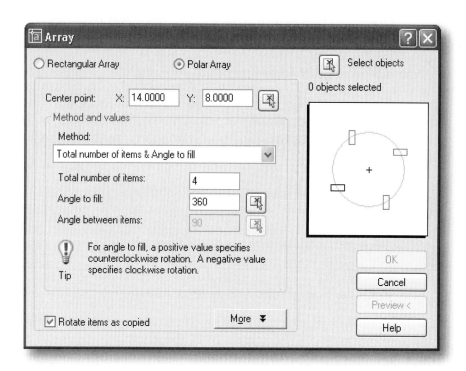

In the **Method and values** section, there are three methods from which to choose: **Total number of items** and **Angle to fill**, **Total number of items** and **Angle between items**, and **Angle to fill** and **Angle between items**. The method you select will determine which two of the three text boxes will be active among the **Total number of items**, the **Angle to fill**, and the **Angle between items** boxes.

The total number of items includes the original item, just as in the rectangular array. Note that a positive angle causes items to array in a counterclockwise direction. A negative angle causes them to array in a clockwise direction.

The array preview window on the right of the dialog box shows a representative example of the number items selected, the total angle of the array, and the angle between items. Choose **Select objects** to return to the drawing screen in order to select the objects to be arrayed. Once one or more objects are selected, the number of objects is shown above the array sample window, and you can preview the resulting array on the screen by choosing **Preview**.

Check the **Rotate items as copied** box to cause each item to be rotated from the original item's orientation the same angle at which the particular item is arrayed.

Choose the **More** button to display more options in the dialog box. You can specify a new reference (base) point relative to the selected objects that will remain at a constant distance from the center point of the array as the objects are arrayed. AutoCAD uses the distance from the array's center point to a base point on the last object selected. The point used is determined by the type of object, as shown in the following table.

Type of Object	Base Point
Arc, circle, ellipse	Center point
Polygon, rectangle	First corner
Donut, line, polyline, 3D polyline, ray, spline	Starting point
Block, paragraph text, single-line text	Insertion point
Construction lines	Midpoint
Region	Grip point

If the **Set to Object's Default** box is set to ON, AutoCAD uses the default base point of the object to position the arrayed object. To specify a different base point, turn OFF the box.

You can use the **Base Point** boxes to set new X and Y base point coordinates. Or you can choose **Pick Base Point** to temporarily close the dialog box and specify a point with your cursor. After you specify a point, the Array dialog box is redisplayed.

Once the settings have been satisfactorily specified, choose **OK**.

The following example of values entered in their respective text boxes draws a rectangular array with 8 items and a 360 'degree symbol' angle to fill, as shown in Figure 10–4.

Rows:	4
Columns:	6
Row offset:	1
Column offset:	1.5
Angle of array:	0

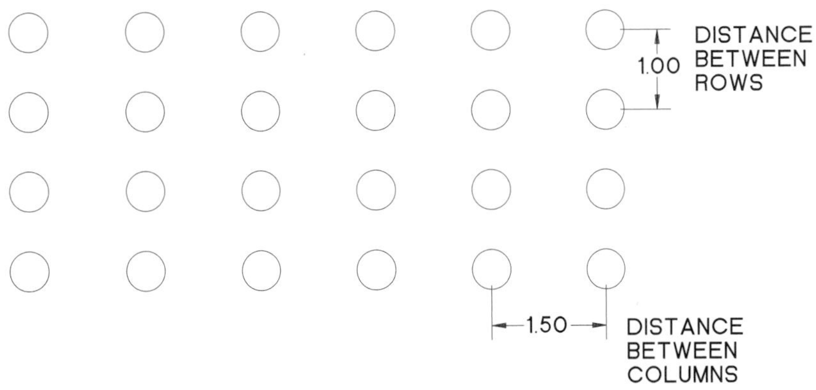

▶ **Figure 10–4**

Using the ARRAY command to place a rectangular array

The following example of values entered in their respective text boxes draws a polar array with 8 items and a 360 'degree symbol' angle to fill, as shown in Figure 10–5.

Center point:	*(specify the center point as shown in Figure 10–5)*
Method	**Total number of items & angle to fill**
Total number of items:	**8**
Angle to fill:	**360**
Rotate items as copied	*(checked to rotate, cleared for non-rotated)*

Figure 10-5

Using the ARRAY command to place rotated and non-rotated polar arrays

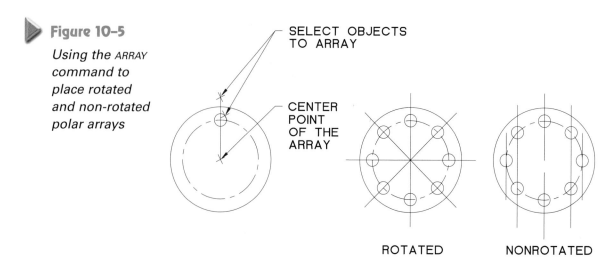

Figure 10–5 shows both non-rotated and rotated polar arrays.

Offset – Parallel Lines/Concentric Circles and Arcs

Figure 10-6

Invoking the OFFSET command from the Modify toolbar

The OFFSET command (see Figure 10–6) is used to create parallel lines, circles, and arcs based on a specified offset distance from existing objects. An offset line object will be created in a similar manner to using the COPY command, but offset circles and arcs will be created with radii that are increased or decreased the value of the specified offset and will have the same centers as their base objects. Figure 10–7 shows parallel lines, parallel curves, and concentric circles created relative to existing objects. Special precautions must be taken when using the OFFSET command to prevent unpredictable results from occurring when using the command on arbitrary curve/line combinations in polylines.

Figure 10-7

Examples created using the OFFSET command

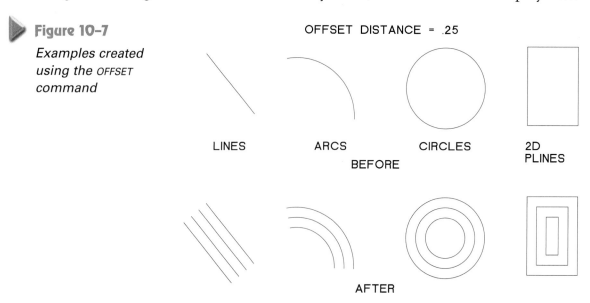

After you invoke the OFFSET command, AutoCAD prompts:

> Specify offset distance or ⬇: *(specify offset distance, or choose one of the available options from the Shortcut menu)*
>
> Select object to offset or ⬇: *(select an object to offset)*
>
> Specify point on side to offset: *(specify a point to one side of the object to offset)*
>
> Select object to offset or ⬇: *(continue selecting additional objects for offset, and specify the side of the object to offset, or press ENTER to terminate the command sequence)*

Instead of specifying the offset distance, select the Through option, and AutoCAD prompts for a through point. Specify a point, and AutoCAD creates an object passing through the specified point.

Valid Objects to Offset

Valid objects include the line, spline curve, arc, circle, and 2D polyline. If you select any other type of object, such as text, you will get the following error message:

> Cannot offset that object.

The object selected for offsetting must be in a plane parallel to the current coordinate system. Otherwise, you will get the following error message:

> Object not parallel with UCS.

Offsetting Miters and Tangencies

The OFFSET command affects single objects in a manner different from a polyline made up of the same objects. Polylines whose arcs join lines and other arcs in a tangent manner are affected differently than polylines with non-tangent connecting points. For example, in Figure 10–8 the seven lines are separate objects. When you specify side to offset as shown, there are gaps and overlaps at the ends of the newly created lines.

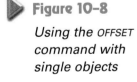

Figure 10-8

Using the OFFSET command with single objects

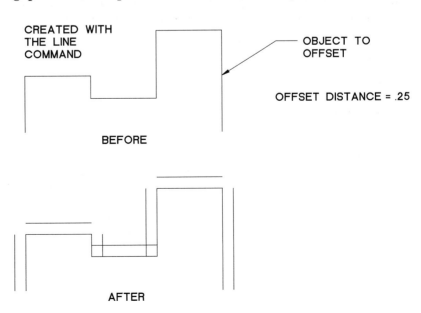

CREATED WITH THE LINE COMMAND

OBJECT TO OFFSET

OFFSET DISTANCE = .25

BEFORE

AFTER

In Figure 10–9, the lines have been joined together (see PEDIT later in this chapter) as a single polyline, and the figure shows how the OFFSET command affects the corners where the new polyline segments join.

Figure 10-9

Using the OFFSET command with polylines

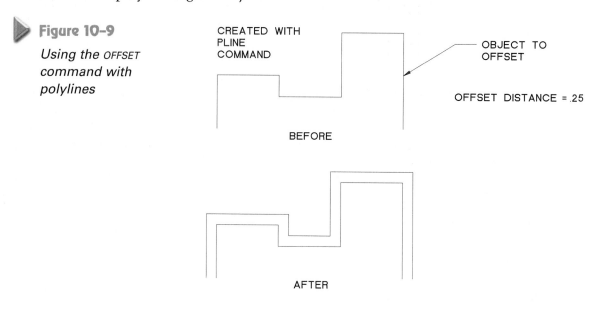

note The results of offsetting polylines with arc segments that connect other arc segments and/or line segments in dissimilar (nontangent) directions might be unpredictable. Examples of offsetting such polylines are shown in Figure 10–10.

Figure 10-10

Using the OFFSET command with non-tangent arc and/ or line segments

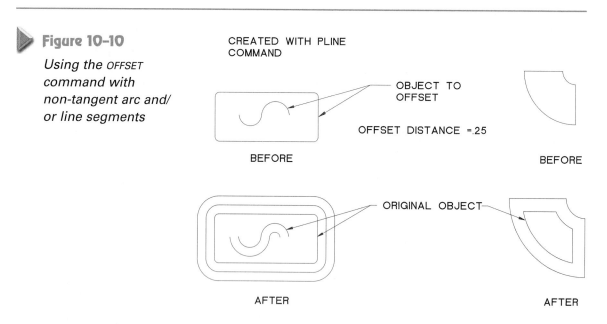

If you are not satisfied with the resulting new polyline configuration, you can use the PEDIT command to edit it. Or, you can explode the polyline and edit the individual segments.

Mirror – Creating a Mirror Copy of Objects

 Figure 10–11

Invoking the MIRROR command from the Modify toolbar

The MIRROR command (see Figure 10–11) is used to create a mirror image of selected objects, that is, a copy mirrored about a specified line. If the objects include text, then you have the option of having the text mirrored in the new location or maintaining the same orientation. The first and second specified points of the mirror line become the endpoints of an invisible line about which the selected objects will be mirrored. In Figure 10–12, the group of selected objects is mirrored about the line produced by specifying points 3 and 4 as shown in the following sequence. AutoCAD prompts:

Select objects: *(specify POINT 1 to specify one corner of a window)*

Second corner: *(specify POINT 2 to specify the opposite corner of the window)*

Select objects: (ENTER)

Specify first point of mirror line: *(specify POINT 3, as shown in Figure 10–12)*

Specify second point of mirror line: *(specify POINT 4, as shown in Figure 10–12)*

Erase source objects? ⬇: (ENTER)

 Figure 10–12

Mirroring a group of objects selected by means of the Window option

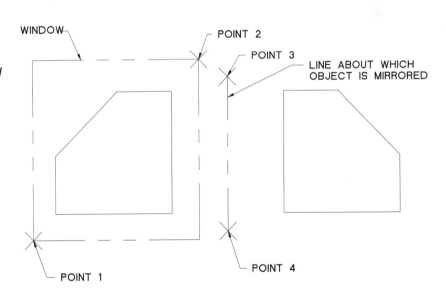

The text as mirrored is located relative to other objects within the selected group. But the text will or will not retain its original orientation, depending on the setting of the system variable called MIRRTEXT. If the value of MIRRTEXT is set to 1, then text items in the selected group will have their orientations mirrored. That is, if their characters were normal and they read left to right in the original group, in the mirrored copy they will read right to left and the characters will be backwards. If MIRRTEXT is set to 0 (zero), then the individual text strings will retain their normal (left-to-right) character appearance.

The MIRRTEXT system variable, like other system variables, is changed by the SETVAR command or by entering **mirrtext** at the On-Screen prompt, as follows:

mirrtext (ENTER)
Enter New Value for MIRRTEXT <1>: **0**

This setting causes mirrored text to retain its readability.

Fillet - Joining Two Objects with or without a Tangent Arc

 Figure 10–13

Invoking the FILLET command from the Modify toolbar

The FILLET command (see Figure 10–13) is used to join two nonparallel lines, arcs, circles, elliptical arcs, polylines, rays, xlines, or splines at their intersection either with or without an arc of specified radius tangent to both objects.

If the TRIMMODE system variable is set to 1 (default), the FILLET command trims the intersecting lines to the endpoints of the fillet arc. And if TRIMMODE is set to 0 (zero), the FILLET command leaves the intersecting lines at the endpoints of the fillet arc. After you invoke the FILLET command, AutoCAD prompts:

Select first object or ⬇: *(select one of the two objects to fillet, or choose one of the available options from the Shortcut menu)*

By default, AutoCAD prompts you to select an object. If you select an object to fillet, then AutoCAD prompts:

Select second object or shift-select to apply corner: *(select the second object to fillet or hold down* SHIFT *and select an object to create a sharp corner)*

AutoCAD joins the two objects with an arc having the default radius. If the objects selected to be filleted are on the same layer, AutoCAD creates the fillet arc on the same layer. If not, AutoCAD creates the fillet arc on the current layer.

AutoCAD allows you to draw a fillet between parallel lines, xlines, and rays. The first selected object must be a line or ray, but the second object can be a line, xline, or ray. The diameter of the fillet arc is always equal to the distance between the lines. The current fillet radius is ignored and remains unchanged.

Setting the Default Radius

The Radius option allows you to change the current fillet radius. The following command sequence sets the fillet radius to 0.5 and draws the fillet between two lines, as shown in Figure 10–14:

Select first object or ⬇: *(choose Radius from the Shortcut menu)*
Specify fillet radius <current>: **0.25**

Select first object or ⬇: *(select one of the lines, as shown in Figure 10–14)*
Select second object: *(select the other line to fillet, as shown in Figure 10–14)*

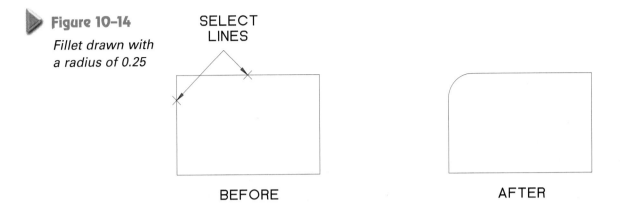

▶ **Figure 10-14**

Fillet drawn with a radius of 0.25

SELECT LINES

BEFORE   AFTER

If you select lines or arcs, AutoCAD extends these lines or arcs until they intersect or trims them at the intersection, keeping the selected segments if they cross.

Drawing a Fillet Arc for a Polyline

With the Polyline option, AutoCAD draws fillet arcs at each vertex of a 2D polyline where two line segments meet. The following command sequence sets the fillet radius to 0.5 and draws the fillet at each vertex of a 2D polyline, as shown in Figure 10–15:

Select first object or ⬇: *(choose Radius from the Shortcut menu)*
Specify fillet radius <current>: **0.5**
Select first object or ⬇: *(choose Polyline from the Shortcut menu)*
Select 2D polyline: *(select the polyline as shown in Figure 10–15)*

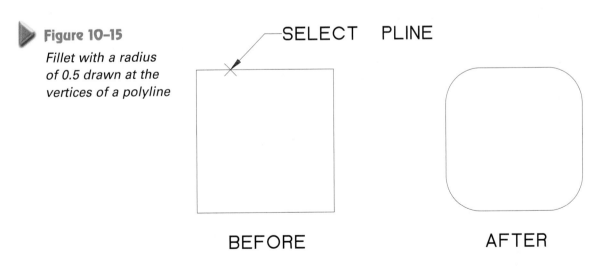

▶ **Figure 10-15**

Fillet with a radius of 0.5 drawn at the vertices of a polyline

SELECT PLINE

BEFORE   AFTER

Setting the Trim Option

The Trim option (Trim/No Trim) controls whether or not AutoCAD trims the selected edges to the fillet arc endpoints. This option is similar to setting the TRIMMODE system variable from 1 to 0 or 0 to 1.

Multiple Selection

The Multiple option, when selected, allows you to specify multiple pairs of objects to be filleted without exiting the FILLET command. To exit the FILLET command, select Enter or Cancel from the Shortcut menu, press ESC, or invoke another command.

Chamfer – Joining Two Objects with a Diagonal Line

 Figure 10-16

Invoking the CHAMFER command from the Modify toolbar

The CHAMFER command (see Figure 10–16) allows you to quickly and easily join two nonparallel lines at their intersection with a diagonal line set back the specified distance(s) from the intersection. The size of the chamfer (setback) is determined by the settings of the first and the second chamfer distances. If it is to be a 45-degree chamfer for perpendicular lines, the two distances are set to the same value.

If the TRIMMODE system variable is set to 1 (default), the CHAMFER command trims the intersecting lines to the endpoints of the chamfer line. And if TRIMMODE is set to 0 (zero), the CHAMFER command leaves the intersecting lines at the endpoints of the chamfer line. After you invoke the CHAMFER command, AutoCAD prompts:

Select first line or ⊡: *(select one of the two lines to chamfer, or choose one of the available options from the Shortcut menu)*

By default, AutoCAD prompts you to select the first line to chamfer. If you select a line to chamfer, AutoCAD prompts:

Select second line or shift-select to apply corner: *(select the second line to chamfer or hold down SHIFT and select an object to create a sharp corner)*

AutoCAD draws a chamfer to the selected lines. If the selected lines to be chamfered are on the same layer, AutoCAD creates the chamfer on the same layer. If not, AutoCAD creates the chamfer on the current layer.

Setting Default Distances for Chamfer

The Distance option allows you to set the first and second chamfer distances. The following command sequence sets the first chamfer and second chamfer distances to 0.5 and 1.0, respectively, and draws the chamfer between two lines, as shown in Figure 10–17.

Select first line or ⊡: *(choose Distance from the Shortcut menu)*
Specify first chamfer distance <0.5000>: **0.5**

Specify second chamfer distance <0.5000>: **1.0**

Select first line or ⊻: *(select the first line, as shown in Figure 10–17)*

Select second line or shift-select to apply corner: *(select the second line, as shown in Figure 10–17)*

Figure 10–17

Chamfer drawn with distances of 0.5 and 1.0

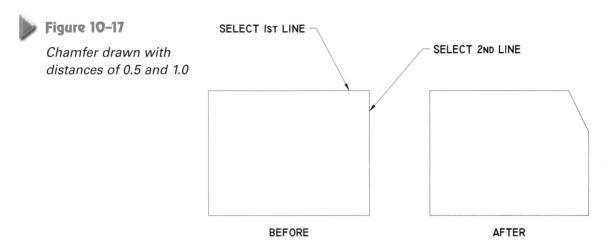

Drawing a Chamfer for a Polyline

With the Polyline option, AutoCAD draws chamfers at each vertex of a 2D polyline where two line segments meet. The following command sequence sets the chamfer distances to 0.5 and draws the chamfer at each vertex of a 2D polyline, as shown in Figure 10–18.

Select first line or ⊻: *(choose Distance from the Shortcut menu)*

Specify first chamfer distance <0.5000>: **0.5**

Specify second chamfer distance <0.5000>: **0.5**

Select first line or ⊻: *(choose Polyline from the Shortcut menu)*

Select 2D polyline: *(select the polyline, as shown in Figure 10–18)*

Figure 10–18

Chamfer with distances of 0.5 drawn on a polyline

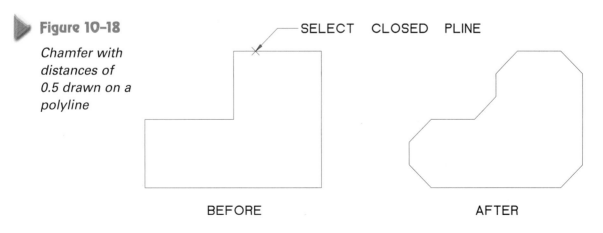

Specifying Angle for Chamfer

The Angle option is similar to the Distance option, but instead of prompting for the first and second chamfer distances, AutoCAD prompts for the first chamfer distance and an angle from the first line. This is another method by which to create the chamfer line.

Selection of a Method for Chamfer

The Method option controls whether AutoCAD uses two distances or a distance and an angle to create the chamfer line.

Setting the Trimming Option

The Trim option (Trim/No Trim) controls whether or not AutoCAD trims the selected edges to the chamfer line endpoints. This option is similar to setting the TRIMMODE system variable from 1 to 0 or from 0 to 1.

---

note   The CHAMFER command set to zero distance operates the same way as the FILLET command operates set to zero radius.

Multiple Selection

The Multiple option, when selected, allows you to specify multiple pairs of objects to be chamfered without exiting the CHAMFER command. To exit the command, select Enter or Cancel from the Shortcut menu, press ESC, or invoke another command.

CUT TO CLEAR, BEAT TO FIT, PAINT TO MATCH

Objects in AutoCAD often need to have changes made to them, either because the requirements have changed, an error has been discovered, or because it might be easier to create the objects one way and then utilize a special modifying command to complete them. AutoCAD allows you to modify objects using the advanced commands such as LENGTHEN, STRETCH, TRIM, BREAK, EXTEND, PROPERTIES, and MATCHPROP.

Lengthen – Changing the Length of Objects

 **Figure 10–19**

Invoking the LENGTHEN command from the Modify menu

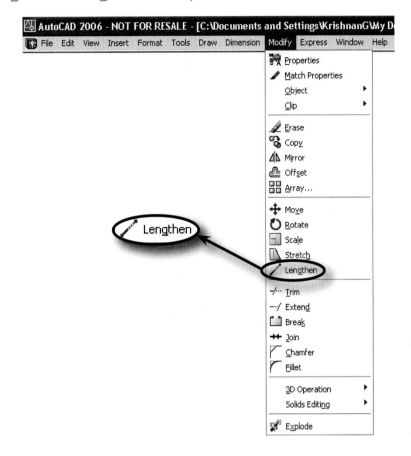

The LENGTHEN command (see Figure 10–19) is used to increase or decrease the lengths of line segments or the included angles of arcs. After you invoke the LENGTHEN command, AutoCAD prompts:

Select an object or ☑: *(select an object or choose one of the available options from the Shortcut menu)*

When you select an object, AutoCAD displays its length in the current units display and, where applicable, the included angle of the selected object.

Incremental Change

The Delta option changes the length or, where applicable, the included angle incrementally from the endpoint of the selected object closest to the pick point. A positive value results in an increase in extension; a negative value results in a trim. When you select the Delta option, AutoCAD prompts:

Select an object or ☑: *(choose Delta from the Shortcut menu)*

Enter delta length or ☑ (current)>: *(specify positive or negative value)*

Select an object to change or ☑: *(select an object, and its length is changed on the end nearest the selection point)*

Select an object to change or ☑: *(select additional objects; when done, press ENTER to exit the command sequence)*

If instead of specifying the delta length you select the Angle option, AutoCAD prompts:

Enter delta length or ⬇ <current>: *(choose Angle from the Shortcut menu)*

Enter delta angle <current>: *(specify positive or negative angle)*

Select an object to change or ⬇: *(select an object, and its included angle is changed on the end nearest the selection point)*

Select an object to change or ⬇: *(select additional objects; when done, press* ENTER *to exit the command sequence)*

The Undo option reverses the most recent change made by the LENGTHEN command.

Percent Change

The Percent option sets the length of an object by a specified percentage of its total length. It will increase the length/angle for values greater than 100 and decrease them for values less than 100. For example, a 12-unit long line will be changed to 15 units by using a value of 125. A 12-unit long line will be changed to 9 units by using a value of 75. When you select the Percent option, AutoCAD prompts:

Select an object or ⬇: *(choose Percent from the Shortcut menu)*

Enter percentage length <current>: *(specify positive nonzero value and press* ENTER*)*

Select an object to change or ⬇: *(select an object, and its length is changed on the end nearest the selection point)*

Select an object to change or ⬇: *(select additional objects; when done, press* ENTER *to exit the command sequence)*

Specifying Total Length

The Total option changes the length/angle of an object to the value specified. When you select the Total option, AutoCAD prompts:

Select an object or ⬇: *(choose Total from the Shortcut menu)*

Specify total length or ⬇ <current)>: *(specify distance, or enter **a**, for angle, and then specify an angle for change)*

Select an object to change or ⬇: *(select an object, and its length is changed to the specified distance on the end nearest the selection point)*

Select an object to change or ⬇: *(select additional objects; when done, press* ENTER *to exit the command sequence)*

Dynamic Change

The Dynamic option changes the length/angle of an object in response to the cursor's final location, relative to the endpoint nearest to where the object is selected. When you select the Dynamic option, AutoCAD prompts:

Select an object or ⬇: *(choose Dynamic from the Shortcut menu)*

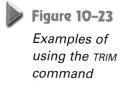 Don't forget to press **ENTER** after selecting
the program will not respond as expected. In fact, TRIM
ting edges until you terminate the cutting edge selecti

Setting the Edge from Trimming

The Edge option determines whether objects that exte
or to an implied intersection are trimmed. AutoCAD pr
option is selected:

> Enter an implied edge extension mode ⬇ <current
> *able options from the Shortcut menu)*

The Extend selection extends the cutting edge along i
object in 3D (implied intersection). The No Extend sele
to be trimmed only at a cutting edge that intersects it i

Reversing the Change

The Undo option reverses the most recent change mad

Setting the Projection Mode

The Project option specifies the projection mode and co
when trimming objects. By default, it is set to the curre

Figure 10–23 shows examples of using the TRIM commar

▶ **Figure 10–23**

*Examples of
using the* TRIM
command

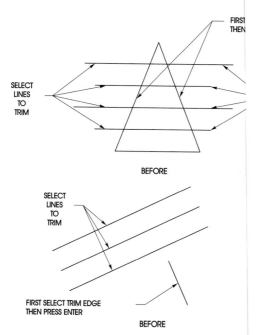

Select an object to change or ⬇: *(select an object to change the endpoint)*

Specify new endpoint: *(specify new endpoint for the selected object)*

Select an object to change or ⬇: *(select additional objects; when done, press* ENTER *to exit the command sequence).*

Stretching Objects

▶ **Figure 10–20**

Invoking the STRETCH
*command from the
Modify toolbar*

The STRETCH command (see Figure 10–20) allows you to stretch the shape of an object without affecting other crucial parts that remain unchanged. A common example is to stretch a square into a rectangle. The length is changed while the width remains the same.

AutoCAD stretches lines, polyline segments, rays, arcs, elliptical arcs, and splines that cross the selection window. The STRETCH command moves the endpoints that lie inside the window, leaving those outside the window unchanged. If the entire object is inside the window, the STRETCH command operates like the MOVE command.

It is possible to make multiple selection sets using the crossing method and AutoCAD will modify all of the selection sets with a single stretch command operation. If you use the pick method of selecting objects the stretch command operates like the move command. After you invoke the STRETCH command from the Modify toolbar, AutoCAD prompts:

> Select objects to stretch by crossing-window or crossing-polygon...
>
> Select objects: *(select the first corner of a crossing window or polygon)*
>
> Specify opposite corner: *(select the opposite corner of the crossing window or polygon)*
>
> Select objects: *(select additional objects; when done, press* ENTER *to complete the selection)*
>
> Specify base point or displacement: *(specify a base point or press* ENTER*)*
>
> Specify second point of displacement or <use first point as displacement>: *(specify the second point of displacement or press* ENTER *to use first point as displacement)*

If you provide the base point and second point of displacement, AutoCAD stretches the selected objects the vector distance from the base point to the second point. If you press ENTER at the prompt for the second point of displacement, then AutoCAD considers the first point as the X,Y displacement value.

Figure 10–21 shows some examples of using the STRETCH command.

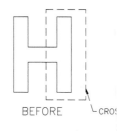

▶ **Figure 10-21**

Examples of using the STRETCH command

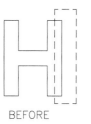

BEFORE ∟CROS

BEFORE

Trimming - Erasing Parts of O Other Objects

 ▶ **Figure 10-22**

Invoking the TRIM command from the Modify toolbar

The TRIM command (see Figure 10-22) is u object(s) that extends past a cutting edge or other objects. Objects that can be trimmed 2D and 3D polylines, xlines, rays, and splines arcs, circles, ellipses, 2D and 3D polylines, splines, and text. After you invoke the TRIM co

> Select objects: *(select the objects and pre the cutting edges)*
>
> Select object to trim or shift-select to exte *choose one of the available options from*

The TRIM command initially prompts you to "S or more cutting edges to establish where to tri Select object to trim: Select one or more objec nate the command. If you press ENTER in resp selecting any objects, by default AutoCAD sele EXTEND command by holding the SHIFT key wh extended to the specified cutting edge instead nation see the section on the EXTEND comman

Break - Erasing Parts of Objects

▶ **Figure 10-24**

Invoking the BREAK command from the Modify toolbar

Break

The BREAK command (see Figure 10-24) is used to remove parts of objects or to split an object in two parts, and it can be used on lines, xlines, rays, arcs, circles, ellipses, splines, donuts, traces, and 2D and 3D polylines. After you invoke the BREAK command, AutoCAD prompts:

> Select object: *(select an object)*
>
> Specify second break point or ⬇: *(specify the second break point, or choose one of the available options from the Shortcut menu)*

See Figure 10-25 for examples of applications of the BREAK command.

▶ **Figure 10-25**

Examples of applications of the BREAK command

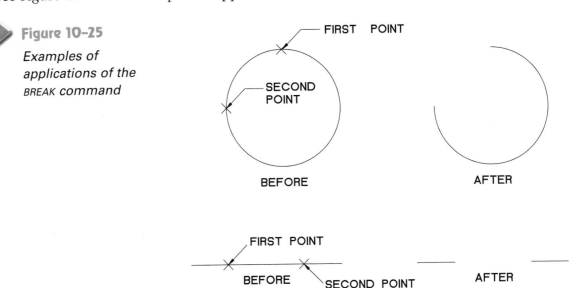

FIRST POINT

SECOND POINT

BEFORE   AFTER

FIRST POINT

BEFORE SECOND POINT   AFTER

AutoCAD erases the portion of the object between the first point (the point where the object was selected) and second point. If the second point is not on the object, AutoCAD selects the nearest point on the object. If you need to erase an object to one end of a line, arc, or polyline, specify the second point beyond the end to be removed.

If instead of specifying the second point, you choose First point from the Shortcut menu, AutoCAD prompts for the first point and then for the second point.

You can split an object into two parts without removing any portion of the object by selecting the same point as the first and second points. You can do so by entering @ to specify the second point. This will not work on a circle. AutoCAD displays the message "Arc cannot be full 360 degrees."

If you select a circle, AutoCAD converts it to an arc by erasing a piece, moving counterclockwise from the first point to the second point. For a closed polyline, the part is removed between two selected points, moving in the direction from the first to the last vertex. And in the case of 2D polylines and traces with width, the BREAK command will produce square ends at the break points.

Extending Objects to Meet Another Object

 Figure 10-26

Invoking the EXTEND command from the Modify toolbar

The EXTEND command (see Figure 10–26) is used to change one or both endpoints of selected lines, arcs, elliptical arcs, open 2D and 3D polylines, and rays to extend to lines, arcs, elliptical arcs, circles, ellipses, 2D and 3D polylines, rays, xlines, regions, splines, text string, or floating viewports. After you invoke the EXTEND command, AutoCAD prompts:

> Select objects: *(select the objects to define the boundary edges and then press* ENTER *to complete the selection)*
>
> Select object to extend or shift-select to trim or ⊥: *(select the object(s) to extend and press* ENTER *to terminate the selection process or choose one of the available options from the Shortcut menu)*

The EXTEND command initially prompts you to Select boundary edges: After selecting one or more boundary edges, press ENTER to terminate the selection process. Then AutoCAD prompts you to Select object to extend: (default option). Select one or more objects to extend to the selected boundary edges. After selecting the required objects to extend, press ENTER to complete the selection process. You can switch to the TRIM command by holding the SHIFT key while selecting the objects that will be trimmed to the specified cutting edge instead of being extended. For detailed explanation, see the section on the TRIM command in this chapter.

The EXTEND and TRIM commands are very similar in the first group of objects selected. With EXTEND you are prompted to select the boundary edge to extend to; with TRIM you are prompted to select a cutting edge. If you press ENTER in response to Select boundary edge(s): without selecting any objects, by default AutoCAD selects all the objects.

Setting the Edge Mode for Extension

The Edge option determines whether objects are extended past a selected boundary or to an implied edge. AutoCAD prompts as follows when the Edge option is selected:

> Enter an implied edge extension mode ⊥ <current>: *(select one of the two available options for extension from the Shortcut menu)*

The Extend selection extends the boundary object along its natural path to intersect another object in 3D space (implied edge). The No Extend selection specifies that the object is to extend only to a boundary object that actually intersects it in 3D space.

Reversing the Change

The Undo option reverses the most recent change made by the EXTEND command.

Setting the Projection Mode

The Project option specifies the projection and AutoCAD uses when extending objects. By default, it is set to the current UCS.

Figure 10–27 shows examples of the use of the EXTEND command.

 Figure 10–27

Examples of applications of the EXTEND command

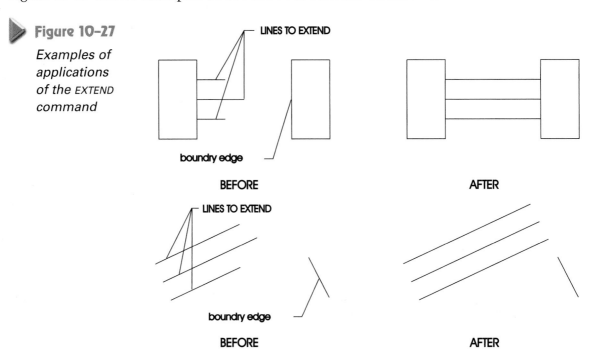

Changing Properties of Selected Objects

 Figure 10–28

Invoking the Properties command from the Standard toolbar

The PROPERTIES command is used to manage and change properties of selected objects by means of the Properties palette. After you invoke the PROPERTIES command (see Figure 10–28), AutoCAD displays the Properties palette, as shown in Figure 10–29.

Figure 10-29

Properties palette

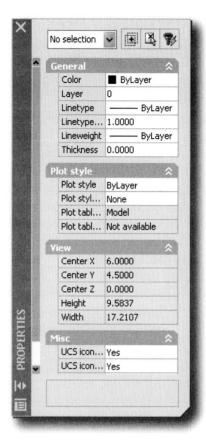

The Properties palette can also be displayed by selecting Properties from the Shortcut menu after selecting one or more objects.

The default position for the Properties palette is docked on the left side of the screen. Its position can be changed by placing the cursor over the double line bar at the top of the window and either double-clicking or dragging the window into the screen area (or across to a docking position on the right side of the screen). Double-clicking causes the Properties palette to become undocked and to float in the drawing area. When the Properties palette in undocked, it can be docked by double-clicking in the title bar (which may be on the left or right side of the window) or by placing the cursor over the title bar and dragging the window all the way to the side you wish to dock it.

When the Properties palette is being displayed and for instance a circle, is selected (make sure PICKFIRST is set to ON), the Properties palette lists all the properties of the selected circle. Not only are properties like color, linetype, and layer listed, but the center X, Y, and Z coordinates, radius, diameter, circumference, and area are also listed. However, if two circles are selected, only the properties that are the same for both circles are listed. Other properties that are not the same but are common properties are listed as *VARIES*. If you specify a value where it is listed as *VARIES*, the new value will be applied. For example, if the centers of the two circles are different and you enter X and Y coordinates in the **Center X** and **Center Y** fields respectively, both circles will be moved to have their centers coincide with the specified center. If different types of objects are in the selection set, the properties listed in the Properties window will include only those common to the selected objects such as color, layer, and linetype. Whenever you change the properties of selected objects in the Properties window, AutoCAD makes the changes immediately in the drawing window.

Matching Properties

 Figure 10–30

Invoking the MATCHPROP *command from the Standard toolbar*

The MATCHPROP command allows you to copy selected properties from one object to one or more other objects located in the current drawing or any other drawing currently open. Properties that can be copied include color, layer, linetype, linetype scale, lineweight, thickness, plot style, and, dimension, text, polyline, viewport, and hatch properties. After you invoke the MATCHPROP command (see Figure 10–30), AutoCAD prompts:

Select source object: *(select the object whose properties you want to copy)*

Select destination object(s) or ⬇: *(select the destination objects and press* ENTER *to terminate the selection, or select Settings from the Shortcut menu)*

Selection of the Settings option displays the Property Settings dialog box, similar to Figure 10–31. Use the Settings option to control which object properties are copied. By default, all object properties in the Property Settings dialog box are set to ON for copying.

 Figure 10–31

Property Settings dialog box

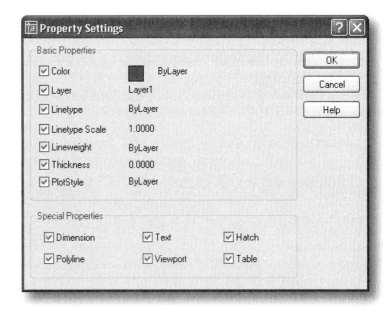

The **Color** selection changes the color of the destination object to that of the source object.

The **Layer** selection changes the layer of the destination object to that of the source object.

The **Linetype** selection changes the linetype of the destination object to that of the source object—available for all objects except attributes, hatches, multiline text, points, and viewports.

The **Linetype Scale** selection changes the linetype scale factor of the destination object to that of the source object—available for all objects except attributes, hatches, multiline text, points, and viewports.

The **Lineweight** selection changes the lineweight of the destination object to that of the source object.

The **Thickness** selection changes the thickness of the destination object to that of the source object—available only for arcs, attributes, circles, lines, points, 2D polylines, regions, text, and traces.

The **PlotStyle** selection changes the plot style of the destination object to that of the source object. If you are working in color-dependent plot style mode (PSTYLEPOLICY is set to 1), this option is unavailable.

The **Dimension** selection changes the dimension style of the destination object to that of the source object—available only for dimension, leader, and tolerance objects.

The **Polyline** selection changes the width and linetype generation properties of the destination polyline to those of the source polyline. The fit/smooth property and the elevation of the source polyline are not transferred to the destination polyline. If the source polyline has variable width, the width property is not transferred to the destination polyline.

The **Text** selection changes the text style of the destination object to that of the source object.

The **Viewport** selection changes the following properties of the destination paper space viewport to match those of the source viewport: on/off, display locking, standard or custom scale, shade plot, snap, grid, and UCS icon visibility and location. The settings for clipping and for UCS per viewport and the freeze/thaw state of the layer are not transferred to the destination object.

The **Hatch** selection changes the hatch pattern of the destination object to that of the source object.

Choosing **Table** changes the table style of the destination object to that of the source object.

After making the necessary changes, choose **OK** to close the Property Settings dialog box. AutoCAD continues with the "Select destination object(s):" prompt. Press ENTER to complete the object selection.

MODIFYING COMPLEX OBJECTS

Polylines, spline curves, and multilines are complex objects. The commands that create them are powerful, and therefore more complicated and harder to learn. Likewise, the commands that modify these objects are more difficult to learn. However, once these editing commands are mastered, AutoCAD can be more fully utilized as a powerful design/drafting tool.

Modifying Polylines

 Figure 10–32

Invoking the PEDIT command from the Modify II toolbar

The PEDIT command allows you to modify polylines. In addition to using such modify commands as MOVE, COPY, BREAK, TRIM, and EXTEND, you can use the PEDIT command to modify polylines. The PEDIT command has special editing features for dealing with the unique properties of polylines and is perhaps the most complex AutoCAD command, with several multi-option submenus totaling some 70 command options. After you invoke the PEDIT command (see Figure 10–32), AutoCAD prompts:

> Select polyline or ⊻: *(select polyline, line, or arc, or choose Multiple from the Shortcut menu)*

If you select a line or an arc instead of a polyline segment, you are prompted as follows:

> Object selected is not a polyline.
>
> Do you want it to turn into one? <Y>

Responding **Y** or pressing ENTER turns the selected line or arc into a single-segment polyline that can then be edited. Normally this is done in order to use the Join option to add other connected segments that, if not polylines, will also be transformed into polylines. It should be emphasized at this time that in order for segments to be joined together into a polyline, their endpoints must coincide. This occurs during line-line, line-arc, arc-line, and arc-arc continuation operations, or by using the endpoint Object Snap mode.

The second prompt does not appear if the first segment selected is already a polyline. It may even be a multi-segment polyline. If you select the Multiple option, AutoCAD allows you to select more than one polyline to modify. After the object selection process, you will be returned to the Multiple option prompt as follows:

> Enter an option ⊻: *(select one of the available options)*

The **Close** option performs in a manner similar to the Close option of the LINE command.

Figure 10–33 shows an example of the application of the Close option.

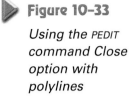

Figure 10-33

Using the PEDIT command Close option with polylines

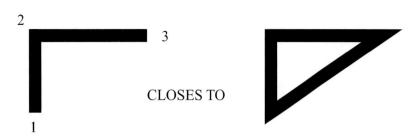

CLOSES TO

The **Open** option deletes the segment that was drawn with the Close option. If the polyline has been closed by drawing the last segment to the first point of the first segment without using the Close option, the Open option will not have a visible effect.

The **Join** option takes selected lines, arcs, or polylines and combines them with a previously selected polyline into a single polyline, if all segments are connected at sequential and coincidental endpoints.

The **Width** option permits uniform or varying widths to be specified for polyline segments.

A vertex is the point where two segments join. When you select the **Edit vertex** option, the visible vertices are marked with an X to indicate which one is to be modified. You can insert a vertex, remove a segment between vertices, relocate a vertex, and attach a tangent direction to the marked vertex for use later in curve fitting.

The **Fit** option draws a smooth curve through the vertices, using any specified tangents.

The **Spline** option provides several ways to draw a curve based on the polyline being edited. These include Quadratic B-spline and Cubic B-spline curves.

The **Decurve** option returns the polyline to the way it was drawn originally. See Figure 10–34 for differences between Fit Curve, Spline, and Decurve.

The **Ltype gen** option controls the display of linetype at vertices. When it is set to ON, AutoCAD generates the linetype in a continuous pattern through the vertices of the polyline. And when it is set to OFF, AutoCAD generates the linetype starting and ending with a dash at each vertex. The Ltype gen option does not apply to polylines with tapered segments.

The **Undo** option reverses the latest PEDIT operation.

The **Exit** option exits the PEDIT command.

Figure 10-34

Comparing PEDIT command Fit Curve, Spline Curve, and Decurve options

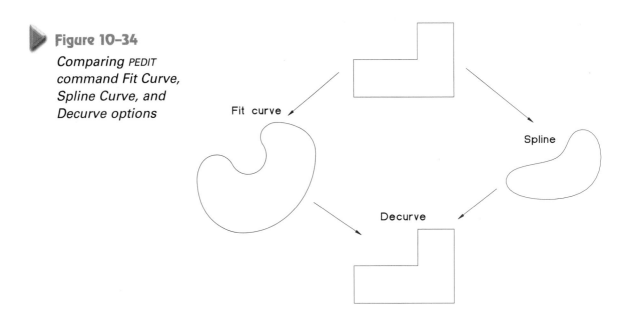

Modifying Spline Curves

Figure 10-35

Invoking the SPLINEDIT command from the Modify II toolbar

Splines created by means of the SPLINE command have numerous characteristics that can be changed with the SPLINEDIT command. These include quantity and location of fit points, end characteristics such as open/close and tangencies, and tolerance of the spline (how near the spline is drawn to fit points).

SPLINEDIT operations on control points (which are different from fit points) of the selected spline, include adding control points (with the Add or the Order option) and changing the weight of individual control points, which determines how close the spline is drawn to individual control points. After you invoke the SPLINEDIT command (see Figure 10–35), AutoCAD prompts:

> Select Spline: *(specify a spline curve)*
> Enter an option [Fit data/Open/Move vertex/Refine/rEverse/Undo]: *(choose one of the available options from the Shortcut menu)*

Control points appear in the grip color, and, if the spline has fit data, fit points also appear in the grip color. If you select a spline whose fit data is deleted, the Fit Data option is not available.

The **Open** option will replace **Close** if you select a closed spline, and vice versa.

The **Fit data** option allows you to edit the selected spline. You can add fit points, close or open the spline, delete selected fit point, relocate the selected fit point, and edit the start and end tangents.

The **Move vertex** option relocates a spline's control vertices.

The **Refine** option of the SPLINEDIT command allows you to fine-tune a spline definition

The **Reverse** option of the SPLINEDIT command reverses the direction of the spline. Reversing the spline does not delete the fit data.

The **Undo** option undoes the effects of the last subcommand.

The **Exit** option terminates the SPLINEDIT command.

Modifying Multilines

Figure 10–36

Invoking the MLEDIT command from the Object flyout of the Modify menu

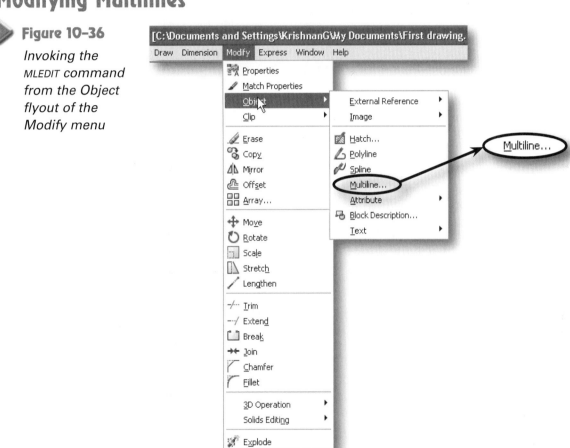

The MLEDIT command (see Figure 10–36) modifies the intersections of two or more multilines and cut gaps in the lines of one multiline. The tools are available for the type of intersection operated on (cross, tee, or vertex) and if one or more elements need to be cut or welded. After you invoke the MLEDIT command, AutoCAD displays the Multiline Edit Tools dialog box (see Figure 10–37) to make it easy to edit multilines.

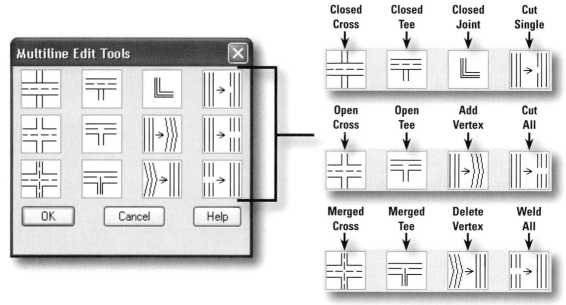

▲ **Figure 10-37**

Multiline Edit Tools dialog box

To choose one of the available options, select one of the image tiles. AutoCAD then prompts for appropriate information.

The first column in the Multiline Edit Tools dialog box works on multilines that cross, the second works on multilines that form a tee, the third works on corner joints and vertices, and the fourth works on multilines to be cut or welded.

Closed Cross

The Closed Cross option cuts all lines that make up the first multiline you select at the point where they cross the second multiline, as shown in Figure 10–38. Choose the Closed Cross image tile (first row and first column in the Multiline Edit Tools dialog box) to invoke the Closed Cross option. AutoCAD then prompts:

Select first mline: *(select the first multiline, as shown in Figure 10–38)*
Select second mline: *(select the second multiline, as shown in Figure 10–38)*

▶ **Figure 10-38**

An example of a closed cross

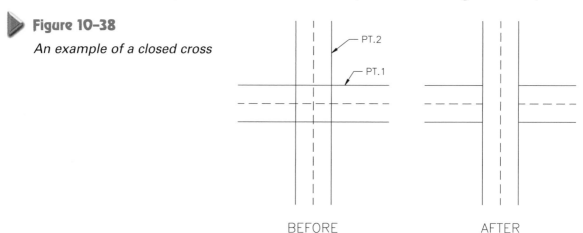

After the closed cross intersection is created, AutoCAD prompts:

> Select first mline or ⬇: *(select another multiline, enter **u**, or press* ENTER*)*

Selecting another multiline repeats the prompt for the second mline. Entering **u** undoes the closed cross just created.

Open Cross

The Open Cross option cuts all lines that make up the first multiline you select and cuts only the outside line of the second multiline, as shown in Figure 10–39. Choose the Open Cross image tile (second row and first column in the Multiline Edit Tools dialog box) to invoke the Open Cross option; AutoCAD prompts:

> Select first mline: *(select a multiline, as shown in Figure 10–39)*
>
> Select second mline: *(select a multiline that intersects the first multiline, as shown in Figure 10–39)*

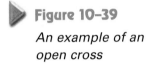

Figure 10–39

An example of an open cross

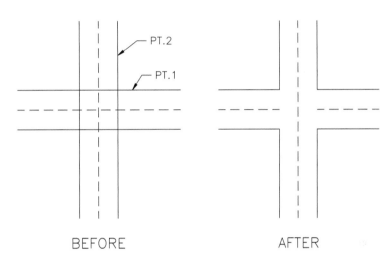

BEFORE   AFTER

After the open cross intersection is created, AutoCAD prompts:

> Select first mline or ⬇: *(select another multiline, enter **u**, or press* ENTER*)*

Selecting another multiline repeats the prompt for the second mline. Entering **u** undoes the open cross just created.

Merged Cross

The Merged Cross option cuts all lines that make up the intersecting multiline you select except the centerlines, as shown in Figure 10–40. Choose the Merged Cross image tile (third row and first column in the Multiline Edit Tools dialog box) to invoke the Merged Cross option; AutoCAD prompts:

> Select first mline: *(select a multiline, as shown in Figure 10–40)*
>
> Select second mline: *(select a multiline that intersects the first multiline, as shown in Figure 10–40)*

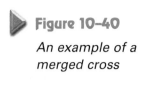

Figure 10–40

An example of a merged cross

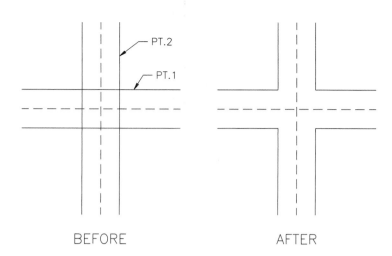

BEFORE   AFTER

After the merged cross intersection is created, AutoCAD prompts:

Select first mline or ⊡: *(select another multiline, enter **u**, or press* ENTER)

Selecting another multiline repeats the prompt for the second mline. Entering **u** undoes the merged cross just created. The order in which the multilines are selected is not important.

Closed Tee

The Closed Tee option extends or shortens the first multiline you identify to its intersection with the second multiline, as shown in Figure 10–41. Choose the Closed Tee image tile (first row and second column in the Multiline Edit Tools dialog box) to invoke the Closed Tee option; AutoCAD prompts:

Select first mline: *(select the multiline to trim or extend, as shown in Figure 10–41)*
Select second mline: *(select the intersecting multiline, as shown in Figure 10–41)*

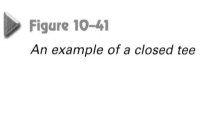

Figure 10–41

An example of a closed tee

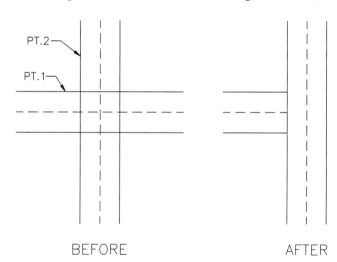

BEFORE   AFTER

After the closed tee intersection is created, AutoCAD prompts:

Select first mline or ⊡: *(select another multiline, enter **u**, or press* ENTER)

Selecting another multiline repeats the prompt for the second mline. Entering **u** undoes the closed tee just created.

Open Tee

The Open Tee option is similar to the Closed Tee option, except that it leaves an open end at the intersecting multiline, as shown in Figure 10–42. Choose the Open Tee image tile (second row and second column in the Multiline Edit Tools dialog box) to invoke the Open Tee option; AutoCAD prompts:

Select first mline: *(select the multiline to trim or extend, as shown in Figure 10–42)*

Select second mline: *(select the intersecting multiline, as shown in Figure 10–42)*

Figure 10–42

An example of an open tee

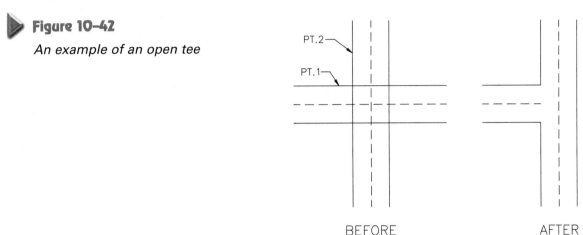

BEFORE   AFTER

After the open tee intersection is created, AutoCAD prompts:

Select first mline or ⊡: *(select another multiline, enter **u**, or press* ENTER*)*

Selecting another multiline repeats the prompt for the second mline. Entering **u** undoes the open tee just created.

Merged Tee

The Merged Tee option is similar to the Open Tee option, except that the centerline of the first multiline is extended to the center of the intersecting multiline, as shown in Figure 10–43. Choose the Merged Tee image tile (third row and second column in the Multiline Edit Tools dialog box) to invoke the Merged Tee option; AutoCAD prompts:

Select first mline: *(select the multiline to trim or extend, as shown in Figure 10–43)*

Select second mline: *(select the intersecting multiline, as shown in Figure 10–43)*

Figure 10–43

An example of a merged tee

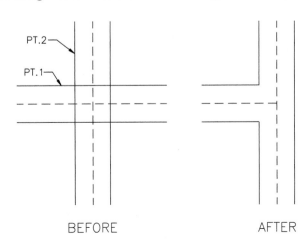

BEFORE   AFTER

After the merged tee intersection is created, AutoCAD prompts:

Select first mline or 📥: *(select another multiline, enter **u**, or press* ENTER*)*

Selecting another multiline repeats the prompt for the second mline. Entering **u** undoes the merged tee just created.

Corner Joint

The Corner Joint option lengthens or shortens each of the two multilines you select as necessary to create a clean intersection, as shown in Figure 10–44. Choose the Corner Joint image tile (first row and third column in the Multiline Edit Tools dialog box) to invoke the Corner Joint option; AutoCAD prompts:

Select first mline: *(select the multiline to trim or extend, as shown in Figure 10–44)*
Select second mline: *(select the intersecting multiline, as shown in Figure 10–44)*

▶ **Figure 10–44**

An example of a corner joint

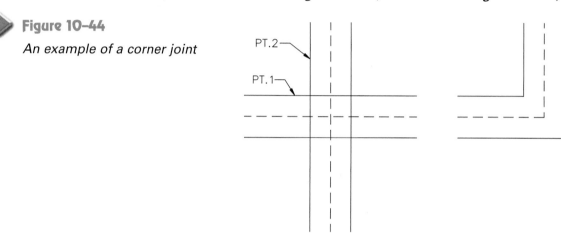

BEFORE   AFTER

After the corner joint intersection is created, AutoCAD prompts:

Select first mline or 📥: *(select another multiline, enter **u**, or press* ENTER*)*

Selecting another multiline repeats the prompt for the second mline. Entering **u** undoes the corner joint just created.

Add Vertex

The Add Vertex option adds a vertex to a multiline, as shown in Figure 10–45. Choose the Add Vertex image tile (second row and third column in the Multiline Edit Tools dialog box) to invoke the Add Vertex option; AutoCAD prompts:

Select mline: *(select a multiline to add a vertex, as shown in Figure 10–45)*

Figure 10–45

An example of adding a vertex

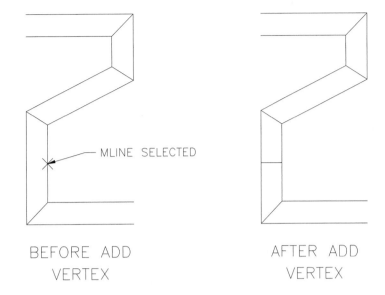

MLINE SELECTED

BEFORE ADD
VERTEX

AFTER ADD
VERTEX

AutoCAD adds the vertex at the selected point and prompts:

Select mline or : *(select another multiline, enter **u**, or press* ENTER*)*

Selecting another multiline allows you to add another vertex. Entering **u** undoes the vertex just created.

Delete Vertex

The Delete Vertex option deletes a vertex from a multiline, as shown in Figure 10–46. Choose the Delete Vertex image tile (third row and third column in the Multiline Edit Tools dialog box) to invoke the Delete Vertex option; AutoCAD prompts:

Select mline: *(select a multiline to delete a vertex, as shown in Figure 10–46)*

Figure 10–46

An example of deleting a vertex

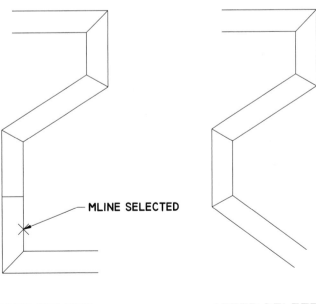

MLINE SELECTED

BEFORE DELETE
VERTEX

AFTER DELETE
VERTEX

AutoCAD deletes the vertex at the selected point and prompts:

Select mline or ⬇: *(select another multiline, enter **u**, or press* ENTER)

Selecting another multiline allows you to delete another vertex. Entering **u** undoes the operation and displays the Select mline: prompt.

Cut Single

The Cut Single option cuts a selected element of a multiline between two cut points, as shown in Figure 10–47. Choose the Cut Single image tile (first row and fourth column in the Multiline Edit Tools dialog box) to invoke the Cut Single option; AutoCAD prompts:

Select mline: *(select a multiline and the selected point becomes the first cut point, as shown in Figure 10–47)*

Select second point: *(select the second cut point on the multiline, as shown in Figure 10–47)*

 Figure 10–47

An example of removing a selected element of a multiline between two cut points

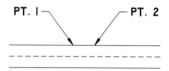

BEFORE   **AFTER**

AutoCAD cuts the multiline and prompts:

Select mline or ⬇: *(select another multiline, enter **u**, or press* ENTER)

Select another multiline to continue, or enter **u** to undo the operation and display the Select mline: prompt.

Cut All

The Cut All option removes a portion of the multiline you select between two cut points, as shown in Figure 10–48. Choose the Cut All image tile (second row and fourth column in the Multiline Edit Tools dialog box) to invoke the Cut All option; AutoCAD prompts:

Select mline: *(select a multiline and the selected point becomes first cut point, as shown in Figure 10–48)*

Select second point: *(select the second cut point on the multiline, as shown in Figure 10–48)*

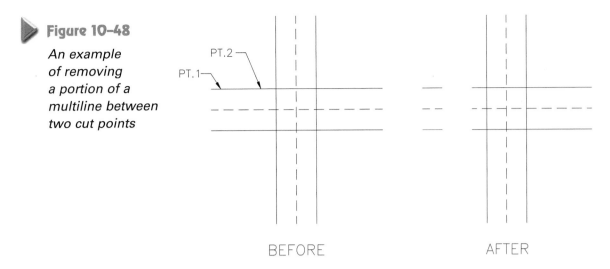

Figure 10-48

An example of removing a portion of a multiline between two cut points

BEFORE   AFTER

AutoCAD cuts all the elements of the multiline and prompts:

> Select mline or ⬇: *(select another multiline, enter **u**, or press* ENTER*)*

Select another multiline to continue, or enter **u** to undo the operation and display the Select mline: prompt.

Weld All

The Weld All option rejoins multiline segments that have been cut, as shown in Figure 10–49. Choose the Weld All image tile(third row and fourth column in the Multiline Edit Tools) to invoke the Weld All option; AutoCAD prompts:

> Select mline: *(select the multiline, as shown in Figure 10–49)*
>
> Select second point: *(select the endpoint on the multiline to be joined, as shown in Figure 10–49)*

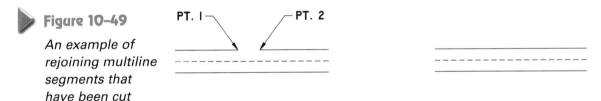

Figure 10-49

An example of rejoining multiline segments that have been cut

PT. 1   PT. 2

BEFORE   AFTER

AutoCAD joins the multiline and prompts:

> Select mline or ⬇: *(select another multiline, enter **u**, or press* ENTER*)*

Select another multiline to continue, or enter **u** to undo the operation and display the Select mline: prompt.

drawing exercises

Open the Exercise Manual PDF file for Chapter 10 on the accompanying CD for discipline-specific exercises.

review questions

1. To create a rectangular array of objects, you must specify:
   a. *The number of items and the distance between them.*
   b. *The number of rows, the number of items, and the unit cell size.*
   c. *The number of rows, the number of columns, and the unit cell size.*
   d. *None of the above.*

2. To create an arc that is concentric with an existing arc, you could use what command?
   a. *ARRAY*
   b. *COPY*
   c. *OFFSET*
   d. *MIRROR*

3. The MATCHPROP command does not allow you to modify an object's:
   a. *Linetype.*
   b. *Fillmode.*
   c. *Color.*
   d. *Layer.*

4. Multiline styles can be saved to an external file, thus allowing their use in multiple drawings.
   a. *True*   b. *False*

5. When filleting multilines, you must specify:
   a. *The radius of the innermost fillet.*
   b. *The radius of the center line.*
   c. *The radius of the outermost fillet.*
   d. *Multilines cannot be filleted.*

8. The _____ command is used to make multiple copies of one or more objects in rectangular or polar patterns.
   a. *ARRAY*   c. *COPY*
   b. *MOVE*   d. *MINSERT*

9. When performing a polar array, the reference point for arraying a block is:
   a. *The upper right corner of the block.*   c. *The lower left corner of the block.*
   b. *The insertion point of the block.*   d. *None of the above.*

10. The command that converts a corner into a beveled corner is:
    a. *FILLET.*   c. *CHAMFER.*
    b. *BEVEL.*   d. *CHANGE.*

11. The best command used to make many copies of an object in a rectangular pattern is:

a. *DUPLICATE.* **c.** *ARRAY.*

b. *PATTERN.* **d.** *REPEAT.*

12. All of the following objects can be filleted or chamfered except:

a. *Lines.* **c.** *Blocks.*

b. *Circles.* **d.** *Arcs.*

13. The ____ command is used to make copies of objects and place them in another location.

a. *OFFSET* **c.** *MIRROR*

b. *MOVE* **d.** *COPY*

14. Filleting two nonparallel, non-intersecting line segments with a zero radius will:

a. *Return an error message.* **c.** *Create a sharp corner.*

b. *Have no effect.* **d.** *Convert the lines to rays.*

15. To reproduce an object seven times in a half circular pattern, use the polar option of the ARRAY command with the number of items and angle of:

a. *8 and 180.* **c.** *7 and 360.*

b. *6 and 360.* **d.** *6 and 180.*

15. Objects are duplicated to form symmetrical shapes using:

a. *The MOVE command.* **c.** *The OFFSET command.*

b. *The COPY command.* **d.** *The MIRROR command.*

16. Which PEDIT option and its description is NOT correct?

a. *Open: Converts and connects non-polyline entities to polylines*

b. *Width: Allows selection of new width for all segments of polylines*

c. *Fit Curve: Constructs a smooth curve using the vertices*

d. *X: Returns to the command prompt*

17. To turn a series of line segments into a polyline, one uses the PEDIT option of:

a. *Spline.* **c.** *Close.*

b. *Join.* **d.** *Edit Curve.*

18. The following are all options of the PEDIT command except:

a. *LType and Color.* **c.** *Width.*

b. *Undo.* **d.** *Arc.*

19. Which command allows you to change the location of the objects and allows a duplicate to remain intact?

a. *CHANGE* **c.** *COPY*

b. *MOVE* **d.** *MIRROR*

20. To efficiently relocate multiple objects with the MOVE command, which selection option would be more efficient?

a. *Last* **c.** *Add*

b. *Window* **d.** *Undo*

21. The MOVE command allows you to:

a. *Move objects to new locations on the screen.*
b. *Move only the objects that are on the current layer.*
c. *Dynamically drag objects on the screen.*
d. *Both a and c.*

22. The ____command allows you to project objects in a drawing to meet a boundary entity.

a. *EXTEND*   c. *SCALE*
b. *ROTATE*   d. *None of the above*

23. The ____ command allows you to define other objects as cutting edges, and then specify the part of the object to be cut.

a. *STRETCH*   c. *TRIM*
b. *SCALE*   d. *None of the above*

24. The MATCHPROP command allows you to copy the following properties:

a. *Color.*   c. *Linetype.*
b. *Layer.*   d. *All of the above.*

25. The LENGTHEN command has the following options available:

a. *Delta.*   c. *Total.*
b. *Percent.*   d. *All of the above.*

26. The STRETCH command will alter the following elements if they reside inside the selection window:

a. *Lines.*   c. *Splines.*
b. *Polylines.*   d. *All of the above.*

27. Which is not an option of the Multiline Line Edit tools?

a. *Merged Cross*   c. *Corner Joint*
b. *Radius*   d. *All of the above*

28. Which is not a multiline justification choice?

a. *Top*   c. *Center*
b. *Zero*   d. *Bottom*

29. Which command brings up the Change Properties dialog box?

a. *CHANGE*   c. *DDCHPROP*
b. *CHPROP*   d. *Both b and c*

advanced object selection

introduction

Chapter 5 explained how to select objects to be modified in AutoCAD. This chapter explains how to use the more advanced object selection features and to enhance the methods of object selection.

After completing this chapter, you will be able to do the following:

- Use advanced object selection methods and modes to modify objects

- Use the QSELECT tool to specify the filtering criteria and how you want AutoCAD to create the selection set from that criteria

- Use FILTER tool to create reusable filters for object selection

- Use the GROUP tool to save a set of objects that you can select and edit together or separately as needed

FINE-TUNING OBJECT SELECTION

AutoCAD provides five modes to tailor the way you select objects to best suit your individual needs. The available modes include Noun/Verb Selection, Use Shift to Add, Press and Drag, Implied Windowing, and Object Grouping. You can also adjust the size of the pick box used for selecting objects.

Object Selection Modes

You can toggle one or more object selection modes ON or OFF from the **Selection Modes** section on the Selection tab of the Options dialog box (see Figure 11–1). You can open the Options dialog box from Shortcut menu that appears in the Graphics window when there are no commands in effect.

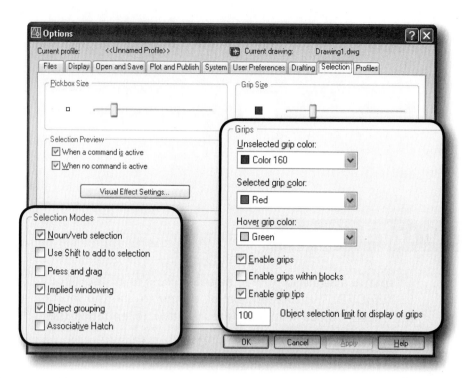

Noun/verb selection, **Implied windowing**, and **Object grouping** are set as defaults.

Noun/Verb Selection

The **Noun/verb selection** feature allows the traditional verb-noun command syntax to be reversed for most modifying commands. When the **Noun/verb selection** is set to ON, you can select the objects first at the Command: prompt and then invoke the appropriate modifying command you want to use on the selection set. For example, instead of invoking the COPY command followed by selecting the objects to be copied, with **Noun/verb selection** set to ON, you can select the objects first and then invoke the COPY command, and AutoCAD skips the object selection prompt.

When **Noun/verb selection** is set to ON, the cursor at the Command: prompt changes to resemble a Running Osnap cursor (see Figure 11-2). Whenever you want to use the **Noun/verb selection** feature, first create a selection set at the Command: prompt. Subsequent modify commands you invoke execute by using the objects in the current selection set without prompting for object selection. To clear the current selection set, press ESC at the Command: prompt. This clears the selection set, so any subsequent editing command will once again prompt for object selection. You can also toggle the **Noun/verb selection** feature ON/OFF by changing the setting PICKFIRST system variable. TRIM, EXTEND, BREAK, CHAMFER, and FILLET are the commands not supported by the **Noun/verb selection** feature.

Use Shift to Add

The **Use Shift to add to selection** feature controls how you add objects to an existing selection set. When **Use Shift to add to selection** is set to ON, it activates an additive selection mode in which SHIFT must be held down while adding more objects to the selection set. For example, if you first pick an object, it is highlighted. If you pick another object, it is highlighted and the first object is no longer highlighted. The only way you can add objects to the selection set is to select objects by holding down the SHIFT key. Similarly, the way to remove objects from the selection set is to select the objects by holding down the SHIFT key.

When **Use Shift to add to selection** is set to OFF (default), objects are added to the selection set by just picking them individually or by using one of the selection options; AutoCAD adds the objects to the selection set. You can also toggle **Use Shift to add to selection** ON/OFF by changing the setting PICKADD system variable.

Press and Drag

The **Press and drag** feature controls the manner by which you draw the selection window with your pointing device. When **Press and drag** is set to ON, you can create a selection window by holding down the pick button and dragging the cursor diagonally while you create the window.

When **Press and drag** is set to OFF (default), you need to use two separate picks of the pointing device to create the selection window. In other words, you need to pick once to define one corner of the selection window, and pick a second time to define its diagonal corner. You can also toggle **Press and drag** ON/OFF by changing the setting PICKDRAG system variable.

Implied Windowing

The **Implied windowing** feature when it is set to ON (default) initiates the drawing of a selection window when you select a point outside an object. Drawing the selection window from left to right selects objects that are entirely inside the window's boundaries. Drawing from right to left selects objects within and crossing the window's boundaries. If **Implied windowing** is set to OFF, you cannot initiate drawing of a window to create a selection set. You can also toggle **Implied windowing** ON/OFF by changing the setting PICKAUTO system variable.

Object Grouping

Selects all objects in a group when you select one object in that group. With the GROUP command, you can create and name a set of objects for selection.

The **Object grouping** feature controls the automatic group selection. If **Object grouping** is set to ON, selecting an object that is a member of a group selects the whole group. If it is set to OFF, only individual objects can be selected instead of a group as a whole. You can also set **Object grouping** to ON by setting the PICKSTYLE system variable to 1. Refer to "Grouping Objects" section for detailed explanation about creating and modifying groups.

The **Associative Hatch** feature controls which objects will be selected when you select an associative hatch. If **Associative Hatch** is set to ON (default), then selecting an associative hatch also selects the boundary objects. Refer to Chapter 12 for a detailed description of hatching.

Pickbox Size

The **Pickbox Size** slider bar controls the display size of the AutoCAD pick box. The pick box is the object selection tool that appears in editing commands. You can also set the pick box by changing PICKBOX system variable in terms of pixels (default is set to 3).

MODIFYING OBJECTS USING GRIPS

The grips feature allows you to edit AutoCAD drawings in an entirely different way than using the traditional AutoCAD modify commands. With grips you can move, stretch, rotate, copy, scale, and mirror selected objects without invoking one of the regular AutoCAD modify commands. When you select an object with grips enabled, small squares appear at strategic points on the object that enable you to edit the selected objects.

Enabling and Setting Grips

To enable the grips feature, first open the Options dialog box and choose the **Selection** tab as shown in Figure 11–1, where AutoCAD displays various options related to grips.

Enable grips controls the display of the grips. If it is set to ON, the grips display is enabled; if it is set OFF, the grips display is disabled.

You can also enable the grips feature by setting the system variable GRIPS to 1.

Enable grips within blocks controls the display of grips on objects within blocks. If it is set to ON, the grips are displayed on all objects within the block; if it is set to OFF, a grip is displayed only on the insertion point of the block.

Enable grip tips, when checked, causes a grip-specific tip to be displayed when the cursor hovers over a grip on a custom object that supports grip tips. Standard AutoCAD objects are not affected by this option.

The **Unselected grip color** box allows you to change the color to the unselected grips.

The **Selected grip color** box allows you to change the color to the selected grips.

The **Hover grip color** box allows you to change the color to the grips in the hover mode.

The **Object selection limit for display of grips** box allows you to specify the number that limits the display of grips when the initial selection set includes more than the specified number of objects. The valid range is 1 to 32,767. The default setting is 20.

The **Grip Size** slider bar allows you to change the size of the grips. To adjust the size of grips, move the slider box left or right. As you move the slider, the size is illustrated to the right of the slider.

After making the necessary settings, choose **OK** to keep the changes and close the Options dialog box.

AutoCAD gives you a visual cue when grips are enabled by displaying a pick box at the intersection of the crosshairs, even when AutoCAD is awaiting a command, as shown in Figure 11–2.

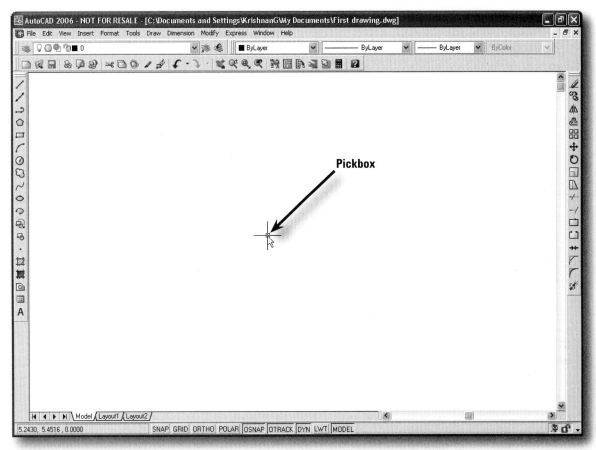

Figure 11-2

The pick box displayed at the intersection of the crosshairs while awaiting a command

note   The pick box is also displayed on the crosshairs when the PICKFIRST (noun/ verb selection) system variable is set to ON.

Applying Grips

To place grips directly from the Command: prompt, select one or more objects you wish to manipulate by any of the selection methods.

Grips appear on the endpoints and midpoint of lines and arcs, on the vertices and endpoints of polylines, on quadrants and the center of circles, on dimensions, text, solids, 3dfaces, 3dmeshes, and viewports, and on the insertion point of a block. Figure 11–3 shows location of the grips on some of the commonly used objects.

Figure 11–3

Locations of grips on commonly used objects

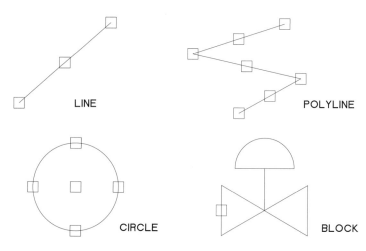

When you move your cursor over a grip, it automatically snaps to the grip point. This allows you to specify exact locations in the drawing without having to use grid, snap, ortho, object snap, or coordinate entry tools.

To use grips, select a grip to act as the base point for the action. You can use multiple grips as the base grips to keep the shape of the object intact between the selected grips. Hold down SHIFT as you select the grips. You can use multiple grips as the base grips to keep the shape of the object intact between the selected grips. Hold down SHIFT as you select the grips. For quadrant grips on circles and ellipses, distance is measured from the center point, not the selected grip.

To clear grips from a selection set, press the ESC key. The grips on the selected objects will be cleared. When you invoke an AutoCAD command, such as LINE or CIRCLE, AutoCAD clears the grips from a selection set.

To use grips to edit the selected objects, as mentioned earlier, specify a grip to act as the base point for the editing operation at the Command: prompt. Then select one of the grip modes. You can cycle through these modes by pressing ENTER or SPACEBAR. You can also use shortcut keys or right-click to see all of the modes as shown in Figure 11–4. To cancel Grip mode, enter **x** (for the mode's eXit option); AutoCAD returns to the Command: prompt. You can also use a combination of the current Grip mode and a multiple copy operation on the selection set.

Figure 11–4

Shortcut menu displaying the Grip modes

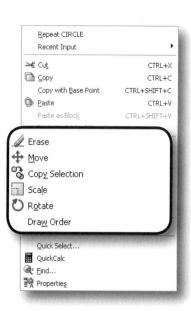

Stretch Mode

The Stretch mode works similar to the STRETCH command. It allows you to stretch the shape of an object without affecting other crucial parts that remain unchanged. When you are in the Stretch mode, the following prompt appears:

Specify stretch point or ⊡:

The Specify stretch point option (the default) refers to the stretch displacement point. As you move the cursor, you see that the shape of the object is stretched dynamically from the base point. You can specify the new point with the cursor or by entering coordinates. The displacement is applied to all selected hot grips.

If necessary, you can change the base point to be other than the base grip by selecting Base Point from the Shortcut menu. Then specify the point with the cursor or enter the coordinates.

To make multiple copies while stretching objects, choose Copy from the Shortcut menu. Then specify destination copy points with the cursor or enter their coordinates.

If you press the SHIFT key while selecting multiple stretch points, the Copy mode is activated, and for subsequent copies the cursor snaps to offset points based on the first two points selected. When the first offset is specified while you hold the SHIFT key, the subsequent second, third and fourth offsets can be specified by moving the cursor until it locks onto the points that have been preset to the same offset as the first one. It you release the SHIFT key, you can place the stretch point at any location.

Move Mode

The Move mode works similar to the MOVE command. It allows you to move one or more objects from their present location to a new one without changing their orientation or size. In addition, you can make copies of the selected objects at the specified displacement, leaving the original objects intact. To invoke the Move mode, cycle through the modes by entering a null response until it takes you to the Move mode, or choose Move from the Shortcut menu. When you are in the Move mode, the following prompt appears:

Specify move point or ⊡:

The Specify move point option refers to the move displacement point. As you move the cursor, AutoCAD moves all the objects in the current selection set to a new point relative to the base point. You can specify the new point with the cursor or by entering the coordinates.

If necessary, you can change the base point to be other than the base grip by selecting Base Point from the Shortcut menu. Then specify the point with the cursor or enter the coordinates.

To make multiple copies while moving objects, choose Copy from the Shortcut menu and then specify the destination copy points with the cursor, or enter their coordinates.

If you press the SHIFT key while moving the object, the Copy mode is activated, and for subsequent copies the cursor snaps to offset points based on the first two points selected. When the first offset is specified while you hold the SHIFT key, the subsequent second, third and fourth offsets can be specified by moving the cursor until it locks onto the points that have been preset to the same offset as the first one. It you release the SHIFT key, you can place the copy at any location.

Rotate Mode

The Rotate mode works similar to the ROTATE command. It allows you to change the orientation of objects by rotating them about a specified base point. In addition, you can make copies of the selected objects and at the same time rotate them about a specified base point.

To invoke the Rotate mode, cycle through the modes by entering a null response until it takes you to the Rotate mode, or choose Rotate from the Shortcut menu. When you are in the Rotate mode, the following prompt appears:

Specify rotation angle or ⬇:

The Specify rotation angle option refers to the rotation angle to which objects are rotated. As you move the cursor, AutoCAD allows you to drag the rotation angle, to position all the objects in the current selection set at the desired orientation. You can specify the new orientation with the cursor or from the keyboard. If you specify an angle by entering a value from the keyboard, this is taken as the angle that the objects should be rotated from their current orientation, around the base point. A positive angle rotates counterclockwise, and a negative angle rotates clockwise. Similar to the ROTATE command, you can use the Reference option to specify the current rotation and the desired new rotation.

If necessary, you can change the base point to be other than the base grip by choosing Base Point from the Shortcut menu. Then specify the point with the cursor or enter the coordinates.

To make multiple copies while rotating objects, choose Copy from the Shortcut menu. Then specify destination copy points with the cursor, or enter the coordinates.

If you press the SHIFT key while rotating the object, the Copy mode is activated, and for subsequent copies the cursor snaps to offset points based on the first two points selected. When the first offset is specified while you hold the SHIFT key, the subsequent second, third and fourth offsets can be specified by moving the cursor until it locks onto the points that have been preset to the same offset as the first one. It you release the SHIFT key, you can place the copy at any location.

Scale Mode

The Scale mode works similar to the SCALE command. It allows you to change the size of objects about a specified base point. In addition, you can make copies of the selected objects and at the same time change the size about a specified base point. To invoke the Scale mode, cycle through the modes by entering a null response until it takes you to the Scale mode, or choose Scale from the Shortcut menu. When you are in the Scale mode, the following prompt appears:

Specify scale factor or ⬇:

The Specify scale factor option refers to the scale factor to which objects are made larger or smaller. As you move the cursor, AutoCAD allows you to drag the scale factor, to change all the objects in the current selection set to the desired size. You can specify the new scale factor with the cursor or from the keyboard. If you specify the scale factor by entering a value from the keyboard, this is taken as a relative scale factor by which all dimensions of the objects in the current selection set are to be multiplied. To enlarge an object, enter a scale factor greater than 1; to shrink an object, use a scale factor between 0 and 1. Similar to the SCALE command, you can use the Reference option to specify the current length and the desired new length.

If necessary, you can change the base point to be other than the base grip by choosing Base Point from the Shortcut menu. Then specify the point with the cursor or enter the coordinates.

To make multiple copies while scaling objects, choose Copy from the Shortcut menu, and then specify the destination copy points with the cursor, or enter their coordinates.

If you press the SHIFT key while scaling the object, the Copy mode is activated. and for subsequent copies the cursor snaps to offset points based on the first two points selected. When the first offset is specified while you hold the SHIFT key, the subsequent second, third and fourth offsets can be specified by moving the cursor until it locks onto the points that have been preset to the same offset as the first one. It you release the SHIFT key, you can place the copy at any location.

Mirror Mode

The Mirror mode works similar to the MIRROR command. It allows you to make mirror images of existing objects. To invoke the Mirror mode, cycle through the modes by entering a null response until it takes you to the Mirror mode, or choose Mirror from the Shortcut menu. When you are in the Mirror mode, the following prompt appears:

Specify second point or ⊡:

Two points are required in AutoCAD to define a line about which the selected objects are mirrored. AutoCAD considers the base grip point as the first point; the second point is the one you specify or enter in response to the Specify second point: prompt.

If necessary, you can change the base point to be other than the base grip by choosing Base Point from the Shortcut menu. Then specify the point with the cursor or enter the coordinates.

To make multiple copies while retaining original objects, choose Copy from the Shortcut menu. Then specify mirror points by specifying the point(s) with the cursor, or enter the coordinates.

If you press the SHIFT key while mirroring the object, the Copy mode is activated, and for subsequent copies the cursor snaps to offset points based on the first two points selected. When the first offset is specified while you hold the SHIFT key, the subsequent second, third and fourth offsets can be specified by moving the cursor until it locks onto the points that have been preset to the same offset as the first one. It you release the SHIFT key, you can place the copy at any location.

SELECTING OBJECTS BY QUICK SELECT

▶ **Figure 11-5**

Invoking the QSELECT command from the Tools menu

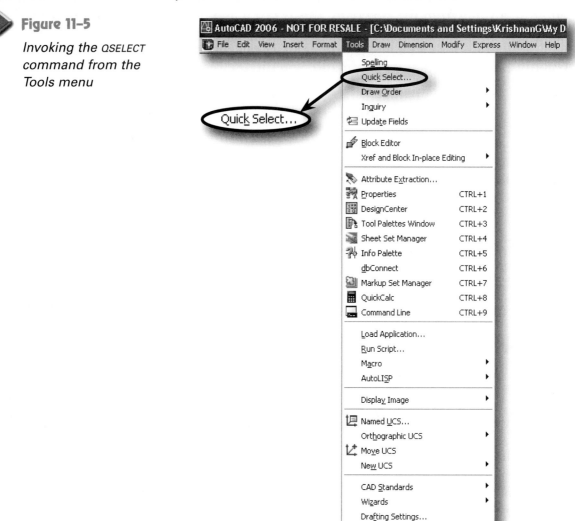

Selecting objects by the QSELECT command is used to create selection sets based on objects that have or do not have similar characteristics or properties as determined by filters that you specify. For example, you can create a selection set of all lines that are equal to or less than 2.5 units long. Or you can create a selection set of all objects that are not text objects on one certain layer. The combinations of possible filters are almost limitless. The selection set created by the QSELECT command replaces or is appended to the current selection set. If you have partially opened the current drawing, QSELECT does not consider objects that you have not loaded. You can cause the created selection set to replace the current selection set or be appended to it.

To create a filtered selection set, choose Quick Select from the Tools menu (see Figure 11–5), and AutoCAD displays the Quick Select dialog box similar to Figure 11–6.

Figure 11-6

Quick Select dialog box

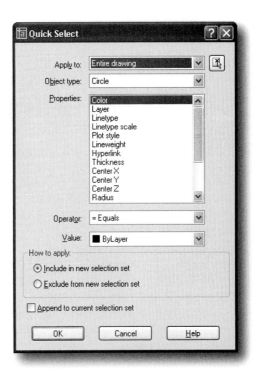

First, use the **Apply to** text box to select whether to apply the specified filters to Current selection or to Entire drawing. If there is a current selection, then Current selection is the default; otherwise, Entire drawing is the default. If the **Append to current selection set** check box is checked, Current selection is not an option. If you wish to create a selection set, choose the **Select Objects** button (located next to the **Apply to** text box). **Select Objects** is available only when **Append to current selection set** is not checked.

The **Object type** box lets you select the type of objects to include in the filtering criteria. If the filtering criteria are being applied to the entire drawing, the **Object type** list includes all object types, including custom. Otherwise, the list includes only the object types of the selected objects.

The **Properties** section lists the properties that can be used for filters for the type of object(s) specified in the **Object type** box. This list includes all searchable properties for the selected object type. The property you select determines the options available in **Operator** and **Value** boxes.

The **Operator** box lets you apply logical operators to specified values. Depending on the selected property, options may include Greater Than, Less Than, Equal To, Not Equal To, and *Wildcard Match. Greater Than and Less Than apply primarily to numeric values, and Wildcard Match applies to text strings.

The **Value** box lets you specify a value to which the operator applies. If you specify the Greater Than Operator to the Value of 1.0 for the Length of lines, then the lines with lengths greater than 1.0 will be filtered to be either included in or excluded from the selection set, depending on which option in the **How to apply** section has been chosen.

The **How to apply** section lets you specify whether you want the new selection set to include or exclude objects that match the specified filtering criteria. Choose **Include in new selection set** to create a new selection set composed only of objects that match the filtering criteria. Choose **Exclude from new selection set** to create a new selection set composed only of objects that do not match the filtering criteria.

Once you have all the required selection criteria set, choose **OK** to close the dialog box. AutoCAD highlights the selected set. You can proceed with the appropriate modification required for the selected objects.

SELECTION SET BY FILTER TOOL

The FILTER command allows you to create reusable filters for object selection based on a list of requirements that an object must meet to be included in a selection set. This is another method of selecting objects.

The new selection set that is created by the FILTER command can be used as the Previous option at the next "Select object:" prompt. If you use the FILTER command transparently, AutoCAD passes the new selection set directly to the command in operation. Named filters can be used at any "Select object:" prompt. This will save you a considerable amount of time.

With the FILTER command, you can select objects based on object properties, such as location, object type, color, linetype, layer, block name, text style, and thickness. For example, you could use the FILTER command to select all the blue lines and arcs with a radius of 2.0 units. You can even name filter lists and save them to a file.

To create a new selection, invoke the FILTER command at the On-screen prompt. AutoCAD displays the Object Selection Filters dialog box, similar to Figure 11–7.

Figure 11–7

Object Selection Filters dialog box

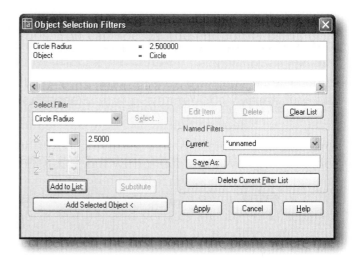

The list box displays the filters currently being used as a selection set. The first time you use the FILTER command in the current drawing, the list box is empty.

Selection of Object Properties to Filter

The **Select Filter** section lets you add filters to the list box based on object properties. Select the object or logical operator from the list box. You can use the grouping operators AND, OR, XOR, and NOT from the list box. The grouping operators must be paired and balanced correctly in the filter list. For example, each Begin OR operator must have a matching End OR operator. If you select more than one filter, AutoCAD by default uses an AND as a grouping operator between each filter.

Choose **Select** to display a dialog box listing all items of the specified type selected in the list box and select items to filter. For example, if you selected the object type

Block, the Select dialog box displays a list of available blocks in the current drawing to select from.

The X, Y, and Z parameters define additional filter parameters depending on the object selected. For example, if you select Center Radius, you can enter the radius in the X box that you want to filter. You can also use filter parameters that contains relative operators such as < (less than) or > (greater than). For example, the following filter selects all circles with radii greater than or equal to 1.5:

Object = Circle

Circle Radius >= 1.5000

Once you have selected the object type and set corresponding parameters if any, then choose **Add to List** to add the current filter selection to the filter list.

You can also add an object in the drawing to the filter list. Choose **Add Selected Object,** and AutoCAD prompts you to select an object and adds the selected object to the filter list. To substitute the selected filter in the filter list with the current filter setting, choose **Substitute** to replace the selected filter with the current one in the **Select Filter** box.

Modifying Object Properties in the Filter List

To modify an existing filter property, first select the filter object in the Filter list box and choose **Edit Item**. AutoCAD moves the selected filter property into the **Select Filter** area for editing. Make necessary changes to the filter properties and then choose **Substitute**. AutoCAD replaces the edited filter to the selected filter property.

To delete an existing filter property listed in the filter list, choose **Delete**. AutoCAD deletes the selected filter property from the current filter.

To clear the filter list and start all over again, choose **Clear List**. AutoCAD deletes all the listed properties from the current filter, and you can build a brand new filter list.

Saving a Filter List

As mentioned earlier, AutoCAD allows you to save the currently displayed filter list to use it again at any "Select objects:" prompt. This will save you a considerable amount of time. To save the current filter list, enter the name for the filter list in the **Save As** field, and choose **Save As** to save the list with the given name. A filter name can contain up to 18 characters, and AutoCAD saves the filter in the *filter.nfl* file.

The **Current** box lists saved filters. Select a filter list to make it current, and AutoCAD loads the named filter and its list of properties from the default file, *filter.nfl*.

To delete a named filter list, first select the named filter in the **Current** list box and choose **Delete Current Filter List**. AutoCAD deletes the selected named filter and all its properties from the default filter file.

Choose **Apply** to close the dialog box and display the "Select objects:" prompt, where you create a selection set. AutoCAD uses the current filter on the objects you select.

If you used the FILTER command as a transparent command during the modification command, AutoCAD will resume the operation after applying filter selection.

GROUPING OBJECTS

A group definition is a saved set of objects you can select and edit together or separately as needed. Groups provide an easy way to combine drawing elements that you need to manipulate as a unit. You can change the components of a group definition as you work by adding or removing objects.

When you create a group definition, you can give the group a name and description. This is similar to creating a block (see Chapter 17). The advantage of using a group instead of a block is that the GROUP command's "selectable" switch can be set to OFF for modifying an individual member without losing its membership in the group. Also, named groups, like blocks, are saved with the drawing and cannot be shared with other drawings.

A named group can be selected for modifying as a group only when its "selectable" switch is set to ON. You can select a group to modify (such as with the MOVE or COPY command) by selecting one of the members of the group or select the group by name. To select a group by name, type **g** at the "Select objects:" prompt; AutoCAD prompts for the group's name. Specify the name of the group to modify. AutoCAD highlights the objects that belong to the selected group for modification.

Creating a Group

To create a new group definition, invoke the GROUP command at the On-screen prompt. AutoCAD displays the Object Grouping dialog box, similar to Figure 11–8.

Figure 11–8

Object Grouping dialog box

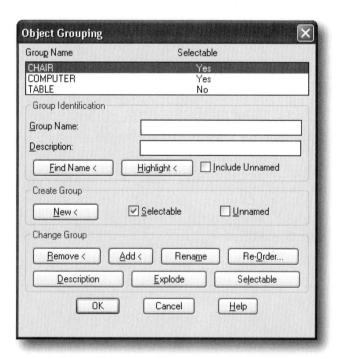

The **Group Name** list box lists the names of the existing groups defined in the current drawing. The **Selectable** column indicates whether a group is selectable. If it is listed as Yes in the **Selectable** column, selecting a single group member selects all the members except those on locked layers. If it is listed as No in the **Selectable** column, selecting a single group member selects only that object.

To create a new group definition, enter the group name and description (optional) in the **Group Name** and **Description** box respectively. Group names can be up to 31 characters long and can include letters, numbers, and the special characters $ and _. Set the **Selectable** check box to ON or OFF, which specifies default Selectable setting for the new group. Choose **New** to create a new group. AutoCAD prompts for the selection of objects. Select all the objects to be included in the new group, and press ENTER to complete the selection. The newly created group will be listed in the **Group Name** list box. Choose **OK** to close the dialog box. To create an unnamed group definition, set the **Unnamed** check box to ON, provide a description (optional) in the **Description** box and choose **New** to create a new group. AutoCAD creates a new group definition and assigns default name, *An, to unnamed group. The n represents a number that increases with each unnamed new group.

Modifying and Deleting a Group

At any time, you can add or remove group members and rename groups.

To add an object to an existing group definition, first select the name of the group in the **Group Name** list box and choose **Add**. AutoCAD prompts for the selection of objects. Select the object(s) to be added to the existing group, and press ENTER to complete the selection. AutoCAD adds the selected object(s) to the group and redisplays the Object Grouping dialog box.

To remove object(s) from an existing group definition, first select the name of the group in the **Group Name** list box and choose **Remove**. AutoCAD prompts for the selection of objects to remove from the selected group. Select the object(s) to remove the selected group, and press ENTER to complete the selection. AutoCAD removes the selected object(s) from the group and redisplays the Object Grouping dialog box.

To lists the groups to which an object belongs, choose **Find Name,** and AutoCAD prompts to select an object. AutoCAD displays the Group Member List dialog box, which lists the group or groups to which the selected object belongs.

To highlight member objects of a selected group, first select the name of the group in the **Group Name** list box and choose **Highlight**. AutoCAD s highlights the member objects of the selected group in the drawing area.

To rename a group definition, first select the name of the group in the **Group Name** list box. The selected group name and description appears in the **Group Name** and **Description** fields respectively. Make the necessary changes in the group name and choose **Rename** to rename the selected group. Similarly, make necessary changes in the group description and choose **Description** to change the description of the selected group. Choosing Reorder displays the Order Group dialog box, in which you can change the numerical order of objects within the selected group.

To change the selectable status of a group, first select the name of the group in the **Group Name** list box and choose **Selectable**. AutoCAD switches the current status either from Yes to No or vice versa.

To delete a group definition, first select the name of the group in the **Group Name** list box and choose **Explode** to delete the definition of the selected group. Objects that were part of the deleted group definition remain in the drawing.

After making the necessary changes to the Object Grouping dialog box, choose **OK** to keep the changes and close the dialog box.

drawing exercises

Open the Exercise Manual PDF file for Chapter 11 on the accompanying CD for discipline-specific exercises.

review questions

1. When editing using grips, you can ____ an object.
   - **a.** *Stretch, rotate, offset, move, or scale.*
   - **b.** *Stretch, move, rotate, scale, or mirror.*
   - **c.** *Scale, trim, stretch, copy, or move.*
   - **d.** *Lengthen, move, rotate, scale, or mirror.*

2. The default setting for the GRIPS system variable is:
   - **a.** *1.*
   - **b.** *0.*
   - **c.** *2.*
   - **d.** *None of the above.*

3. What is the default value for GRIPSIZE?
   - **a.** *4*
   - **b.** *3*
   - **c.** *7*
   - **d.** *None of the above*

4. Which is not an option in the Options dialog box (Grips section)?
   - **a.** *Enable Grips*
   - **b.** *Grip Size*
   - **c.** *Grip Colors*
   - **d.** *Grip Locations*

5. Grips appear ____ on selected objects.
   - **a.** *Randomly*
   - **b.** *At Key Points*
   - **c.** *At End Points only*
   - **d.** *At Mid Points only*

6. After activating grips on an object, AutoCAD will ____ when you press the right mouse button.
   - **a.** *Display cursor menu*
   - **b.** *Exit grip mode*
   - **c.** *Scale objects to a specified number*
   - **d.** *Ask for a start point*

7. Which command allows you to select multiple objects by specifying criteria?
   - **a.** FILTER
   - **b.** GROUP
   - **c.** SELECT
   - **d.** WINDOW

8. A group can be:
   - **a.** *Named.*
   - **b.** *Renamed.*
   - **c.** *Selectable or not.*
   - **d.** *All of the above.*

9. What is a set of objects with an assigned name and description called?
   - **a.** *Block*
   - **b.** *Shape*
   - **c.** *Group*
   - **d.** *Entity*

10. Object grips are small squares that appear on a selected object.
    - **a.** *True*
    - **b.** *False*

11. When the GRIPS system variable is set to 1, grips are turned off, while a setting of 0 turns grips on.

a. *True*   **b.** *False*

12. To activate grips on an object, you can select the object whenever any commands are in use.

a. *True*   **b.** *False*

13. Pressing ESC will cancel grips and delete them from the display.

a. *True*   **b.** *False*

14. Filter lists can be named and saved on a disk.

a. *True*   **b.** *False*

15. You must include a description when creating a new group.

a. *True*   **b.** *False*

16. Like a block, a group is one object until it is exploded.

a. *True*   **b.** *False*

17. Which key enables you to cycle through the objects available for selection?

a. *ALT*   **c.** *CTRL*
b. *SHIFT*   **d.** *SHIFT & CTRL*

18. Which of the following is not a valid object selection option?

a. *Box*   **c.** *Entity*
b. *Fence*   **d.** *Last*

19. Which option deletes objects from the selection set and is a switch from the Add mode?

a. *Erase*   **c.** *Undo*
b. *Delete*   **d.** *Remove*

20. Which system variable allows you to set the noun/verb selection?

a. *PICKAUTO*   **c.** *SELECTIONSET*
b. *PICKFIRST*   **d.** *NOUNVERB*

21. Which key, along with the pick button, allows you to remove an object from the selection set?

a. *CTRL*   **c.** *DELETE*
b. *SHIFT*   **d.** *ALT*

22. If the **Press and drag** option is set to off, how many points must be specified on an object to select it?

a. *1*   **c.** *3*
b. *2*   **d.** *4*

23. Which selection method allows you to pick a point and drag the cursor across the screen by holding the pick button down?

a. *Hold*   **c.** *Pick & Drag*
b. *Press & Drag* **d.** *Drag*

24. What command allows you to name a selection set of objects in order to always select them together?

   a. *all*   **c.** *group*

   b. *block*   **d.** *name*

25. Using the Shift to Add mode allows you to:

   a. *Only add objects to a selection set.*

   b. *Only subtract objects from a selection set.*

   c. *Add and subtract objects in a selection set.*

   d. *Erase objects as you pick them.*

hatching, gradients and boundaries

introduction

This chapter introduces some of the basic commands and concepts in AutoCAD that can be used to create repeating patterns, called hatching, along with boundaries that may or may not include internal hatching.

After completing this chapter, you will be able to do the following:

- Use the HATCH command to draw the hatch patterns

- Use the GRADIENT command to draw gradient fills

- Modify hatch patterns through the HATCHEDIT command

- Control the visibility of hatch patterns and gradient fills

WHAT IS HATCHING AND GRADIENT FILL?

Drafters and designers use repeating patterns, called hatching, to fill regions in a drawing for various purposes (see Figure 12–1). In a cutaway (cross-sectional) view, hatch patterns help the viewer differentiate between components of an assembly and indicate the material of each. In surface views, hatch patterns depict material and add to the readability of the view. In general, hatch patterns greatly help the drafter/designer in communicating information. Because drawing hatch patterns is a repetitive task, it is an ideal application of computer-aided drafting.

You can use patterns that are supplied in an AutoCAD support file called *acad.pat* or patterns in files available from third-party custom developers, or you can create your own custom hatch patterns. AutoCAD allows you to fill with a solid color in addition to a hatch pattern. AutoCAD creates an associative hatch, which updates when its boundaries are modified, or a non-associative hatch, which is independent of its boundaries. Before AutoCAD draws the hatch pattern, it allows you to preview the hatching and to adjust the definition if necessary.

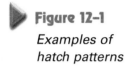 **Figure 12-1**

Examples of hatch patterns

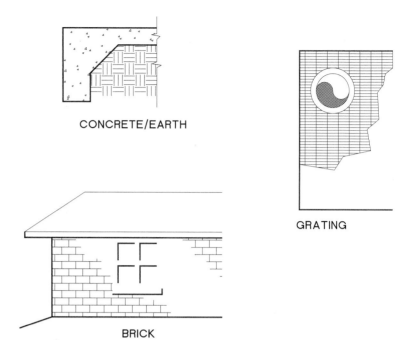

CONCRETE/EARTH

GRATING

BRICK

The hatch pattern behaves as one object; if necessary, you can separate it into individual objects with the EXPLODE command. Once it is separated into individual objects, the hatch pattern will no longer be associated with the boundary object.

Hatch patterns are stored with the drawing, so they can be updated, even if the pattern file containing the hatch is not available. You can control the display of the hatch pattern with the FILLMODE system variable. If FILLMODE is set to OFF, then the patterns are not displayed, and regeneration calculates only the hatch boundaries. By default, FILLMODE is set to ON.

The hatch pattern is drawn with respect to the current coordinate system, current elevation, current layer, color, linetype, and current snap origin.

A gradient fill is a solid hatch fill that gives the blended-color effect of a surface with light on it. You can use gradient fills to suggest a solid form in two-dimensional drawings. The color in a gradient fill makes a smooth transition from light to dark, or from dark to light, and back. You select a predefined pattern (for example, linear, spherical, or radial sweep) and specify an angle for the pattern. In a two-color gradient fill, the transition is both from light to dark and from the first color to the second.

Gradient fills are applied to objects in the same way solid fills are and can be associated with their boundaries or not. An associated fill is automatically updated when the boundary changes. The Hatch and Gradient dialog box applies to both hatch and gradient patterns settings.

 note   You cannot use plot styles to control the plotted color of gradient fills.

Defining the Hatch or Gradient Boundary

A region of the drawing may be filled with a hatch pattern or gradient fill if it is enclosed by a boundary of connecting lines, circles, or arc objects. Overlapping boundary objects can be considered as terminating at their intersections with other boundary objects. If there are any gaps between boundary objects, they must be within the limits set by the **gap tolerance**. Figure 12–2 illustrates variations of objects and the potential boundaries that might be established from them.

Note how in Figure 12–2 the enclosed regions are defined by their respective boundaries. A boundary might include all or part of one or more objects. In addition to lines, circles, and arcs, boundary objects can include 2D and 3D polylines, 3D faces, and viewports. Boundary objects should be parallel to the current UCS. You can also hatch Block References that have been inserted with unequal X and Y scale factors.

▶ **Figure 12–2**

Allowed hatching boundaries made from different objects

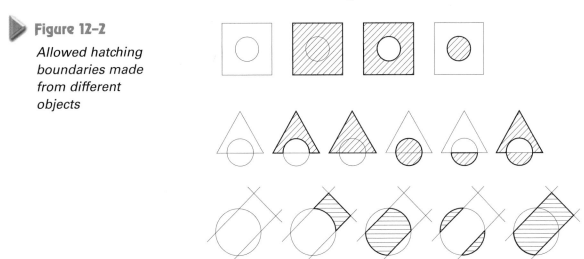

SELECTING OBJECTS versus PICKING A POINT

AutoCAD provides two methods of determining the area to be filled with a hatch or gradient pattern; selecting the object(s) that define(s) a closed boundary or picking a point in a closed area so AutoCAD can define the boundary. Using the select object method (see Figure 12–3), select the four lines via the Window option. They connect at their endpoints (no overlap or gaps). Instead of using the Window option of the object selection process, you can select the four lines individually. This may be desirable if there are unwanted objects within a window used to select them.

▶ **Figure 12–3**

Using the select objects method to define the boundary

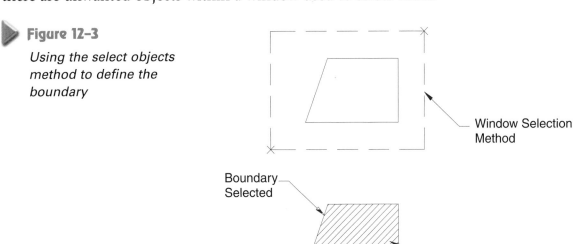

Window Selection Method

Boundary Selected

Resulting Hatch

In Figure 12–4, using the pick point method, select a point in the region enclosed by the four lines. Then AutoCAD creates a polyline with vertices that coincide with the intersections of the lines. There is also an option that allows you to retain or discard the boundary when the hatching is complete.

▶ **Figure 12–4**

Using the pick points method to create the boundary

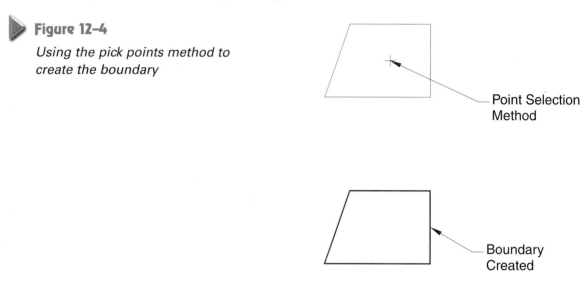

Point Selection
Method

Boundary
Created

If the four lines shown in Figure 12–3 had been segments of a closed polyline, then you could have selected that polyline by picking it with the cursor. Otherwise, all objects enclosing the region to be hatched must be selected and those objects must be connected at their endpoints or, as mentioned above, they must be within the limits set by the **gap tolerance**. If you select objects whose endpoints overlap, the results are unpredictable. For example, to use the HATCH command for the region in Figure 12–5, you would need to draw three lines (from 1 to 2, 2 to 3, and 3 to 4) and an arc from 4 to 1, select the four objects or join them together as a polyline, and select them (or it) to be the boundary.

Using the pick points method permits you to select a point in the region and have AutoCAD automatically create the needed polyline boundary.

▶ **Figure 12–5**

A Region bounded by three lines and an arc

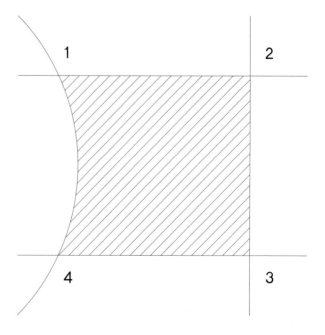

note For users of previous versions of AutoCAD, the BHATCH command has been renamed HATCH. If you enter **bhatch**, you will invoke the HATCH command and the Hatch and Gradient dialog box will be displayed. If you enter **–bhatch**, the prompts will be displayed at the On-Screen input box and on the command line.

The dialog box used by the HATCH or GRADIENT command provides a variety of easy-to-select options, including a means to preview the pattern or fill before completing the command. This saves time. Consider the variety of effects possible, such as areas to be hatched, angle, spacing between segments in a pattern, and even the pattern selected. The Preview option lets you make necessary changes without having to start over.

HATCH PATTERNS AND GRADIENT FILLS

Figure 12–6

Invoking the HATCH command from the Draw toolbar

After you invoke the HATCH or GRADIENT command (see Figure 12–6), AutoCAD displays the Hatch and Gradient dialog box, similar to Figure 12–7. The Hatch and Gradient dialog box can be expanded or contracted in size by choosing the arrow at the lower right corner of the dialog box.

Figure 12–7

Hatch and Gradient dialog box (expanded) with the Hatch tab selected

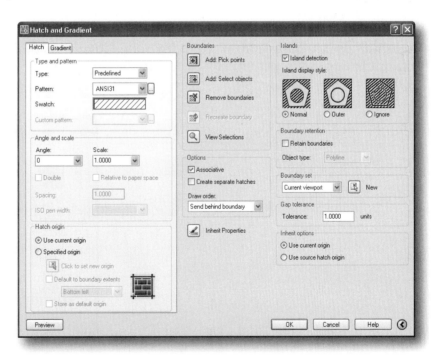

The majority of the Hatch and Gradient dialog box applies to both hatch and gradient patterns such as the sections on **Boundaries**, **Options**, **Islands**, **Boundary retention**, **Boundary set**, **Gap tolerance**, and **Inherit options**. The last five sections listed are displayed only in the expanded dialog box. There are separate **Hatch** and **Gradient** tabs that have controls and options that apply to hatch patterns and gradient fills, respectively.

Hatch related settings

The **Hatch** tab of the Hatch and Gradient dialog box enables you to specify the type of hatch pattern to be applied and the angle, scale and origin of the pattern.

The **Type** text box in the **Type and pattern** section lets you select the type of pattern from **Predefined**, **User defined**, or **Custom** hatch patterns.

When the **Predefined** pattern type is chosen, the **Pattern** list box lets you select a pattern from those defined in the *acad.pat* file. Or, you can select one of the available patterns by choosing the button located at the right of the **Pattern** text box or clicking in the **Swatch** sample box. This causes the Hatch Pattern Palette to be displayed (see Figure 12–8). There are four tabs from which to select predefined patterns: **ANSI**, **ISO**, **Other Predefined**, and **Custom**. Each tab displays icons representing a selected pattern group. To select one of the patterns, choose it and then choose **OK** or double-click on the icon. An example of the selected pattern is displayed in the **Swatch** box in the **Hatch** tab of the dialog box. To create a solid fill in an enclosed area, select the **Solid** pattern (located in the **Other Predefined** section of the Hatch Pattern Palette). The solid fill is drawn with the current color settings, and all pattern properties are disabled, such as scale, angle, and spacing. The selected pattern becomes the value of the HPNAME system variable.

The **User-defined** pattern type allows you to define a simple pattern using the current linetype, on the fly. You can specify a simple pattern of parallel lines or two groups of parallel lines (crossing at 90 degrees) at the spacing and angle desired. Specify the angle and spacing for the userdefined pattern in the **Angle** and **Spacing** text boxes in the **Angle and scale** section. To draw a second set of lines at 90 degrees to the original lines, select **Double**.

When the **Custom** pattern type is chosen, the **Custom Pattern** text box lets you select a custom pattern from a *.pat* file other than the *acad.pat* file. Or you can select one of the available patterns by choosing the image tile located at the right end of the **Custom Pattern** text box.

 Figure 12–8

Hatch Pattern Palette dialog box with the ANSI tab selected

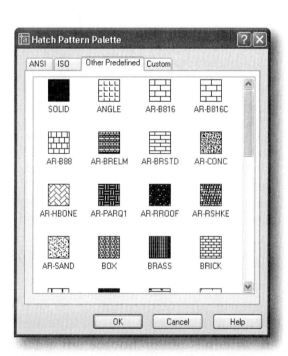

The **Angle and scale** section controls the angle and relative size of the pattern drawn. The **Angle** text box, lets you determine the angle at which the lines in the hatch pattern are drawn relative to the angle at which they were defined. The default angle is set to 0 degrees. The angle 0 (zero) corresponds to the direction of the positive X axis of the current UCS. The **Scale** text box lets you change the scale of the pattern. This affects the spacing between the lines of the pattern relative to the spaces at which they were defined. The default scale is set to 1. These settings can be changed to suit the desired appearance, as shown in Figure 12–9.

Figure 12–9

Hatch pattern with different scale and angle values

SCALE=1   SCALE=2   SCALE=1
ANGLE=0   ANGLE=0   ANGLE=90

The angle and scale specified in the **Angle** and **Scale** text boxes refer to the selected hatch pattern in the **Pattern** list box and shown in the **Swatch** sample box.

Choosing **Double** causes a user-defined hatch pattern to be drawn as defined in the **Angle**, **Scale,** and **Spacing** text boxes in the **Angle and scale** section and then repeated at 90 degrees to the original definition. This can only be used when a user-defined hatch pattern has been selected.

Choosing **Relative to paper space** allows you to scale the hatch pattern relative to the units in Paper Space. This can only be used in layout mode.

The **Spacing** text box allows you to specify the spacing between lines of a user-defined hatch pattern. This can only be used when a user-defined hatch pattern has been selected.

The **ISO pen width** box allows you to specify ISO-related pattern scaling based on the selected pen width. This option is available only if a predefined ISO hatch pattern is selected.

The **Hatch origin** section controls the origin of the pattern drawn. Changing the origin causes the lines in the hatch pattern to be offset by the distance between 0,0 and the specified origin. This is sometimes necessary when you wish to offset the whole pattern of lines for visual effects. For example, if you wish to use the same hatch pattern in adjacent boundaries but do not want their lines to coincide, you can use different origins for the two patterns. Or, if a hatch pattern, such as those for brick and masonry, needs to begin at a certain point, you can specify that point as the origin.

Selecting **Use current origin** causes the hatch pattern to use the current setting of the hatch origin which is stored in the HPORIGIN system variable. By default it is set to 0,0 in the current UCS. If it has been specified as another point, then selecting **Use current origin** uses the current setting.

Choosing **Specified origin** lets you specify a different origin than the current one by utilizing one of the available methods. Choosing **Click to set new origin** lets you specify the new origin on the screen with the pointing device or by entering the new coordinates. Choosing **Default to boundary extents** lets you specify one of the four corners of the rectangular extents of the boundary as the new origin. Choosing

Store as default origin lets you store the newly specified origin in the HPORIGIN system variable. The **origin preview** sample box shows the newly specified location of the origin.

Gradient related settings

The **Gradient** tab of the Hatch and Gradient dialog box (see Figure 12–10) lets you specify the type of gradient pattern to be applied and the color, orientation, and angle of the gradient.

The **Color** section controls the color of the gradient drawn.

Selecting **One color** specifies a fill that uses a smooth transition between darker shades and lighter tints of one color. When **One color** is selected (see Figure 12–10), AutoCAD displays a color swatch with a browse button and a **Shade and Tint** slider. The color swatch specifies the color for the gradient fill, and the **Shade and Tint** slider specifies the tint (the selected color mixed with white) or shade (the selected color mixed with black) of a color to be used for a gradient fill of one color.

Selecting **Two color** specifies a fill that uses a smooth transition between two colors. When **Two color** is selected, AutoCAD displays a color swatch with a browse button for **Color 1** and for **Color 2**. The Color swatch specifies the color for the gradient fill.

Figure 12–10

Hatch and Gradient dialog box with the Gradient tab selected

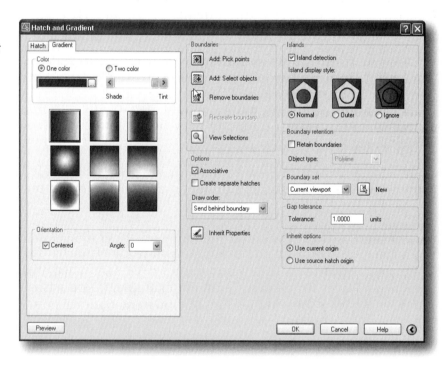

The **Shade and Tint** slider lets you specify the tint (amount of white mixed in) of the selected color or the shade (the amount of black mixed in) of the selected color for a one-color gradient fill.

The **Orientation** section controls the base location and angle of the gradient drawn. Choosing **Centered** causes the gradient to be drawn with the gradation from the center of the boundary outward. When it is not selected, the gradient fill is shifted up and to the left, creating the illusion of a light source to the left of the object.

The **Angle** text box lets you specify the angle (relative to 0) that the gradation is drawn. Valid values are 0 through 360 degrees and the specified angle is relative to the current UCS. This option is independent of the angle specified for hatch patterns.

Settings for Both Hatching and Gradient Fill Patterns

The options and controls that are applicable to both hatching and gradient fill patterns include: **Boundaries**, **Options**, **Islands**, **Boundary retention**, **Boundary set**, **Gap tolerances**, and **Inherit options**.

The **Boundaries** section provides two methods by which you can select the objects that determine the boundary for drawing hatch patterns: **Add Pick points** and **Add Select objects**.

Choosing **Add Pick points** lets you determine a boundary from existing objects that form an enclosed area around the specified point. If you set **Island Detection** (available in the **Islands** section) to ON, objects that enclose areas within the outermost boundary are detected as islands. How HATCH or GRADIENT detects objects using this option depends on which island detection method (**Normal, Outer,** or **Ignore**) you select in the **More Options** area of the dialog box. If it is set to OFF, then AutoCAD draws an imaginary line from the selected point to the nearest object and traces the boundary in the counterclockwise direction. If it cannot trace a closed boundary, AutoCAD will return to the drawing without hatching the object.

For example, in Figure 12–11, point A is valid and point B is not when **Island Detection** is not chosen. The object nearest to point A is the line that is part of a potential boundary (the square) of which point A is inside and AutoCAD considers the square as the hatch boundary. Conversely, point B is nearest a line that is part of a potential boundary (the triangle) of which point B is outside and AutoCAD displays an error message with the selected point as outside the boundary.

 Figure 12–11

Selecting points for hatching or gradient

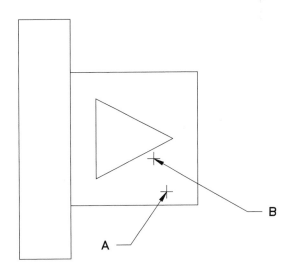

To define a hatch or gradient boundary from existing objects, choose **Add Pick points** and the dialog box closes temporarily, and you are prompted to pick a point as follows:

Select internal point or ⊡: *(specify a point within the area to be hatched)*

Select internal point or ⊡: *(specify a point, enter **u** to undo the selection, or press* ENTER *to end point specification)*

AutoCAD redisplays the dialog box after choosing the internal points. See Figure 12–12 for an example of hatching by specifying a point inside a boundary. While picking internal points, you can right-click in the drawing area at any time to display a shortcut menu that contains several options (see Figure 12–13). You can undo the last selection or all selections, change the boundary selection method, change the island detection style, or preview the hatch or gradient fill.

Figure 12–12

Hatching by specifying a point

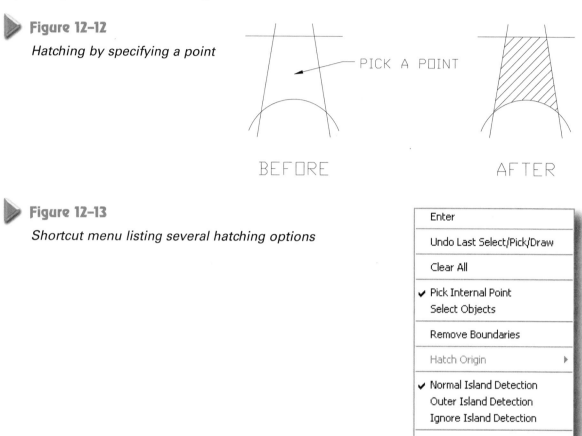

PICK A POINT

BEFORE   AFTER

Figure 12–13

Shortcut menu listing several hatching options

| Enter |
| Undo Last Select/Pick/Draw |
| Clear All |
| ✔ Pick Internal Point |
| Select Objects |
| Remove Boundaries |
| Hatch Origin ▸ |
| ✔ Normal Island Detection |
| Outer Island Detection |
| Ignore Island Detection |
| Preview |

Choosing **Add Select objects** lets you select specific objects for a pattern or fill that determine a boundary to form an enclosed area. The dialog box closes temporarily, and you are prompted to select objects as follows:

Select objects or ⬇: *(select the object(s) by one of the standard methods, and press* ENTER *to terminate object selection)*

When you use the **Add Select objects** option, HATCH does not detect interior objects automatically. You must select the objects within the selected boundary to hatch or fill those objects according to the current island detection style. While selecting objects, you can right-click at any time in the drawing area to display a shortcut menu. You can undo the last selection or all selections, change the selection method, change the island detection style, or preview the hatch or gradient fill.

The **Remove boundaries** and **Recreate boundary** options are not available during creation of a new pattern/fill. They are used on existing patterns and fills with internal boundaries. See their explanation in the section on the Hatch Edit dialog box.

Choosing **View Selections** causes AutoCAD to highlight the defined boundary set.

Caution must be observed when hatching over dimensioning. Dimensions are not affected by hatching as long as the DIMASSOC dimension variable (short for "associative dimensioning") is set to ON when the hatching is created and the dimension has not been exploded. The DIMASSOC system variable toggles between associative and nonassociative dimensioning. If the dimensions are drawn with DIMASSOC set to OFF (or exploded into individual objects), then the lines (dimension and extension) have an unpredictable (and undesirable) effect on the hatching pattern. Therefore, selection in this case, should be done by specifying the individual objects on the screen.

Blocks are hatched as though they are separate objects. Note however, that when you select a block, all objects that make up the block are selected as part of the group to be considered for hatching.

If the selected items include text objects, shapes, or attribute objects, AutoCAD does not hatch through these items if identified in the selection process. AutoCAD leaves an unhatched area around the text objects so they can be clearly viewed, as shown in Figure 12–14. Using the **Ignore** (Island display style) style will negate this feature so that the hatching is not interrupted when passing through the text, shape, and attribute objects.

Figure 12–14

Hatching in an area where there is text

note When you select objects individually (after choosing the **Select Objects** button in the Hatch and Gradient dialog box), AutoCAD no longer automatically creates a closed border. Therefore, any objects selected that will be part of the desired border must be either connected at their endpoints or a closed polyline.

When a filled solid or trace with width is selected in a group to be hatched, AutoCAD does not hatch inside that solid or trace. However, the hatching stops right at the outline of the filled object, leaving no clear space around the object as it does around text objects, shapes, and attributes.

The **Options** section of the Hatch and Gradient dialog box controls several commonly used hatch or fill options.

Associative controls whether the hatch or fill is associative or nonassociative. Choosing **Associative** causes the hatch pattern elements or gradient fills to be associated with the objects that make up the boundary. For example, if the object is stretched, the hatch pattern expands to fill the new size. Figure 12–15 shows examples of associative and nonassociative hatch patterns.

 **Figure 12-15**

Examples of associative and nonassociative hatch pattern, when an object is stretched

ASSOCIATIVE HATCH

NON-ASSOCIATIVE HATCH

Create Separate Hatches controls whether a single hatch object or multiple hatch objects are created when several separate closed boundaries are specified.

The **Draw Order** list box lets you determine the order in which a pattern or fill is drawn relative to other objects. You can place a hatch or fill behind all other objects, in front of all other objects, behind the hatch boundary, or in front of the hatch boundary.

Inherit Properties allows you to apply the hatch pattern settings, such as pattern type, pattern angle, and pattern scale, from an existing pattern to another area to be hatched. The dialog box closes temporarily, and you are prompted to select hatch objects as follows:

Select hatch object: *(select a hatch pattern)*

Select objects or ⊥: *(select a closed area to hatch)*

Select objects or ⊥: *(select a closed area to hatch or press* ENTER *to complete the selection)*

The **Islands** section specifies the method used to hatch or fill objects within the outermost boundary. If no internal boundaries exist, specifying an island detection style has no effect.

Island detection lets you control whether internal closed boundaries, called islands, are detected. Choose one of the three Island display styles: **Normal**, **Outer**, and **Ignore**.

Choosing **Normal** hatches or fills between alternate areas, starting with the outermost area.

Choosing **Outer** hatches or fills only the outermost area and leaves the internal structure blank.

Choosing **Ignore** hatches or fills the entire area enclosed by the outermost boundary, regardless of how you select the object, as long as its outermost objects comprise a closed polygon and are joined at their endpoints.

For example, in Figure 12–16, specifying the point shown in the upper-right image in response to the **Add Pick points** option, results in hatching for the **Normal** style, as shown in the upper-right; **Outer** style, as shown in the lower-left; and **Ignore** style, as shown in the lower-right.

Figure 12–16

Examples of hatching by specifying a point for Normal, Outer, and Ignore styles

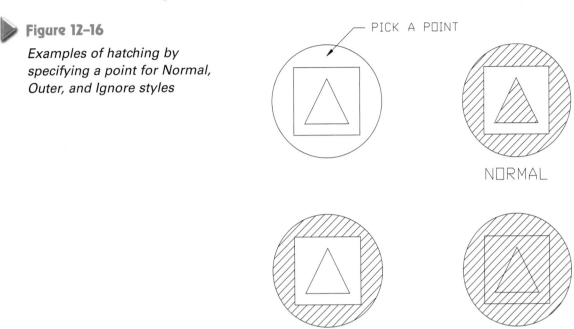

The **Boundary retention** section specifies whether to retain boundaries as objects and allows you to select the object type AutoCAD applies to those objects.

Choosing **Retain boundaries** specifies whether the boundary defining the area to be hatched/filled is to be retained. If it is to be retained, then the adjacent selection box lets you select whether the area is to be determined by a Polyline boundary or by a Region. The **Object type** list box lets you specify whether the object is a polyline or a region.

The **Boundary set** section allows you to select a set of objects (called a boundary set) that AutoCAD analyzes when defining a boundary from a specified point. The selected boundary set applies only when you use the **Add Pick points** selection to create a boundary to draw hatch patterns. By default, when you use **Add Pick points** to define a boundary, AutoCAD analyzes all objects visible in the current viewport.

The **Boundary set** list box lets you select whether the boundary set will be selected from the **Current Viewport** or **Existing Set**. Selecting **Current Viewport** causes the boundary set to be defined from everything in the current viewport extents. Selecting this option discards any current boundary set. Selecting **Existing Set** causes the boundary set to be defined from the objects that you selected with **New**. If you have not created a boundary set with **New**, the **Existing Set** option is not available.

Choosing **New** causes AutoCAD to clear the dialog box and return you to the drawing area to select objects from which a new boundary set will be defined. AutoCAD creates a new boundary set from those objects selected that are hatchable; existing boundary sets are abandoned. If hatchable objects are selected, they remain as a boundary set until you define a new one or exit the HATCH command. Defining a boundary set will be helpful when you are working on a drawing that has too many objects to analyze to create a boundary for a pattern/fill.

It is important to distinguish between a boundary set and a boundary. A boundary set is the group of objects from which AutoCAD creates a boundary. As explained earlier, a boundary set is defined by selecting objects in a manner similar to selecting objects for some modify commands. The objects (or parts of them) in the group are used to define the subsequent boundary. A boundary is created by AutoCAD after it has analyzed the objects (the boundary set) you have selected. It is the boundary that determines where the hatching begins and ends. The boundary consists of line/arc segments, which can be considered to be a closed polygon with segments that connect at their endpoints. If objects in the boundary set overlap, then in creating the boundary AutoCAD uses only the parts of objects in the boundary set that lie between intersections with other objects in the boundary set.

The **Gap tolerance** section treats a set of objects that almost encloses an area as a closed hatch boundary.

The **Tolerance** text box lets you specify a value, in drawing units, from 0 to 5000 to set the maximum size of gaps that can be ignored when the objects serve as a hatch boundary. Any gaps equal to or smaller than the value you specify are ignored, and the boundary is treated as closed. The default value, 0, specifies that the objects enclose the area, with no gaps.

The **Inherit options** section of the Hatch and Gradient dialog box lets you determine the origin the hatch pattern uses.

Choosing **Use current origin** causes the hatch pattern to use the origin stored in the HPORIGIN system variable.

Choosing **Use source hatch origin** causes the hatch pattern to use the origin of the hatch pattern from which properties are to be inherited.

Choose **Preview** to see an example of the hatch pattern to be drawn as specified for the selected objects, AutoCAD displays the currently defined boundaries with the current hatch settings. After previewing the hatch, press ESC to return to the dialog box. If necessary, make any changes to the settings, choose **OK** to apply the hatch pattern, or choose **Cancel** to disregard the selection.

DRAGGING PATTERNS/FILLS FROM A TOOL PALETTE

You can also hatch a closed shape by dragging a hatch pattern from a tool palette (see Figure 12–17).

 **Figure 12–17**

Tool Palettes window with Imperial Hatches tab selected

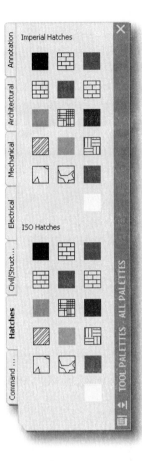

Tool palettes are tabbed areas within the Tool Palettes window, and hatches that reside on a tool palette are called tools. Several tool properties, including scale, rotation, and layer, can be set for each tool individually. To change the tool properties, right-click a tool and select Properties on the Shortcut menu. Then you can change the tool's properties in the Tool Properties dialog box as shown in Figure 12–18. The Tool Properties dialog box has two categories of properties—the Pattern properties category, which controls object-specific properties such as scale, rotation, and angle, and the General properties category, which overrides the current drawing property settings such as layer, color, and linetype.

 Figure 12–18

Tool properties of a selected tool

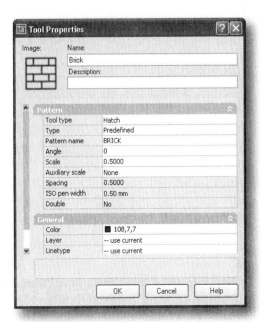

You can place hatches that you use often on a tool palette by dragging the hatch pattern from DesignCenter by opening the *acad.pat* file. See Chapter 19 for a detailed explanation about using the DesignCenter.

CHANGING A HATCH PATTERN OR ITS SETTINGS

 Figure 12-19

Invoking the HATCHEDIT *command from the Modify II toolbar*

The HATCHEDIT command (see Figure 12-19) allows you to modify hatch patterns and gradient fills or choose a new pattern for an existing hatch. In addition, it allows you to change the pattern style of an existing pattern. After you invoke the HATCHEDIT command, AutoCAD prompts:

Select hatch object: *(select an associative hatch object)*

AutoCAD displays the Hatch Edit dialog box, similar to Figure 12-20.

 **Figure 12-20**

Hatch Edit dialog box

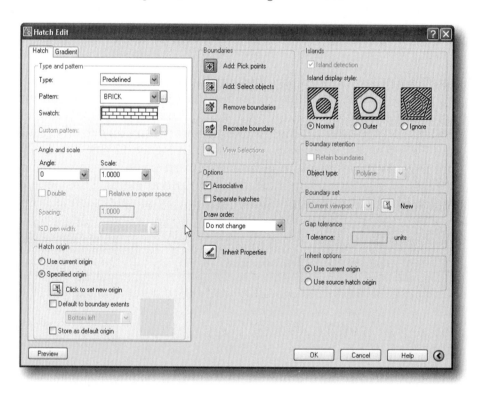

The Hatch Edit dialog box is similar to the Hatch and Gradient dialog box. After the hatch object is selected, the dialog box will be displayed with the **Hatch** tab selected. To edit a gradient fill, invoke the HATCHEDIT command and when prompted to select a hatch object, select a gradient fill and the dialog box will be displayed with the **Gradient** tab selected. The hatch pattern or gradient fill selected can be modified in the same manner as specifying a new pattern or fill.

If a hatch pattern or gradient fill is created using the normal island display style, you can cause inner objects that are used as boundaries to no longer be effective with the **Remove boundaries** option located in the **Boundaries** section. This does not physically remove the objects. It only removes their being used as boundaries. The **Recreate boundary** option can be applied only to boundaries that have been removed with the **Remove boundary** option.

Choosing **Remove boundaries** causes AutoCAD to prompt you to select a boundary set to be removed from the defined boundary set. You cannot remove the outermost boundary.

Choosing **Recreate boundary** creates a polyline or region around the selected hatch or fill, with the option to associate the hatch object with it. Choosing **Recreate Boundary** causes the dialog box to close temporarily, and AutoCAD prompts:

> Select objects: *(select the object(s) by one of the standard methods, and press* ENTER)
>
> Enter type of boundary object *(Enter **r** to create a region or **p** to create a polyline)*
>
> Reassociate hatch with new boundary? ⊡ *(Enter **y** or **n**)*

You can also display and change the current properties for hatch or fill objects with the **Properties** palette. Open the **Properties** palette and view and change the settings for all properties of the selected hatch or fill objects (see Figure 12–21). The **Properties** palette lists the current settings for properties of the selected objects. You can modify any property that can be changed by specifying a new value. The **Properties** palette also enables you to view the area of a hatch (see Figure 12–21). If you select multiple hatch objects, you can now view their cumulative area.

Figure 12–21

Properties Palette

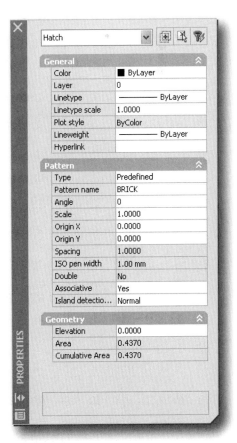

CONTROLLING THE VISIBILITY OF HATCH PATTERNS

The fill command controls the visibility of hatch patterns in addition to the filling of multilines, traces, solids, and wide polylines. After you invoke the fill command by typing **fill** at the Command: prompt, AutoCAD prompts:

Enter mode [ON/OFF] <current>: *(specify* **ON** *to display the hatch pattern and* **OFF** *to turn off the display of hatch pattern)*

You have to invoke the REGEN command after changing the setting of FILL to see the effect.

drawing exercises

Open the Exercise Manual PDF file for Chapter 12 on the accompanying CD for discipline-specific exercises.

review questions

1. AutoCAD will ignore text within a crosshatching boundary.
   a. *True*   **b.** *False*

2. The BHATCH command allows you to create an associative hatch pattern that updates when its boundaries are modified.
   a. *True*   **b.** *False*

3. By default, hatch patterns are drawn at a 45-degree angle.
   a. *True*   **b.** *False*

4. All of the following may be used as boundaries of the HATCH command, except:
   a. *ARC.*   **d.** *CIRCLE.*
   b. *LINE.*   **e.** *Polyline.*
   c. *BLOCK.*

5. The following are all valid AutoCAD commands except:
   a. *ANGLE.*   **d.** *ELLIPSE.*
   b. *POLYGON.*   **e.** *MULTIPLE.*
   c. *HATCH.*

6. When using the HATCH command with a named hatch pattern, one can change:
   a. *The color and scale of the pattern.*
   b. *The angle and scale of the pattern.*
   c. *The angle and linetype of the pattern.*
   d. *The color and linetype of the pattern.*
   e. *The color and angle of the pattern.*

7. The AutoCAD hatch feature:
   a. *Provides a selection of numerous hatch patterns.*
   b. *Allows you to change the color and linetype.*
   c. *Hatches over the top of text when the text is contained inside the boundary.*
   d. *All of the above.*

8. The BHATCH command will allow you to create a polyline around the area being hatched and to retain that polyline upon completion of the command.
   a. *True*   **b.** *False*

9. Hatch patterns created with the BHATCH command can be non-associative and associative.
   a. *True*   **b.** *False*

10. The _____ command automatically defines the nearest boundary surrounding a point you have specified.

 a. _HATCH_   **c.** _BOUNDARY_
 b. _FILL_   **d.** _PTHATCH_

11. The HATCH and Hatch Edit dialog boxes look the same.

 a. _True_   **b.** _False_

12. From what default file does AutoCAD use to load hatch patterns?

 a. _ACAD.mnu_ **c.** _ACAD.dwg_
 b. _ACAD.pat_ **d.** _ACAD.hat_

13. You can terminate the HATCH command before applying the hatch pattern by pressing ESC.

 a. _True_   **b.** _False_

14. AutoCAD allows the properties of a non-associative hatch pattern to be inherited.

 a. _True_   **b.** _False_

15. The ____ command requires you to specify manually each segment of the boundary to be hatched.

 a. _HATCH_   **c.** _BOUNDARY_
 b. _-BHATCH_   **d.** _Both a and b_

16. The hatch command will place a hatch pattern within the boundaries of all of the following:

 a. _Polylines._ **c.** _Circles._
 b. _Lines._   **d.** _All of the above._

17. Which is not a valid hatch boundary style?

 a. _Outer_   **c.** _Ignore_
 b. _Inner_   **d.** _Normal_

18. The default setting of the fillmode system variable in a hatch pattern is:

 a. _Off._   **c.** _Fill._
 b. _On._   **d.** _1._

19. The default angle for hatch patterns is:

 a. _30._   **c.** _0._
 b. _60._   **d.** _15._

all about styles

introduction

This chapter introduces commands and concepts that can be used to create and modify styles. A style is a collection of system variable settings, or in the case of dimensions, dimension variable settings. The collection of settings can be saved as a style. It is given a unique name so that when the same combination of settings and characteristics needs to be applied, it can be recalled by its name and made current.

After completing this chapter, you will be able to do the following:

* Create and modify text styles

* Create and modify multiline styles

* Create and modify dimension styles

* Create and modify table styles

CREATING AND MODIFYING TEXT STYLES

 Figure 13–1

Invoking the STYLE command from the Text toolbar

The Style option of the TEXT and MTEXT commands (in conjunction with the STYLE command) lets you determine how text characters and symbols appear, other than adjusting the usual height, slant, and angle of rotation. In order for a text style to be specified through the Style option of the TEXT and MTEXT commands, it must have been defined by using the STYLE command. In other words, the STYLE command (see Figure 13–1) creates a new style or modifies an existing style. The Style option under the TEXT or MTEXT command allows you to choose a specific style from the styles available.

There are three things to consider in creating a new style with the STYLE command:

First, you must name the newly defined style. Style names can contain up to 255 characters, numbers, and special characters ($, –, and _). Names like "title block," "notes," and "bill of materials" can remind you of the purpose for which the particular style was designed.

Second, you can apply a particular font to a style. The font that AutoCAD uses as a default is called TXT. It has blocky looking characters, which are economical to store in memory. But the *txt.shx* font, made up entirely of straight-line (noncurved) segments, is not considered as attractive or readable as others. Other fonts offer many variations in characters, including those for foreign languages. All fonts that are available in AutoCAD are stored for use in files of their font name with an extension of *.shx*. The most effective way to get a distinctive appearance in text strings is to use a specially designed font. You can also use TrueType fonts and Type 1 postscript fonts (which must first be compiled into an AutoCAD shape file). If necessary, you can buy additional fonts from third-party vendors.

The third consideration is how AutoCAD treats general physical properties of the characters, regardless of the font that is selected. The properties includes the height, width-to-height ratio, obliquing angle, backwards, upside down, and orientation (horizontal/vertical) options. After you invoke the STYLE command, AutoCAD displays the Text Style dialog box, similar to Figure 13–2.

Figure 13-2

Text Style dialog box

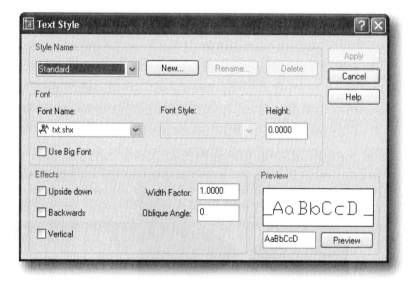

Choose **New** to create a new style. AutoCAD displays the New Text Style dialog box shown in Figure 13–3. Enter the appropriate name for the text style and choose **OK** to create the new style.

To rename an existing style, first select the style from the **Style Name** box in the Text Style dialog box, and then choose **Rename**. AutoCAD displays the Rename Text Style dialog box. Make the necessary changes in the name of the style, and choose **OK** to rename the text style.

Figure 13-3

New Text Style dialog box

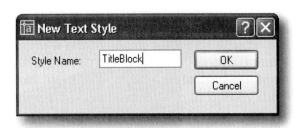

To delete an existing style, first select the style from the **Style Name** list box in the Text Style dialog box, and choose **Delete**. AutoCAD displays the AutoCAD Alert dialog box to confirm the deletion of the selected style. Choose **Yes** to confirm the deletion or **No** to cancel the deletion of the selected style.

> **note**   You cannot rename nor delete the Standard style name.

You can assign a font to the selected text style from the **Font Name** box. Similarly, select a font style for the selected font from the **Font Style** box. The font style specifies the font character formatting, such as italic, bold, or regular.

The **Use Big Font** check box specifies an Asian-language Big Font file, and it is available only if you specify an *.shx* file under Font Name. Only *.shx* files are valid file types for creating Big Fonts.

Specify the text height in the **Height** box for the selected font. If you set the height to 0 (zero), when you use this style in the TEXT or MTEXT command, you are given an opportunity to change the text height with each occurrence of the command. If you set it to any other value, that value will be used for this style, and you will not be allowed to change the text height.

The **Upside down** and **Backwards** check boxes control whether the text is drawn upside down (left to right) or right to left (with the characters backward), respectively as shown in Figure 13-4.

Figure 13-4

Examples of backward text and upside down text

The **Vertical** check box controls the display of the characters aligned vertically. The Vertical option is available only if the selected font supports dual orientation. See Figure 13-5 for an example of vertically oriented text.

▶ **Figure 13–5**

Example of vertically oriented text

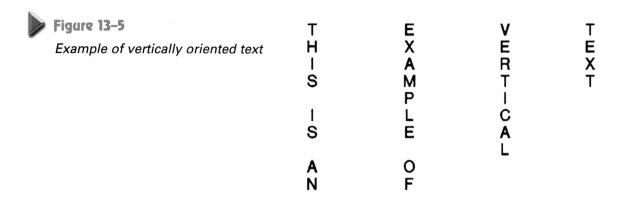

The **Width Factor** box sets the character width relative to text height. If it is set to more than 1.0, the text widens; if set to less than 1.0, it narrows.

The **Oblique Angle** box sets the obliquing angle of the text. If it is set to 0 (zero) degrees, the text is drawn upright (or in AutoCAD, 90 degrees). A positive value slants the top of the characters toward the right, or in the clockwise direction. A negative value slants the characters in the counterclockwise direction. See Figure 13–6 for examples of oblique angle settings applied to a text string.

▶ **Figure 13–6**

Example of oblique angle settings applied to a text string

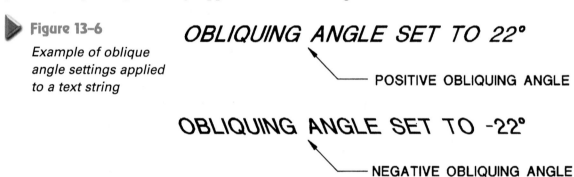

The **Preview** section of the Text Style dialog box displays sample text that changes dynamically as you change fonts and modify the effects. To change the sample text, type characters in the box below the larger preview image.

After making the necessary changes in the Text Style dialog box, choose **Apply** to create or modify an existing text style. All the existing text (of the text style being modified) in the drawing will reflect the changes immediately made to the text style. The default text style will be set to the newly created text style. Choose **Close** to close the Text Style dialog box.

CREATING AND MODIFYING MULTILINE STYLES

 **Figure 13-7**

Invoking the Multiline Style command from the Format menu

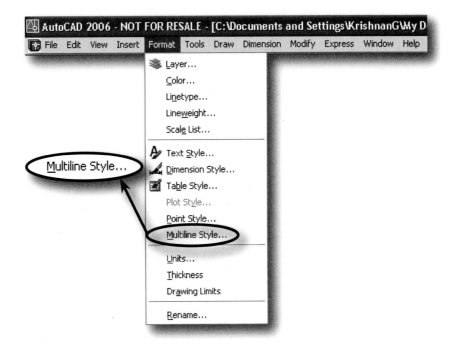

The MLSTYLE command is used to create a new Multiline Style or modify an existing one. You can define a multiline style comprised of up to 16 lines, called elements. The style controls the number of elements and the properties of each element. In addition, you can specify the background color and the end caps of each multiline.

The Style option of the MLINE command allows you to choose a specific style from the available styles to draw multilines. After you invoke the MLSTYLE command (see Figure 13–7), AutoCAD displays the Multiline Styles dialog box, as shown in Figure 13–8.

 Figure 13-8

Multiline Styles dialog box

Creating a New Multiline Style

AutoCAD displays the name of the current multiline style. The current multiline style is the style that will be used for all new multilines.

The **Styles** list box allows you to choose from the available multiline styles loaded in the current drawing.

The **Description** area displays a description (if available) of the selected multiline style.

The **Preview of** area displays the name and an image of the selected multiline style.

Set Current is used to make the selected style current.

Selecting **New** causes the Create New Multiline Style dialog box to be displayed (see Figure 13–9).

 Figure 13-9

Create new Multiline Style dialog box

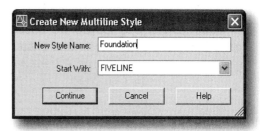

Specify a new multiline style name in the **New Style Name** text box. The **Start With** box lets you choose the existing Multiline style that you would like to start with in creating your new Multiline style. To save time, choose a multiline style that is similar to the one that you want to create. Choose **Continue** to display the New Multiline Style dialog box (see Figure 13–10).

 Figure 13-10

New Multiline Style dialog box

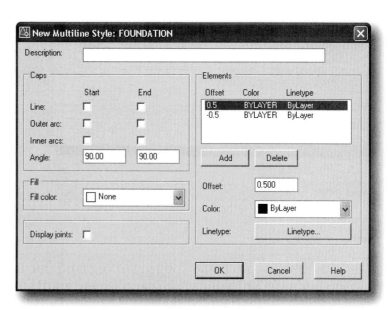

The **Description** text box allows you to add a description of up to 255 characters, including spaces. The description entered here will be displayed in the **Description** text box when this style is selected in the Multiline Styles dialog box.

The **Elements** section sets element properties, such as the offset, color, and linetype, of new and existing multiline elements. AutoCAD lists all the elements in the current multiline style. Choose **Add** to add a new element to the multiline style. Each new element in the style is defined by its offset, its color (default is set to ByLayer), and its linetype (default is set to ByLayer). The **Offset**, **Color**, and **Linetype** fields allow you to specify an element's offset distance from the middle of the multiline, color, and linetype. Elements are always displayed in descending order of their offsets. To delete the selected element, choose **Delete**.

The **Caps** section allows you to specify the appearance of multiline start and end caps. The **Line** check boxes control the display of the start and end caps, by adding a straight line to the start or end of a multiline (see Figure 13–11). The **Outer arc** check boxes control the display of the start and end caps by connecting the ends of the outermost elements with a semicircular arc (see Figure 13–12). The **Inner arcs** check boxes control the display of the start and end caps by connecting the ends of the innermost elements with a semicircular arc (see Figure 13–13). For a multiline with an odd number of elements, the center element is not connected. For an even number of elements, connected elements are paired with elements that are the same number from each edge. For example, the second element from the outer-left will be connected to the second element from the outer-right, the third to the third, and so forth. The **Angle** edit field sets the angle of endcaps. Figure 13–14 shows the display of the end caps with an angular cap.

▶ **Figure 13–11**

Display of the line for start and end cap

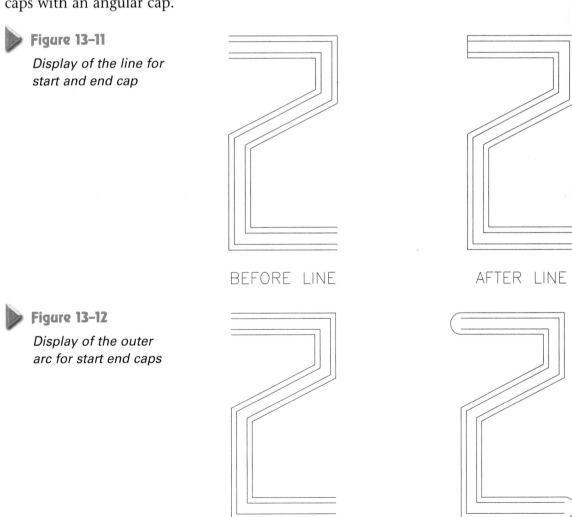

BEFORE LINE

AFTER LINE

▶ **Figure 13–12**

Display of the outer arc for start end caps

BEFORE OUTER
ARC

AFTER OUTER
ARC

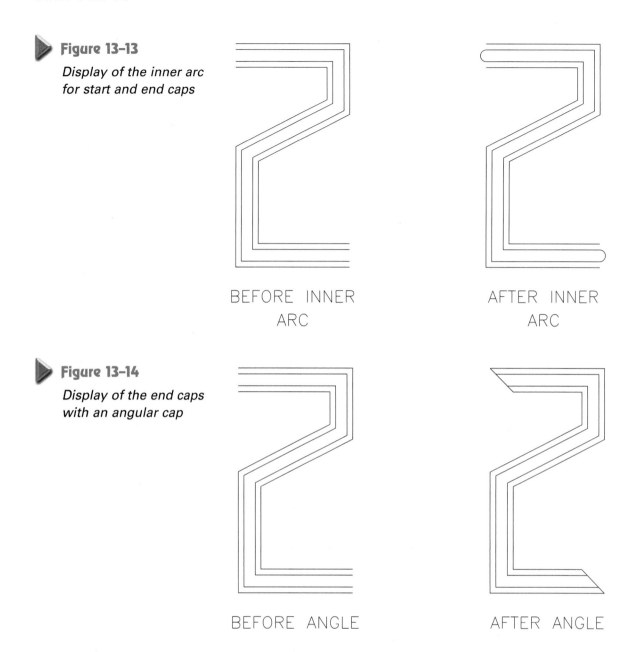

Figure 13-13

Display of the inner arc for start and end caps

BEFORE INNER ARC

AFTER INNER ARC

Figure 13-14

Display of the end caps with an angular cap

BEFORE ANGLE

AFTER ANGLE

The **Fill** section contains a **Fill color** list box which allows you to include a specified background color for the multiline style. Choosing **Select Color** displays the Select Color dialog box from which you can specify a non-standard color for the background fill.

Selecting **Display joints** causes a line (miter) to be drawn at the vertices of each multiline segment (see Figure 13–15).

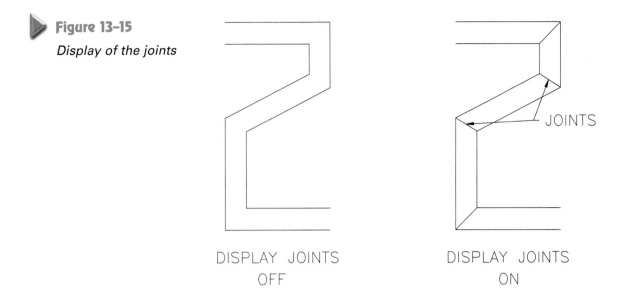

Figure 13-15

Display of the joints

JOINTS

DISPLAY JOINTS
OFF

DISPLAY JOINTS
ON

Choose **OK** to create the selected multiline style.

In the Multiline Style dialog box, choose **Modify** to change the element properties of an existing multiline style. AutoCAD displays the Modify Multiline Style dialog box which is similar to the New Multiline Style dialog box (see Figure 13–10).

Choose **Rename** to change the name of the selected multiline style.

Choose **Delete** to delete the selected multiline style.

Choose **Load** to load multiline style from a specified *.mln* file.

Choose **Save** to save a multiline style to a multiline library (*.mln*) file. If you specify an *.mln* file that already exists, the new style definition is added to the file and existing definitions are not erased. The default file name is *acad.mln*.

Choose **OK** to close the Multiline Style dialog box to create your new multiline style.

Modifying an Existing Multiline Style

Select a multiline style name in the **Current** box of the Multiline Styles dialob box to modify the properties. If there are multiple styles, the name of the current style is selected.

---

note You cannot edit the element and multiline properties of the Standard multiline style or any multiline style that is being used in the drawing. If you try to edit the options in either the Element Properties dialog box or the Multiline Properties dialog box, the options are unavailable. To edit an existing multiline style, you must do so before you draw any multilines in that style.

To rename an existing multiline style, first select the style from the **Current** box and the selected style name will be displayed in the **Name** box. Rename the style displayed in the **Name** box and choose **Rename**. To change the Element properties, choose **Element Properties** and make necessary changes. Do the same for Multiline properties by choosing **Multiline Properties**.

CREATING AND MODIFYING DIMENSION STYLES

Figure 13–16

*Invoking the
DIMSTYLE command
from the
Dimension toolbar*

Each time a dimension is drawn, it conforms to the settings of the dimensioning system variables in effect at the time. The entire set of dimensioning system variable settings can be saved in their respective states as a dimension style, with a name by which it can be recalled for application to a dimension later in the drawing session or in a subsequent session. Some dimensioning system variables affect every dimension. For example, every time a dimension is drawn, the DIMSCALE setting determines the relative size of the dimension. But DIMDLI, the variable that determines the offsets for baseline dimensions, comes into effect only when a baseline dimension is drawn. However, when a dimension style is created and named, all dimensioning system variable settings (except DIMASO and DIMSHO) are recorded in that dimension style, whether or not they will have an effect. The DIMASO and DIMSHO dimensioning variable settings are saved in the drawing separate from the dimension styles.

Dimension Style Manager Dialog Box

AutoCAD provides a comprehensive set of dialog boxes accessible through the Dimension Style Manager dialog box for creating new dimension styles and managing existing ones. In turn, these dialog boxes compile and store dimensioning system variable settings. Creating dimension styles through use of the DIMSTYLE command's dialog boxes allows you to make the desired changes to the appearance of dimensions without having to search for or memorize the names of the dimensioning system variables in order to change the settings directly. After you invoke the DIMSTYLE command (see Figure 13–16), AutoCAD displays the Dimension Style Manager dialog box, shown in Figure 13–17.

Figure 13–17

*Dimension Style
Manager dialog box*

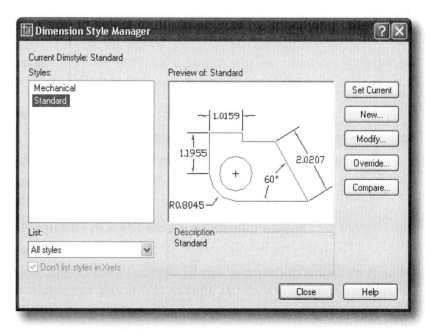

The **Current Dimstyle** heading shows the name of the dimension style to which the next drawn dimension will conform. The name of the current Dimstyle is recorded as the value of the DIMSTYLE dimensioning system variable.

The **Styles** section displays the name(s) of the current style and any other style(s) available depending on the option chosen in the **List** selection box.

From the **List** selection box you can list all of the styles available in the drawing displayed or only those in use. The Standard style cannot be deleted. Styles from externally referenced drawings can be listed, although they are not changeable in the current drawing.

The **Preview of** window shows how dimensions will appear when drawn using the current Dimstyle.

The **Description** section shows the difference(s) between the Dimstyle chosen in the **Styles** section and the current Dimstyle. For example, if the current Dimstyle is Standard and the Dimstyle chosen in the **Styles** section is named Harnessing, the Description section might say "Standard + Angle format = 1, Fraction format = 1, Length units = 4" to indicate that the Angle format is in Degrees/Minutes/Seconds, the Fraction format is in Diagonal, and the Length units is in Architectural, these settings being different from the corresponding settings for the Standard Dimstyle.

Choosing **Set Current** causes the Dimstyle chosen in the **Styles** section to become the current Dimstyle.

Creating a New Dimension Style

Choose **New** to create a new dimension style. AutoCAD displays Create New Dimension Style dialog box, shown in Figure 13–18.

Figure 13–18

Create New Dimension Style dialog box

Specify the name of the new style in the **New Style Name** box. The **Start With** box lets you choose the existing Dimstyle that you would like to start with in creating your new Dimstyle. The dimensioning system variables will be the same as those in the **Start With** list box until you change their settings. In many cases you may wish to change only a few dimensioning system variable settings. The **Use For** box lets you choose the type(s) of dimensions to which to apply the new Dimstyle you will be creating.

Choose **Continue** to display the New Dimension Style dialog box, similar to Figure 13–19, in which you define the new style properties. There are six tabs to choose from: **Lines, Symbols and Arrows**, **Text**, **Fit**, **Primary Units**, **Alternate Units**, and **Tolerances**. The dialog box initially displays the properties of the dimension style that you selected to start the new style in the Create New Dimension Style dialog box.

Setting Format and Properties for Lines

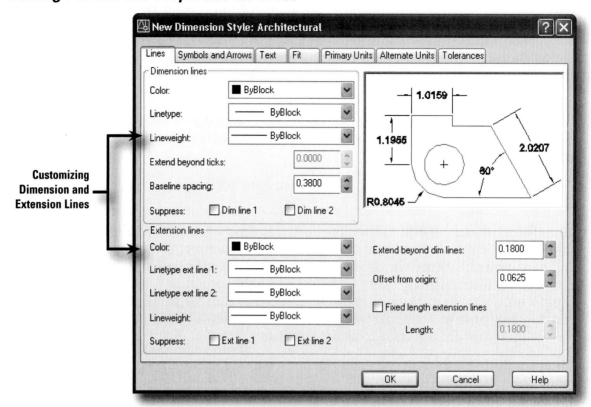

Customizing Dimension and Extension Lines

Figure 13-19

New Dimension Style dialog box with Lines and Arrows tab chosen

The **Lines** tab allows you to modify the geometry of the various elements that make up the dimensions, such as dimension lines and extension lines. There is a preview showing how dimensions will appear when drawn using the settings as you change them.

The **Dimension Lines** and **Extension Lines** sections allow you set the dimension line and extension lines properties and appearance.

Choose the color for dimension lines and extension lines from the **Color** box. ByLayer causes the color of the line to match that of its layer. ByBlock causes the color of the line to match that of its block reference if it is part of a block reference. You can also choose one of the standard colors to match.

The **Linetype** list box let you choose how the linetype of Dimension and Extension Lines will be determined. ByLayer causes the linetype of the line to match that of its layer. ByBlock causes the linetype of the line to match that of its block reference if it is part of a block reference. You can also choose one of the standard linetypes or other linetype to load through the Select Linetype dialog box.

Choose the lineweight for dimension lines and extension lines from the **Lineweight** box. ByLayer causes the lineweight of the line to match that of its layer. ByBlock causes the lineweight of the line to match that of its block reference if it is part of a block reference. You can also choose one of the standard lineweights or specify a value.

Choose one or more of the Suppress boxes to suppress one or more of the dimension or extension lines when the dimension is drawn.

In the **Dimension Lines** section only:

Specify the distance that the dimension line will be drawn past extension line if the arrowhead type is a tick (small diagonal line) in the **Extend beyond ticks** box.

In the **Baseline spacing** box:

Specify the distance that AutoCAD uses between dimension lines when they are drawn using the Baseline Dimension command.

Selecting Suppress lets you determine if one or more of the Dimension Lines is suppressed when the dimension is drawn. The value is recorded in DIMSD1 and DIMSD2 system variables.

In the **Extension Lines** section only:

Specify the distance that extension lines will be drawn past dimension lines in the **Extend beyond dim lines** box.

Specify the distance that AutoCAD uses between the extension line and the origin point specified when drawing an extension line in the **Offset from origin** box.

Selecting **Fixed length extension lines** causes dimension extension Lines to be drawn to the length specified in the **Length** text box.

Selecting Suppress let you determine if one or more of the Extension Lines is suppressed when the dimension is drawn. The value is recorded in DIMSE1 and DIMSE2 system variables.

Setting Format and Properties for Symbols and Arrows

The **Symbols and Arrows** tab lets you specify the appearance, location and display of symbols and arrowheads (see Figure 13–20).

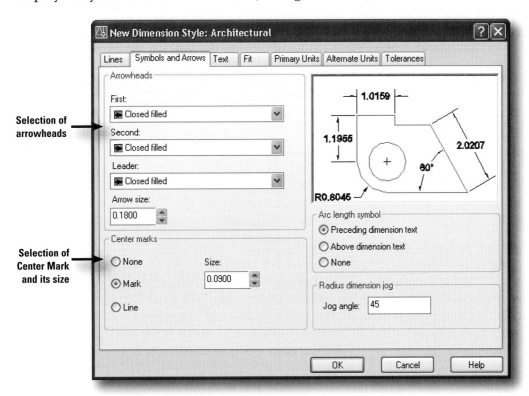

▲ Figure 13–20

Modify dimension Style dialog box with Symbols and Arrows tab chosen

The **Arrowheads** section controls the appearance of the arrowheads.

The **First, Second**, and **Leader** list boxes let you determine if and how arrowheads will be drawn at the terminations of the Extension Lines or the start point of a Leader. Unless you specify a different type for the second line termination, it will be the same type as the first line termination. The type of arrowheads is recorded in DIMBLK if both the first and second are the same or in DIMBLK1 and DIMBLK2 if they are different.

The **Arrow size** text box lets you enter the distance that AutoCAD uses for the length of an arrowhead when drawn in a dimension or leader. The value is recorded in the DIMASZ system variable.

Some of the arrowhead types included in the standard library are **None, Closed, Dot, Closed Filled, Oblique, Open, Origin Indication**, and **Right-Angle**. Figure 13–21 presents an example of each.

Figure 13–21

Arrowhead types

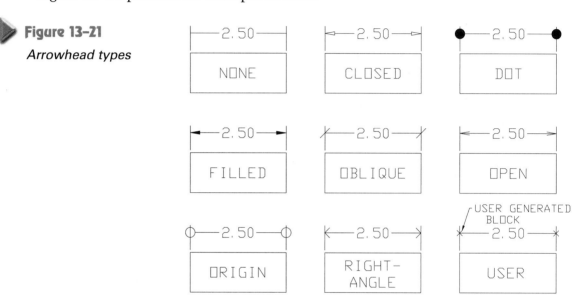

Choosing **User Arrow** lets you use a previously saved block definition by entering its name. Figure 13–22 shows the Select Custom Arrow Block dialog box. The block definition should be created as though it were drawn for the right end of a horizontal Dimension Line, with the insertion point at the intersection of the Extension Line and the Dimension Line.

Figure 13–22

Select Custom Arrow Block dialog box

The **Center Marks** section controls the appearance of center marks and centerlines for diameter and radial dimensions.

Choose if and what type of center mark is drawn when a diameter dimension or radius dimension is drawn from the **Type** list box. You can choose from **None**, **Mark**, and **Line**. If Mark is chosen, a center mark will be drawn with its length determined by the value in the **Size** box. If Line is chosen, the center marks will be drawn as lines that extend outside the circle a distance determined by the value in the **Size** text box. Figure 13–23 shows an example of each.

Figure 13-23

Circles with various center marks

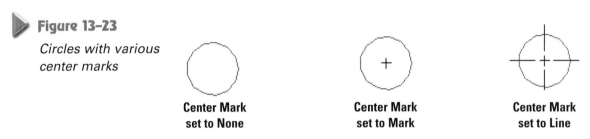

Center Mark
set to None

Center Mark
set to Mark

Center Mark
set to Line

The **Arc length symbol** section controls the appearance of the symbol used for the DIMARC command. Selecting **Preceding dimension text** causes the arc length symbol to be drawn before the dimension text. Selecting **Above dimension text** causes the arc length symbol to be drawn above the dimension text. Selecting **None** suppresses the arc length symbol (none will be drawn).

The **Radius dimension jog** section controls the appearance of the line used to dimension long radius arcs whose center is located out of the drawing area. The **Jog Angle** lets you specify the angle that is drawn in the shortened dimension line for the long radius arc. Figure 13–24 shows an example of a line with a jog in it representing a long radius arc being dimensioned.

Figure 13-24

A long radius arc dimensioned with jog

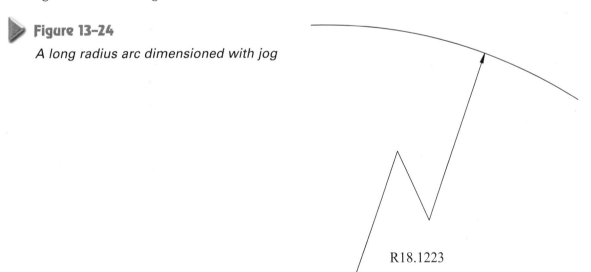

R18.1223

Setting Format, Placement, and Alignment for Dimension Text

The **Text** tab allows you to modify the appearance, location, and alignment of dimension text that is included when a dimension is drawn. There is a preview showing how dimensions will appear when drawn using the settings as you change them (see Figure 13–25).

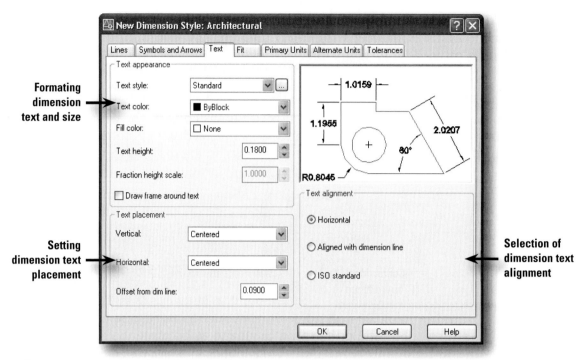

Formating dimension text and size

Setting dimension text placement

Selection of dimension text alignment

Figure 13–25

New Dimension Style dialog box with Text tab chosen

The **Text Appearance** section controls the dimension text format and size.

Choose the text style to which the dimension text will conform from the **Text style** box. The value is recorded in the DIMTXSTY system variable.

---

note   Do not confuse text style with dimension style. Dimensions are drawn in accordance with the current Dimstyle, which has as part of its configuration a text style to which dimension text will conform.

Choose the color for dimension text from the **Text color** box. ByLayer selection causes the color of the text to match that of its layer. ByBlock selection causes the color of the text to match that of its block reference if it is part of a block reference. You can also choose one of the standard colors for the text color to match. The value is recorded in the DIMCLRT system variable.

Choose the fill color for dimension text from the **Fill color** box. ByLayer causes the fill color of the text to match that of its layer. ByBlock causes the fill color of the text to match that of its block reference if it is part of a block reference. You can also choose one of the standard colors for the fill color to match.

Specify text height for dimension text in the **Text height** box. The value is recorded in the DIMTXT system variable.

Specify the scale of fractions relative to dimension text in the **Fraction height scale** box. This scale is the ratio of the text height for normal dimension text to the height of the fraction text. The value is recorded in the DIMTFAC system variable. This option is available only when Fractional is selected as the **Unit format** on the **Primary Units** tab.

Set the **Draw frame around text** box to ON to draw the dimension text inside a rectangular frame and to OFF to draw the dimension text without a rectangular frame (see Figure 13–26). The value is recorded in the DIMGAP system variable as a negative value.

The **Text Placement** section controls the placement of dimension text.

Choose one of the available options to determine how the dimension text will be drawn in relation to the dimension line in the **Vertical** box. The available options include Centered, Above, Outside, and JIS. The value is recorded in the DIMTAD system variable. Figure 13–26 shows an example of each.

Figure 13–26

Dimensioning Text with various Text appearance selections

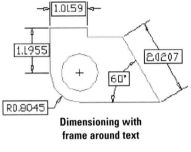

Dimensioning with frame around text

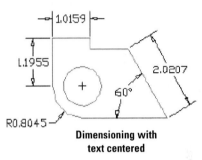

Dimensioning with text centered

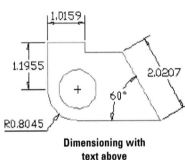

Dimensioning with text above

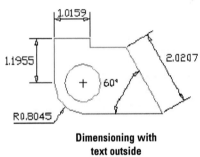

Dimensioning with text outside

Choose one of the available options to determine how the dimension text will be drawn in relation to the extension lines in the **Horizontal** box. The options include: **Centered**, **At Ext Line 1**, **At Ext Line 2**, **Over Ext Line 1**, and **Over Ext Line 2**. The value is recorded in the DIMJUST system variable. Figure 13–27 shows an example of each.

Figure 13–27

Dimensioning Text with various Text placement (horizontal) selections

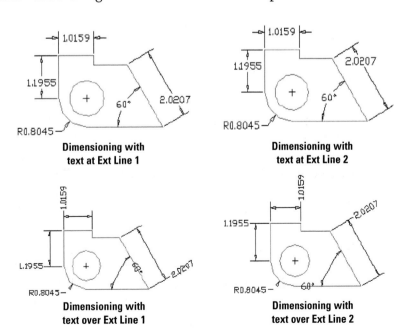

Dimensioning with text at Ext Line 1

Dimensioning with text at Ext Line 2

Dimensioning with text over Ext Line 1

Dimensioning with text over Ext Line 2

Specify the distance from the dimension line that AutoCAD uses when drawing dimension text as part of the dimension in the **Offset from dim line** box. The value is recorded in the DIMGAP system variable.

The **Text Alignment** section controls the orientation (horizontal or aligned) of dimension text. The value is recorded in the DIMTIH and DIMTOH system variables.

Choose the **Horizontal** option to place text in a horizontal position. Choose the **Aligned with dimension line** option to align text with the dimension line. Choose the **ISO Standard** option to align text with the dimension line when text is inside the extension lines, but aligns it horizontally when text is outside the extension lines.

Controlling the Placement of Various Elements of Dimensions

The **Fit** tab lets you to determine the arrangement of the various elements of dimensions when a dimension is drawn. There is a preview showing how dimensions will appear when drawn using the settings as you change them (see Figure 13–28).

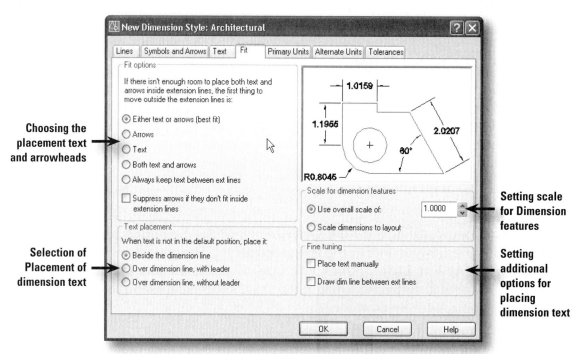

▲ **Figure 13-28**

New Dimension Style dialog box with Fit tab chosen

The **Fit Options** section lets you choose which of the text or arrowheads will be drawn between the extension lines.

Choose **Either the text or the arrows, whichever fits best** to place the text and arrowheads between the extension lines if there is enough space. Otherwise, AutoCAD places either the text or the arrowheads based on the best fit. Choose **Arrows** to place the text and arrowheads between the extension lines if there is enough space. Otherwise, if there is not enough space AutoCAD places text only inside the extension lines, and the arrowheads outside. Choose **Text** to place the text and arrowheads between the extension lines if there is enough space.

Otherwise, if there is not enough space AutoCAD places arrowheads only inside the extension lines, and the text outside. Choose **Both text and arrows** to place text and arrowheads outside extension lines when enough space is not available inside extension lines. Choose **Always keep text between ext lines** to place text between extension lines.

Set the **Suppress arrows if they don't fit inside the extension lines** box to ON to suppress arrowheads if not enough space is available inside the extension lines (see Figure 13–29). The value is recorded in the DIMSOXD system variable.

▶ **Figure 13–29**

Dimensioning examples with various Fit options

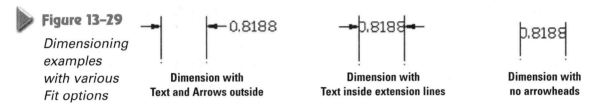

Dimension with Text and Arrows outside   **Dimension with Text inside extension lines**   **Dimension with no arrowheads**

The **Text Placement** section sets the placement of dimension text when it is moved from the default position, that is, the position defined by the dimension style. The value is recorded in the DIMTMOVE system variable.

Choose **Beside the dimension line** to place dimension text beside the dimension line. Choose **Over the dimension line, with a leader** to draw a leader connecting the text to the dimension line if text is moved away from the dimension line. Choose **Over the dimension line, without a leader** to keep the dimension line in the same place when text is moved. Text that is moved away from the dimension line is not connected to the dimension line with a leader. See Figure 13–30 for examples of text placement options.

▶ **Figure 13–30**

Dimensioning example with various Text placement options

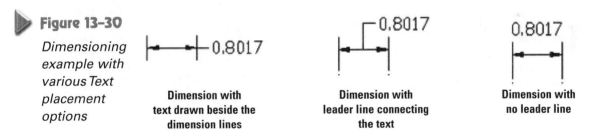

Dimension with text drawn beside the dimension lines   **Dimension with leader line connecting the text**   **Dimension with no leader line**

The **Scale for Dimension Features** section sets the overall dimension scale value or the paper space scaling.

Choose **Use overall scale of** to activate the text box that lets you specify the number AutoCAD uses as a ratio of the true scale to drawn dimensions. This is useful when you wish to create different parts of the drawing at different scales but have the dimension elements uniform in size. This does not include distances, coordinates, angles, or tolerances. The value is recorded in the DIMSCALE system variable.

Choose **Scale dimensions to layout (paperspace)** to cause dimensions to be drawn with elements scaled to layout. When you work in paper space, but not in a model space viewport, or when TILEMODE is set to 1, the default scale factor of 1.0 is used for the DIMSCALE system variable.

The **Fine Tuning** section sets additional fit options.

> Set **Place text manually when dimensioning** to ON to dynamically specify where dimension text is placed when the dimension is drawn. The value is recorded in the DIMUPT system variable.

> Set **Always draw dim line between ext lines** to ON to draw dimension line between the extension lines regardless of the distance between extension lines. The value is recorded in the DIMTOFL system variable.

Setting the Format and Precision of Primary Dimension Units

The **Primary Units** tab lets you to determine the appearance and format of numerical values of distances and angles when they are drawn with dimensions. There is a preview showing how dimensions will appear when drawn using the settings as you change them (see Figure 13–31).

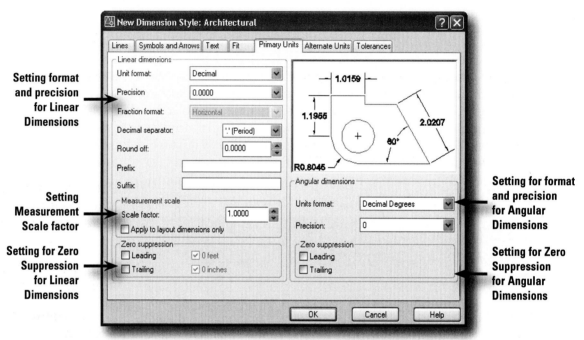

Figure 13-31

New Dimension Style dialog box with Primary Units tab chosen

The **Linear Dimensions** section sets the format and precision for linear dimensions.

> Choose one of the available formats of the units that AutoCAD uses when drawing dimension text as part of the dimension from the **Unit format** box. The available options include Scientific, Decimal, Engineering, Architectural, Fractional, and Windows Desktop. The value is recorded in the DIMLUNIT system variable. Figure 13–32 shows an example of each.

▶ **Figure 13-32**

Dimensioning example with various dimensioning unit formats

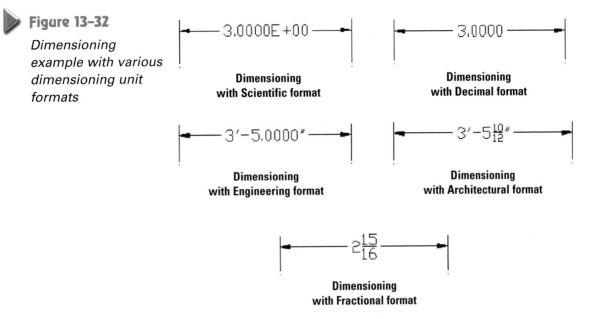

Dimensioning
with Scientific format

Dimensioning
with Decimal format

Dimensioning
with Engineering format

Dimensioning
with Architectural format

Dimensioning
with Fractional format

Set the number of decimal places in the dimension text from the **Precision** box. The value is recorded in the DIMDEC system variable.

Set the format for fractions from the **Fractional format** box. This option is available only when Architectural or Fractional format is selected. The value is recorded in the DIMFRAC system variable.

Choose the separator for decimal formats from the **Decimal separator** box. The available options include Period, Comma, or a Space if the selected format is in Scientific, Decimal, or Windows Desktop. The value is recorded in the DIMDSEP system variable.

Specify a value to which distances will be rounded in the **Round off** box. For example, a value of 0.5 causes dimensions to be rounded to the nearest 0.5 units.

Specify a prefix if any for the dimension text in the **Prefix** box (see Figure 13–33). The prefix text will override any default prefixes, such as those used in radius (R) dimensioning. The value is recorded in the DIMPOST system variable.

Specify suffix if any for the dimension text in the **Suffix** box (see Figure 13–33). If you specify tolerances, AutoCAD includes the suffix in the tolerances, as well as in the main dimension.

▶ **Figure 13-33**

Dimensioning example with prefix, suffix, and alternate units

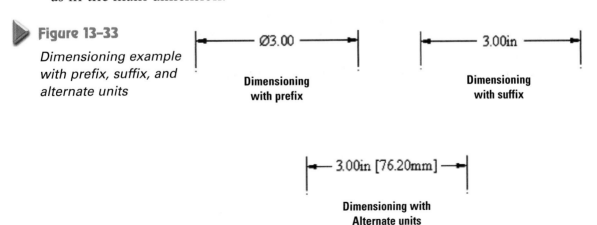

Dimensioning
with prefix

Dimensioning
with suffix

Dimensioning with
Alternate units

The **Measurement Scale** section defines measurement scale options

Specify the scale factor in the **Scale factor** box, which is the ratio of the true dimension distances to drawn dimensions distances, or the linear scale factor for the linear measured distances of a dimension without affecting the components, angles, or tolerance values. For example, you are drawing with the intention of plotting at the quarter-size scale, or 3" = 1'-0". You have scaled up a detail by a factor of 4 so that it will plot to full scale. If you wish to dimension it after it has been enlarged, you can set the linear scale factor in this section of the Primary Units dialog box to .25 so that dimensioned distances will represent the dimension of the object features before it was scaled up. This method keeps the components at the same size as dimensions created without a scale change. This is useful when you wish to create different parts of the drawing at different scales but have the dimension elements uniform in size.

Set **Apply to layout dimensions only** to ON to apply the ratio only to layout dimensions.

The **Zero Suppression** section controls the suppression of leading and trailing zeros, and of feet and inches that have a value of zero. The value is recorded in the DIMZIN system variable.

Set **Leading** to ON to suppress leading zeros in all decimal dimensions. For example: .700 is displayed instead of 0.700.

Set **Trailing** to ON to suppress trailing zeros in all decimal dimensions. For example: 7 or 7.25 is displayed instead of 7.000 or 7.250, respectively.

Set **0 Feet** to ON to suppress the feet portion of a feet-and-inches dimension when the distance is less than one foot. For example: 7" or 7 1/4" is displayed instead of 0'-7" or 0'-7 1/4".

Set **0 Inches** to ON to suppress the inches portion of a feet-and-inches dimension when the distance is an integral number of feet. For example: 7' is displayed instead of 7'-0".

The **Angular Dimensions** section sets the current angle format for angular dimensions. The value is recorded in the DIMAUNIT system variable.

Choose a unit format for the angular Dimension Text from the **Units format** box. The available option includes Decimal Degrees, Degrees/Minutes/Seconds, Grads, and Radians.

Choose the number of decimal places for angular dimensions from the **Precision** box. The value is recorded in the DIMADEC system variable.

The **Zero Suppression** section suppresses leading and trailing zeros.

Set **Leading** to ON to suppress leading zeros in angular decimal dimensions and **Trailing** to ON to suppress the trailing zeros in angular decimal dimensions. The value is recorded in the DIMAZIN system variable.

Setting the Format and Precision of Alternate Dimension Units

The **Alternate Units** tab (see Figure 13–34) is similar to the **Primary Units** tab, except for the **Display alternate units** box, which lets you enable or disable alternate units, and there is no **Angular Dimensions** section. There is a **Placement** section with **After primary value** and **Before primary value** options that let you determine where to place the alternate units.

Setting the Display for alternate units

Setting format and precision for alternate dimensions

Setting for Zero Suppression for alternate Dimensions

Setting for the placement of alternate units

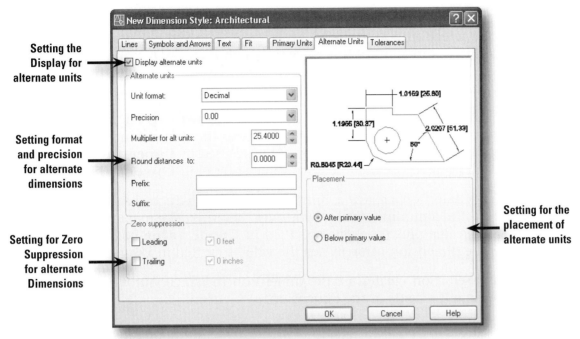

 Figure 13–34

New Dimension Style dialog box with Alternate Units tab chosen

Setting the Format and Precision of Tolerances

The **Tolerances** tab (see Figure 13–35) controls the display and format of dimension text tolerances.

Selection of Tolerance Method and corresponding precision

Setting for Zero Suppression for tolerance

Setting for the Zero suppression for Alternate Unit Tolerance

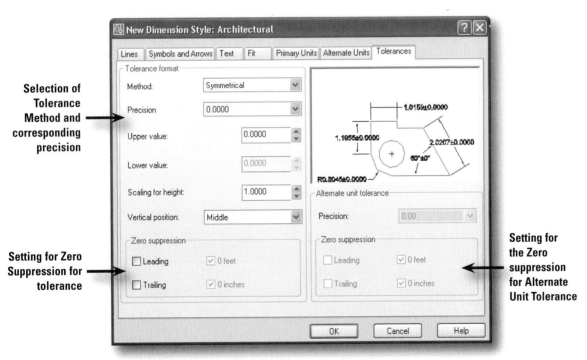

 Figure 13–35

New Dimension Style dialog box with Tolerances tab chosen

The **Tolerance Format** section controls the tolerance format.

Choose one of available methods for calculating the tolerance from the **Method** box.

The None selection does not add a tolerance. The Symmetrical selection adds a plus/minus expression of tolerance in which AutoCAD applies a single value of variation to the dimension measurement. A ± appears after the dimension. Specify the tolerance value in **Upper value** box.

The Deviation selection adds a plus/minus tolerance expression. AutoCAD applies different plus and minus values of variation to the dimension measurement. A plus sign (+) precedes the tolerance value specified in **Upper value** box, and a minus sign (–) precedes the tolerance value entered in **Lower value** box.

The Limits selection creates a limit dimension in which AutoCAD displays a maximum and a minimum value, one over the other. The maximum value is the dimension value plus the value specified in **Upper value** box. The minimum value is the dimension value minus the value specified in **Lower value** box.

The Basic selection creates a basic dimension in which AutoCAD draws a box around the full extents of the dimension.

Choose the number of decimal places in the **Precision** text box.

Specify the tolerance text height as a ratio of the tolerance height to the main dimension text height in the **Scaling for height** box.

Set the text justification for symmetrical tolerances and deviation tolerances in the **Vertical position** box.

Alternate Unit Tolerance sets the precision and zero suppression rules for alternate tolerance units.

Once the necessary changes are made for the appropriate settings, choose **OK** to close the New Dimension Style dialog box to create the new dimensional style.

Modifying a Dimension Style

Select one of the named dimension styles in the Dimension Style list in the Dimension Style Manager, and choose **Modify**.

AutoCAD displays Modify Dimension Style dialog box as shown in Figure 13–36.

Figure 13-36

Modify Dimension Style dialog box

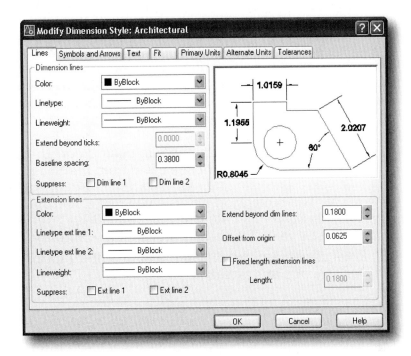

The content of the Modify Dimension Style dialog boxes is identical to the New Dimension Style dialog box explained earlier, although you are modifying an existing dimension style rather than creating a new one.

Overriding The Dimension Feature

Figure 13-37

Invoking the DIMOVERRIDE command from the Dimension menu

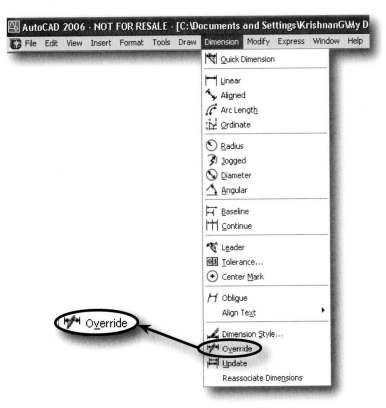

The DIMOVERRIDE command allows you to change one of the features in a dimension without having to change its dimension style or create a new dimension style. For example, you may wish to have one leader with the text centered at the ending horizontal line, rather than over the ending horizontal line in the manner that the current dimension style may call for. By invoking the DIMOVERRIDE command, you can respond to the prompt with the name of the dimensioning system variable (DIMTAD in this case) and set the value to 0 rather than 1. Then you can select the dimension leader you wish to have overridden. After you invoke the DIMOVERRIDE command (see Figure 13–37), AutoCAD prompts:

dimoverride

Enter dimension variable name to override or ⬇: *(specify the dimensioning system variable name or choose Clear from the Shortcut menu to clear overrides)*

If you specify a dimensioning system variable, you are prompted:

Enter new value for dimension variable <current>: *(specify the new value)*
Enter dimension variable name to override: *(specify another dimension system variable or press* ENTER*)*

If you press ENTER, you are prompted:

Select objects: *(select the dimension objects)*

The dimensions selected will have the dimensioning system variable settings overridden in accordance with the value specified.

If you choose Clear, you are prompted:

Select objects: *(select the dimension to clear the overrides)*

The dimensions selected will have the dimensioning system variable setting overrides cleared.

Updating Dimensions

 Figure 13–38

Invoking Dimension Update from the Dimension menu

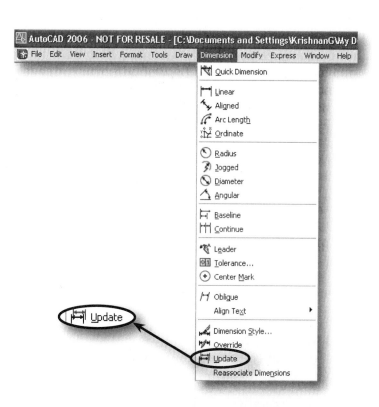

The UPDATE command permits you to make selected existing dimension(s) conform to the settings of the current dimension style. After you invoke the UPDATE command (see Figure 13–38), AutoCAD prompts:

Select objects: *(select any dimension(s) whose settings you wish to have updated to conform to the current dimension style)*

CREATING AND MODIFYING TABLE STYLES

Figure 13-39

Invoking the TABLESTYLE command from the Format menu

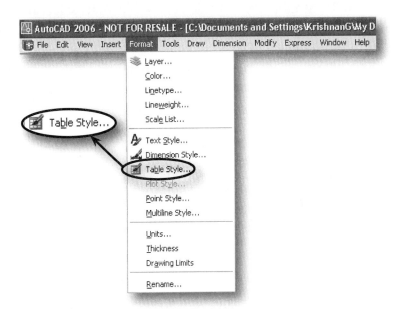

The TABLESTYLE command is used to create a new table style or modify an existing one. The appearance of the table is controlled by its table style. The table style can specify a different justification and appearance for the text and gridlines in each type of row. The Insert Table dialog box has a **Style** menu that lists available table styles from which you can select table style to create a table. After you invoke the TABLESTYLE command (see Figure 13–39), AutoCAD displays the Table Style dialog box, as shown in Figure 13–40.

Figure 13-40

Table Style dialog box

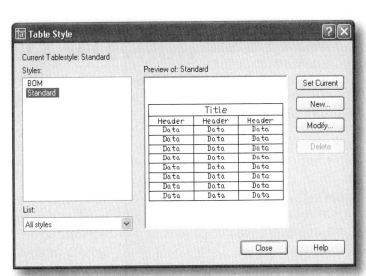

The **Current Tablestyle** heading shows the name of the table style to which the next drawn table will conform.

The **Styles** section displays the name(s) of the current style and any other styles available depending on the option chosen in the **List** selection box.

From the **List** selection box you can list all of the styles available in the drawing displayed or only those in use. The Standard style cannot be deleted.

The **Preview of** window shows how table will appear when drawn using the current table style.

Choosing **Set Current** causes the table style chosen in the **Styles** section to become the current table style. To delete a table style, select the style in the **Styles** section and choose **Delete**. The Standard style cannot be deleted.

Creating a New Table Style

Choose **New** to create a new table style. AutoCAD displays the Create New Table Style dialog box similar to Figure 13–41. Specify an existing table style whose settings are the default for the new table style.

Figure 13–41

Create New Table Style dialog box

Specify the name of the new style in the **New Style Name** box. The **Start With** list box lets you choose the existing table style whose settings are the default for the new table style. Choose **Continue** to display the New Table Style dialog box similar to Figure 13–42, in which you define the new style properties. There are three tabs to choose from: **Data**, **Column Heads**, and **Title**. Options on each tab set the appearance of the data cells, the column heads, or the table title. The dialog box initially displays the properties of the table style that you selected as the **Start With** style.

Figure 13–42

New Table Style dialog box with the Data tab selected

Depending on which tab is active, the **Cell properties** section allows you to set the text style, text height, text color, fill color, and alignment for the text in the applicable cell.

Depending on which tab is active, the **Border properties** section allows you to set the **Grid lineweight** and **Grid color** for the applicable cells in the applicable cell.

The **General** section allows you specify whether the table has the title at the top (for the Down option) or the bottom (for the Up option) and determines the direction in which additional rows are added.

The **Cell margins** section allows you to specify the size of the horizontal and vertical margins for the applicable cells.

The preview window shows how a table conforming to the selected table style will appear. After setting up the Table Style and Cell and Border properties, choose **OK** to create a new table style and close the New Table Style dialog box.

Modifying a Table Style

Select one of the named table styles in the **Styles** list in the Table Style dialog box and choose **Modify**.

AutoCAD displays Modify Table Style dialog box. The content of the Modify Table Style dialog box is identical to the New Table Style dialog box explained earlier, although you are modifying an existing table style rather than creating a new one.

drawing exercises

Open the Exercise Manual PDF file for Chapter 13 on the accompanying CD for discipline-specific exercises.

review questions

1. The elements of a multiline can have different colors.
   a. *True*
   b. *False*

2. To use incline lettering so that the vertical portions of the characters point to two o'clock for a horizontal line of text, the obliquing angle should be:
   a. *–60.*
   b. *–30.*
   c. *0.*
   d. *30.*
   e. *60.*

3. In order to select a different font for use in the TEXT command, a text style must be created.
   a. *True*
   b. *False*

4. Multiline styles can be saved to an external file, thus allowing their use in multiple drawings.
   a. *True*
   b. *False*

5. Which of the following are not valid options when creating a text style?
   a. *Width factor*
   b. *Upside down*
   c. *Vertical*
   d. *Backwards*
   e. *None of the above (all answers are valid)*

6. How many different elements can a multiline style contain?
   a. *40*
   b. *8*
   c. *10*
   d. *16*

7. Which part of a dimension will the DIMCLRD system variable assign color to?
   a. *Arrowheads*
   b. *Text*
   c. *Extension lines*
   d. *All of the above*

8. Which system variable controls the size of dimension line arrowheads?

    a. *dimasz*

    b. *dimaso*

    c. *dimaltz*

    d. *dimaltu*

9. MULTILINE allows the user to draw multiple parallel lines at one time.

    a. *True*   **b.** *False*

10. A multiline's elements may be of different line types.

    a. *True*   **b.** *False*

11. A multiline consists of parallel lines that act as one object.

    a. *True*   **b.** *False*

12. The joints of a multiline can be turned off.

    a. *True*   **b.** *False*

13. Multilines can be edited if they are exploded.

    a. *True*   **b.** *False*

14. Multiline vertices can be deleted.

    a. *True*   **b.** *False*

15. To set the size of the most commonly used DIM variables such as the size of the arrowheads, select _____ in the DDIM dialog box.

    a. *Format*

    b. *Annotation*

    c. *Geometry*

    d. *None of the above*

16. The default gap width between the extension line and the object is:

    a. *1/8 inch.*

    b. *1/2 inch.*

    c. *1/4 inch.*

    d. *None of the above.*

17. To turn the associative dimensioning variable on, use the _____ system variable.

    a. *DIMASO*

    b. *DIMCIA*

    c. *Both a and b*

    d. *None of the above*

18. With the associative mode variable set to ON, changing the size of the object will cause:

    a. *The dimension text to change accordingly.*

    b. *The dimension length to automatically change.*

    c. *Both a and b.*

    d. *None of the above.*

19. To change the size of an arrowhead, use the _____ dialog box.

   a. *Geometry*
   b. *Format*
   c. *Annotation*
   d. *None of the above*

20. To change the dimension text height, use the _____ dialog box.

   a. *Geometry*   *c. Annotation*
   b. *Format*   *d. None of the above*

21. To change the linear tolerance settings, use the _____ dialog box.

   a. *Geometry*   *c. Annotation*
   b. *Format*   *d. None of the above*

SECTION

4

The Package is Complete

(Not the End of the World)

14

introduction

Chapter 8 explained how to draw the fundamental linear, aligned, radius, diameter, angular, and ordinate dimensions. This chapter introduces the more advanced dimensioning features available in AutoCAD, such as baseline, continue, quick, leader and oblique dimensioning. Also covered is tolerance symbology, which communicates how much the "as-built" size and locations of the manufactured object can vary from the dimensions on the drawing and still be acceptable.

After completing this chapter, you will be able to do the following:

- Draw baseline and continue dimensioning

- Draw quick dimensioning

- Draw oblique dimensioning

- Edit dimensioning text

- Edit dimensioning with grips

ADVANCED DIMENSIONING COMMANDS

AutoCAD offers advanced commands to make dimensioning easier and help conform to industry standards. These include baseline, continue, quick, and oblique dimensions.

Baseline Dimensioning

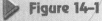 **Figure 14–1**

Invoking the Baseline Dimension command from the Dimension toolbar

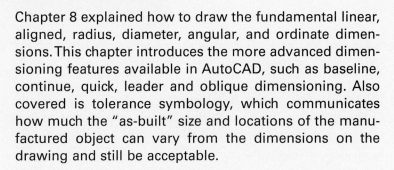

Baseline dimensioning (sometimes referred to as parallel dimensioning) is used to draw dimensions to multiple points from a single datum baseline, as shown in Figure 14–2. Before the Baseline dimension feature is invoked, there must be an existing dimension in the drawing. By default AutoCAD uses the first extension line origin of the last dimension created as the base line. If you wish to have a different dimension extension line origin be the base line, right-click and choose Select from the Shortcut menu when prompted to "Specify a second extension line origin or ⊥:".

The first extension line origin of the initial dimension (it can be a linear, angular, or ordinate dimension) establishes the base from which the baseline dimensions are drawn. That is, all of the dimensions in the series of baseline dimensions will share a common first extension line origin. AutoCAD automatically draws a dimension line/ arc beyond the initial (or previous baseline) dimension line/arc. The location of the new dimension line/arc is an offset distance established by the DIMDLI (for dimension line increment) dimensioning variable. Invoke the DIMBASELINE command (see Figure 14–1), AutoCAD prompts:

> Specify a second extension line origin or ⊥: *(specify a point or choose one of the available options from the Shortcut menu)*

After you specify a point for the second extension line origin, AutoCAD will use the first extension line origin of the previous linear, angular, or ordinate dimension as the first extension line origin for the new dimension, and the prompt is repeated. Like the LINE command, the Baseline Dimension command continues to prompt for additional dimensions until you exit the command. To exit the command, choose Enter from the Shortcut menu.

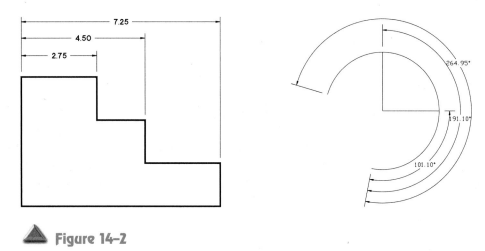

▲ **Figure 14–2**

Examples of applications of baseline dimensioning

Continue Dimensioning

▶ **Figure 14–3**

Invoking the Continue Dimension command from the Dimension toolbar

Continue dimensioning is used for drawing a string of dimensions, each of whose second extension line origin coincides with the next dimension's first extension line origin. Before the Continue dimension feature is invoked, there must be an existing dimension in the drawing. In the example shown in Figure 14–4, the dimension from A to B must first be drawn using the Aligned dimensioning command. Invoke the Continue Dimension command (see Figure 14–3), select points C, D, E, and F in sequence to cause the proper dimensions to be drawn. AutoCAD prompts:

Specify a second extension line origin or ⊡: *(specify point C)*

Dimension text = *(measured angle or dimension)*

... (continue to select points D, E, and F in sequence and then exit by pressing ENTER*)*

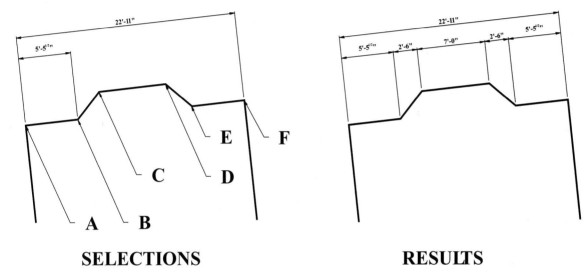

SELECTIONS   **RESULTS**

 Figure 14–4

Continue dimensioning

After you specify a point for the second extension line origin, AutoCAD will use the second extension line origin of the previous linear, angular, or ordinate dimension as the first extension line origin for the new dimension, and the prompt is repeated. Like the LINE command, the Continue Dimension command continues to prompt for additional dimensions until you exit the command. To exit the command, choose Enter from the Shortcut menu.

Quick Dimensioning

▷ **Figure 14–5**

Invoking the Quick Dimension command from the Dimension toolbar

Quick Dimensioning, or the QDIM command, is used to draw a string of dimensions between all of the end and center points of the selected object(s). In the example shown in Figure 14–6, the Quick Dimension command can be used to draw the five vertical dimensions at one time. Invoke the QDIM command (see Figure 14–5), AutoCAD prompts:

> Select geometry to dimension: *(select one or more objects and then press* ENTER*)*

After selecting the object(s), drag the cursor and select the location of the dimension lines to cause the proper dimensions to be drawn.

> Specify dimension line position or ⊻: *(specify location for dimension line or select one of the available options from the Shortcut menu)*

If you specify a location for a dimension line, AutoCAD will draw continuous dimensioning between all end or center points of the objects selected horizontally or vertically, depending on where the dimension line location is specified. You can choose one of the options from the Shortcut menu that will draw the type of dimension chosen.

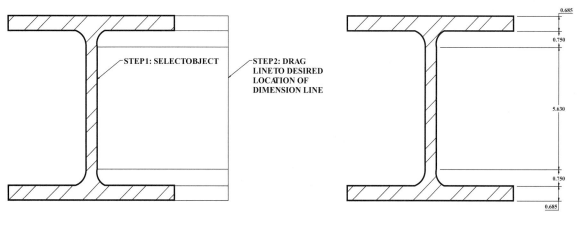

SELECTION   **RESULTS**

 Figure 14–6

An example of quick dimensioning

Leader with Annotations

▶ Figure 14–7

Invoking the QLEADER *command from the Dimension toolbar*

The QLEADER command creates a leader and leader annotation and assumes that you wish to enter multiline text. The QLEADER command minimizes the steps required to draw the text with the line(s) and arrowhead pointing to an object (or feature on an object) for annotations and callouts used to describe them. Invoke the QLEADER command (see Figure 14–7), AutoCAD prompts:

Specify first leader point : *(specify a point for the arrowhead end of the leader line or press enter to open the Leader Settings dialog box)*

Specify next point: *(specify another point for the end of the first leader segment opposite the arrowhead)*

Specify text width <0.0000>: *(specify a point to determine the maximum width of the multiline text)*

Enter first line of annotation text <Mtext>: *(enter the first line of annotation text or right-click for the Multiline Text Editor dialog box)*

Enter next line of annotation text: *(enter the next line of annotation text or press* ENTER *to complete the* QLEADER *command)*

---

note The text width that is specified causes the text to wrap to the next line when the text on the current line exceeds the specified width. If no width is specified, the entire text is drawn on one line.

If you select Settings when prompted to specify the first leader point, AutoCAD displays the Leader Settings dialog box, shown in Figure 14–8.

Figure 14-8

Leader Settings dialog box with Annotation tab chosen

Annotation

The **Annotation** tab allows you to set the annotation type, select multiline text options, and determine how annotations are reused.

Choose one of the five options available for the type of annotation that will be used in the **Annotation Type** section. The options include **MText**, **Copy an Object**, **Tolerance**, **Block Reference**, and **None**. Choosing **MText** causes AutoCAD to use the MTEXT command when the annotation is called for. For a detailed explanation about MTEXT command usage, refer to Chapter 4.

Choosing **Copy an Object** causes AutoCAD to prompt you to Select an object to copy: when the annotation is called for. If you have not chosen a text object, AutoCAD will prompt: Please select an mtext, text, block reference, or tolerance object. The text object you select will become the text for leader annotation.

Choosing **Tolerance** causes the Tolerance option to be used (as described in the following section) when the annotation is called for.

Choosing **Block Reference** causes AutoCAD to use the INSERT command (for inserting a block) when the annotation is called for.

Choosing **None** causes AutoCAD to draw the leader and exit the LEADER command without prompting for or placing text or another object.

The **MText options** section lets you specify how MText will be formatted and is active only if **MText** has been selected in the **Annotation Type** section. The options include **Prompt for width**, **Always left justify**, and **Frame text**.

Choosing **Prompt for width** causes AutoCAD to prompt you for the width of the MText box when the annotation is called for.

Choosing **Always left justify** causes the annotation text to be left justified whether the leader is to the left or the right of the annotation. When this box is checked, the **Prompt for width** check box becomes disabled.

Choosing **Frame text** causes AutoCAD to draw a rectangle around the annotation text.

The **Annotation Reuse** section lets you specify whether or not AutoCAD reuses the next or the current annotation text for subsequent leader annotation text. The options include **None**, **Reuse Next**, and **Reuse Current**.

Choosing **None** causes AutoCAD to not reuse annotation text and to prompt you for annotation text to be used.

Choosing **Reuse Next** causes AutoCAD to reuse the next annotation text used for subsequent annotation text.

Choosing **Reuse Current** causes AutoCAD to reuse the current annotation text for subsequent annotation text.

Leader Line & Arrow

Figure 14-9

Leader Settings dialog box with Leader Line & Arrow tab chosen

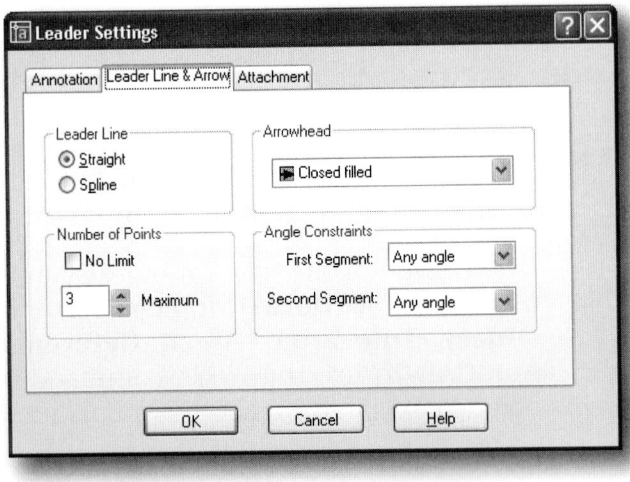

The **Leader Line & Arrow** tab, as shown in Figure 14–9, allows you to modify the geometry of the lines and arrowheads that make up the leader, along with specifying the number of points permitted (which limits the number of leader segments that may be drawn) and the angle constraints of the first and second leader lines. The options are as follows:

The **Leader Line** section sets the leader line format. Options include **Straight** and **Spline**.

Choosing **Straight** causes leaders to be drawn with straight segments.

Choosing **Spline** causes leaders to be drawn with curved segments using the same input as in the SPLINE command.

The **Number of Points** section sets the number of leader points that QLEADER prompts you to specify before prompting for the leader annotation.

Choosing **No Limit** lets you put in any number of leader segments.

If **No Limit** is not checked, the number of points you are can specify to draw leader segments is limited to the number (between 2 and 999) in the **Maximum** box. The number of segments will be one fewer than the number of points specified.

The **Arrowhead** section defines the leader arrowhead. From the list box you can select one of the many available arrowhead shapes. Select None to have no arrowhead or select User Arrow and name a user-defined block to be used as an arrowhead.

The **Angle Constraints** section sets angle constraints for the first and second leader lines.

The **First Segment** list box lets you specify if the first segment can be drawn at any angle or will be constrained to one of the following angles: 90, 45, 30, or 15 degrees.

The **Second Segment** list box lets you specify if the second segment can be drawn at any angle or will be constrained to one of the following angles: 90, 45, 30, or 15 degrees.

Attachment

Figure 14–10

Leader Settings dialog box with Attachment tab chosen

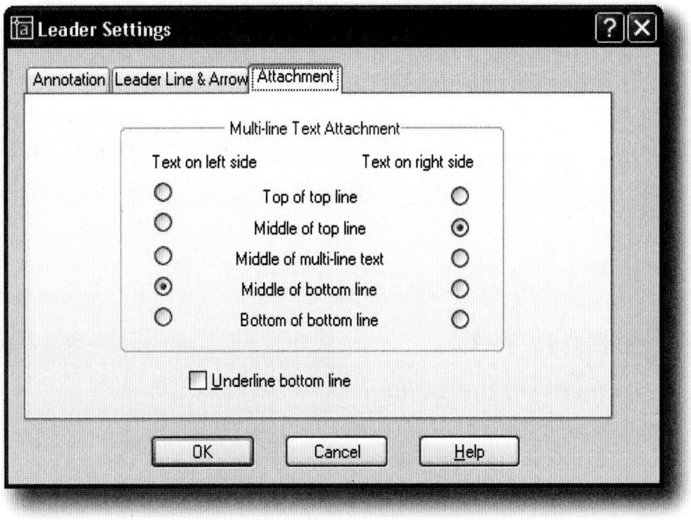

The **Attachment** tab, as shown in Figure 14–10, allows you to specify which part of Multi-line text will be lined up with the annotation end of the leader. The attachment can be specified to line up one way when the text is to the left of the leader and a different way when it is to the right of the leader. An option in the **Multi-line Text Attachment** section is specified by selecting one of the following five options: **Top of top line**, **Middle of top line**, **Middle of multi-line text**, **Middle of bottom line**, and **Bottom of bottom line**.

The **Attachment** tab has a check box that, when checked, causes the leader to terminate at the bottom line of text and continue as an underline of the text. This overrides any of the first four attachment options mentioned above.

Drawing Cross Marks for Arcs or Circles

 Figure 14–11

Invoking the
DIMCENTER *command*
from the Dimension
toolbar

The DIMCENTER command is used to draw the cross marks that indicate the center of an arc or circle as shown in Figure 14–12.

 Figure 14–12

Circles with center cross marks

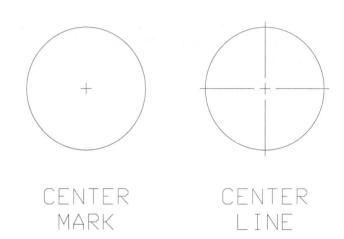

The DIMCENTER command is used to draw the cross marks that indicate the center of an arc or circle as shown in Figure 14–12.

Invoke the DIMCENTER command (see Figure 14–11), AutoCAD prompts:

    Select arc or circle: *(select arc or circle)*

After you select an arc or circle, AutoCAD draws the cross marks in accordance with the setting of the DIMCEN dimensioning variable.

Oblique Dimensioning

 Figure 14–13

Invoking the Oblique
option of the DIMEDIT
command from the
Dimension toolbar

The OBLIQUE command allows you to align the extension lines of a linear dimension to a specified angle. The dimension line will follow the extension lines, retaining its original direction. This is useful for having the dimension stay clear of other dimensions or objects in your drawing. It is also a conventional method of dimensioning isometric drawings. After you invoke the Oblique option of the DIMEDIT command (see Figure 14–13) AutoCAD prompts:

> Enter type of dimension editing [Home/New/Rotate/Oblique]: *(choose Oblique)*
> Select objects: *(select the dimension(s) for obliquing)*
> Enter obliquing angle (press ENTER for none): *(specify an angle or press ENTER)*

---

note If you select the OBLIQUE command from the Dimension menu it is actually the Dimension Style Edit command in which AutoCAD automatically chooses the Oblique option for you. All you have to do is select a dimension.

The extension lines of the selected dimension(s) will be aligned to the specified angle.

Tolerances

Tolerance symbology and text can be included with the dimensions that you draw in AutoCAD. AutoCAD provides a special set of subcommands for the two major methods of specifying tolerances. One set is for lateral tolerances; the other is for geometric tolerances show acceptable deviations of form, profile, orientation, location, and runout of a feature.

Lateral tolerances specify the amount by which dimensions can vary.

Geometric tolerances specify the amount a shape may vary.

---

note Lateral tolerance symbols and text are accessible from the Dimension Style Manager dialog box. Geometric tolerance symbols and text are accessible from the Dimension toolbar, from the Dimension menu, and at the Command: prompt.

Lateral Tolerance

Lateral tolerancing draws the traditional symbols and text for Limit, Plus or Minus (unilateral and bilateral), Single Limit, and Angular tolerance dimensioning. Lateral tolerances will appear with the dimension text in accordance with how they are set up in the Tolerances tab of the Modify Dimension Style dialog box described below.

Lateral tolerance is the range from the smallest to the greatest that a dimension is allowed to deviate and still be acceptable. For example, if a dimension is called out as 2.50 ± 0.05, then the tolerance is 0.1 and the feature being dimensioned may be anywhere between 2.45 and 2.55. This is the symmetrical plus-or-minus convention. If the dimension is called out as $2.50 {}^{+0.10}_{-0.00}$, then the feature may be 0.1 greater than 2.50, but may not be smaller than 2.50. This is referred to as unilateral. The Limits tolerance dimension are shown as $\frac{2.55}{2.45}$, wherein the minimum and maximum values for a dimension are shown directly.

To set the tolerance method, first open the Dimension Style Manager dialog box. Select one of the dimension styles to modify and choose **Modify**. AutoCAD displays the Modify Dimension Style dialog box. Select the **Tolerances** tab, and select the tolerance method from the **Method** option menu as shown in Figure 14–14.

To set the tolerance method, first open the Dimension Style Manager dialog box by typing **dimstyle** (for a discussion of the other features of the Dimension Style manager refer to Chapter 13). Choose **Modify**, and AutoCAD displays the Modify Dimension Style dialog box. Select the **Tolerances** tab and select the **Tolerance** method from the **Method** option menu. Figure 14–14 shows all the options available in the **Tolerance Format** section and the **Alternate Unit Tolerance** section of the **Tolerances** tab. Selecting a method (other than **None**) in the **Method** text box applies only to lateral tolerancing.

Figure 14–14

Modify Dimension Style dialog box with the Tolerances tab selected

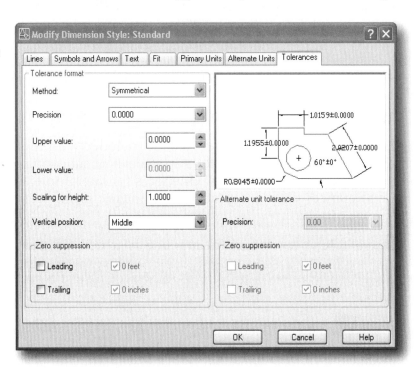

Choosing the **Method** list box in the **Tolerance Format** section of the **Tolerances** tab displays the following options: None, Symmetrical, Deviation, Limits, and Basic. Selecting a method (other than None) in the **Method** box applies only to lateral tolerancing. See Table 14–1 for examples of the various options.

Table 14–1 *Examples of Lateral Tolerance Methods*

Tolerance Method	Description	Example
Symmetrical	Only the **Upper value** box is usable. Only one value is required.	1.00 ± 0.05
Deviation	Both **Upper value** and **Lower value** boxes are active. A 0.00 value may be entered in either box, indicating that the variation is allowed in only one direction.	$1.00 \, {}^{+0.07}_{-0.03}$
Limits	Both **Upper value** and **Lower value** boxes are active. In this case, there is no base dimension. The Justification box is not active, because the values are not tolerance values but are actual dimensions to be drawn in full height text. Using an upper value of 0.0500 and a lower value of 0.0250 will cause annotation of a 1.000 basic dimension to be ${}^{1.0500}_{0.9750}$.	1.070 0.930
Basic	The dimension value is drawn in a box, indicating that it is the base value from which a general tolerance is allowed. The general tolerance is usually given in other notes or specifications on the drawing or other documents.	1.00

note   For a preview of how your tolerance will appear in relation to the dimension, see the format set on the *Primary Units* tab of the Modify Dimension Style dialog box.

In the **Tolerance Format** and **Alternate Unit Tolerance** sections, the **Precision** box lets you determine how many decimals the text will be shown in decimal units. The **Upper value** and **Lower value** boxes allow you to preset the values where they apply. The **Scaling for height** box lets you set the height of the tolerance value text. The **Vertical position** selection box determines if the tolerance value that is drawn after the dimension will be at the top, middle, or bottom of the text space.

In the **Alternate Unit Tolerance** section, the **Zero suppression** subsection lets you specify whether or not zeros are displayed, as follows:

Selecting **Leading** causes zeros ahead of the decimal point to be suppressed. Example: .700 is displayed instead of 0.700.

Selecting **Trailing** causes zeros behind the decimal point to be suppressed. Example: 7 or 7.25 is displayed instead of 7.000 or 7.250, respectively.

Selecting **0 Feet** causes zeros representing feet to be suppressed if the dimension text represents inches and/or fractions only. Example: 7" or 7 1/4" is displayed instead of 0'-7" or 0'-7 1/4".

Selecting **0 Inches** causes zeros representing inches to be suppressed if the dimension text represents feet only. Example: 7' is displayed instead of 7'-0".

Geometric Tolerance

Figure 14–15

Invoking the Geometric Tolerance command from the Dimension toolbar

Geometric tolerancing draws a Feature Control Frame for use in describing standard tolerances according to the geometric tolerance conventions. Geometric tolerancing is applied to forms, profiles, orientations, locations, and runouts. Forms include squares, polygons, planes, cylinders, and cones. After you invoke the Geometric Tolerance command (see Figure 14–15), AutoCAD displays the Geometric Tolerance dialog box, shown in Figure 14–16.

Figure 14–16

Geometric Tolerance dialog box

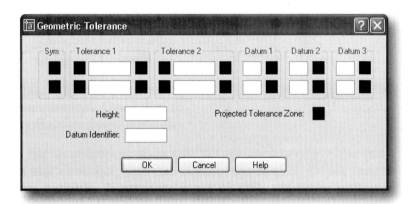

The conventional method of expressing a geometric tolerance for a single dimensioned feature is in a Feature Control Frame, which includes all necessary tolerance information for that particular dimension. A Feature Control Frame has the Geometric Characteristic Symbol box and the Tolerance Value box. Datum reference/material condition datum boxes may be added where needed. The Feature Control Frame is shown in Figure 14–17. An explanation of the Characteristic Symbols is given in Figure 14–18. The supplementary material conditions of datum symbols are shown in Figure 14–19.

Once the symbol button located in the **Sym** column in the Geometric Tolerance dialog box is chosen, AutoCAD displays the Symbol dialog box, as shown in Figure 14–20. Select one of the available symbols and specify appropriate tolerance values in the **Tolerance** field.

Figure 14–17

Feature control frame

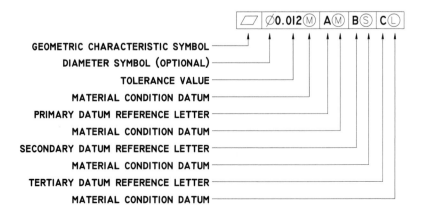

GEOMETRIC CHARACTERISTIC SYMBOL
DIAMETER SYMBOL (OPTIONAL)
TOLERANCE VALUE
MATERIAL CONDITION DATUM
PRIMARY DATUM REFERENCE LETTER
MATERIAL CONDITION DATUM
SECONDARY DATUM REFERENCE LETTER
MATERIAL CONDITION DATUM
TERTIARY DATUM REFERENCE LETTER
MATERIAL CONDITION DATUM

Maximum Materials Condition (MMC) means that a feature contains the maximum material permitted within the tolerance dimension, that is, the minimum hole size, maximum shaft size.

Least Material Condition (LMC) means a feature contains the least material permitted within the tolerance dimension, that is, maximum hole size, minimum shaft size.

Regardless of Feature Size (RFS) applies to any size of the feature within its tolerance. This is more restrictive. For example, RFS does not allow the tolerance of the center-to-center dimension of a pair of pegs fitting into a pair of holes greater leeway if the peg diameters are smaller and/or the holes are bigger, whereas MMC does.

The diameter symbol, Ø, is used in lieu of the abbreviation DIA.

 Figure 14-18

Geometric characteristics symbols

CHARACTERISTIC SYMBOLS			
FORM		▱	FLATNESS
		▭	STRAIGHTNESS
		◯	ROUNDNESS
		⌒⌿	CYLINDRICITY
		⌒	LINE PROFILE
		◠	SURFACE PROFILE
		∠	ANGULARITY
		//	PAREALLELISM
		⊥	PERPENDICULARITY
LOCATION		◎	CONCENTRICITY
		⊕	POSITION
		⊜	SYMMETRY
RUNOUT		↗	CIRCULAR RUNOUT
		↗↗	TOTAL RUNOUT

 **Figure 14-19**

Material conditions of datum symbols

MATERIAL CONDITIONS OF DATUM SYMBOLS	
Ⓜ	MAXIMUM MATERIAL CONDITIONS (MMC)
Ⓢ	REGARDLESS OF FEATURE SIZE
Ø	DIAMETER

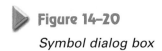 **Figure 14-20**

Symbol dialog box

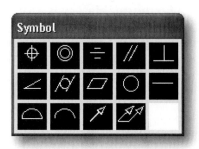

Geometric tolerancing is becoming widely accepted. It is highly recommended that you study the latest drafting texts concerning the significance of the various symbols, so that you will be able to apply geometric tolerancing properly.

EDITING DIMENSION TEXT

AutoCAD allows you to edit dimensions with modify commands and grip editing modes. In addition, AutoCAD provides two additional Dimension commands specifically designed to work on dimension text objects, DIMEDIT and DIMTEDIT. You can also edit dimensions with grips.

DIMEDIT Command

 Figure 14-21

Invoking the DIMEDIT command from the Dimension toolbar

The options that are available in the DIMEDIT command allow you to replace the dimension text with new text, rotate the existing text, move the text to a new location, and if necessary, restore the text to its home position, which is the position defined by the current style. In addition, these options allow you to change the angle of the extension lines (normally perpendicular) relative to the direction of the dimension line (by means of the Oblique option). Invoke the DIMEDIT command (see Figure 14-21), AutoCAD prompts:

> **dimedit**
>
> Enter type of dimension editing [Home/New/Rotate/Oblique]<Home> *(press* ENTER *for the Home option or choose one of the available options from the Shortcut menu)*

The **Home** option returns the dimension text to its default position. AutoCAD prompts:

> Select objects: *(select the dimension objects and press* ENTER*)*

The **New** option allows you to change the original dimension text to the new text. AutoCAD opens the Multiline Text Editor dialog box. Enter the new text and choose **OK**. AutoCAD then prompts:

> Select objects: *(select the dimension objects for which the existing text will be replaced by the new text)*

The **Rotate** option allows you to change the angle of the dimension text. AutoCAD prompts:

Specify angle for dimension text: *(specify the rotation angle for text)*
Select objects: *(select the dimension objects for which the dimension text has to be rotated)*

The **Oblique** option adjusts the obliquing angle of the extension lines for linear dimensions. This is useful for keeping the dimension parts from interfering with other objects in the drawing. Also, it is an easy method by which to generate the slanted dimensions used in isometric drawings. AutoCAD prompts:

Select objects: *(select the dimension objects)*
Enter obliquing angle (press ENTER for none): *(specify the angle or press ENTER)*

DIMTEDIT Command

Figure 14–22

Invoking the
DIMTEDIT *command*
from the
Dimension toolbar

The DIMTEDIT command is used to change the location of dimension text (with the Left/Right/Center/Home options) along the dimension line and its angle (with the Rotate option). Invoke the DIMTEDIT command (see Figure 14–22), AutoCAD prompts:

Select dimension: *(select the dimension object to modify)*

A preview image of the dimension selected is displayed on the screen, with the text located at the cursor. You will be prompted:

Specify new location for dimension text or ⊡: *(specify a new location for the dimension text or select one of the available options from the Shortcut menu)*

By default, AutoCAD allows you to position the dimension text with the cursor, and the dimension updates dynamically as it drags.

The **Center** option will cause the text to be drawn at the center of the dimension line.

The **Left** option will cause the text to be drawn toward the left extension line.

The **Right** option will cause the text to be drawn toward the right extension line.

The **Home** option returns the dimension text to its default position.

The **Angle** option changes the angle of the dimension text. AutoCAD prompts:

Specify angle for dimension text: *(specify the angle)*

The angle specified becomes the new angle for the dimension text.

EDITING DIMENSIONS WITH GRIPS

If the Grips feature is set to ON, you can select an associative dimension object, and its grips will be displayed at strategic points. The grips will be located at the object ends of the extension lines, the intersections of the dimension and extension lines, and at the insertion point of the dimension text. In addition to the normal grip editing of the dimension as a group (rotate, move, copy, etc.), each grip can be selected for editing the configuration of the dimension as follows: Moving the object end grip of an extension line will move that specified point, making the value change accordingly. Horizontal and vertical dimensions remain horizontal and vertical. Aligned dimensions follow the alignment of the relocated point. Moving the grip at the intersection of the dimension line and one of the extension lines causes the dimension line to be nearer to or farther from the object dimensioned. Moving the grip at the insertion point of the text does the same as the intersection grip and also permits you to move the text back and forth along the dimension line.

CHANGING ASSOCIATION

The status of a dimension's associativity can be undone and redone by the DIMDISASSOCI-ATE and DIMREASSOCIATE commands respectively.

DIMDISASSOCIATE removes associativity from selected dimensions. It filters the selection set to include only associative dimensions that are not on locked layers, and that are not in a space different from the current space (for example, if model space is active, associative dimensions in paper space are excluded). DIMDISASSOCIATE then disassociates these dimensions and reports the number of dimensions that are filtered out and the number that are disassociated. Invoke the DIMDISASSOCIATE command by typing at the On-Screen cursor. AutoCAD prompts:

Select objects: *(select the dimension objects and press* ENTER *to complete the selection)*

AutoCAD removes the associativity of the selected dimensions.

DIMREASSOCIATE associates selected dimensions to geometric objects. With DIMREASSOCIATE, a non-associative dimension can be associated to geometric objects, or the existing associations in an associative dimension can be changed. Each selected dimension is highlighted in turn, and prompts for association points appropriate for the selected dimension are displayed. A marker is displayed for each association point prompt. If the definition point of the current dimension is not associated to a geometric object, the marker appears as an X; if the definition point is associated, the marker appears as an X inside a box. Press ESC to terminate the command without losing the changes that were already specified. Use UNDO to restore the previous state of the changed dimensions. Invoke the DIMREASSOCIATE command by typing at the On-Screen cursor. AutoCAD prompts:

Select objects: *(select the dimension objects and press* ENTER *to complete the selection)*

Follow through the prompts by selecting the points to re-associate the dimensions.

drawing exercises

Open the Exercise Manual PDF file for Chapter 14 on the accompanying CD for discipline-specific exercises.

review questions

1. Dimension types available in AutoCAD include:
   a. *Linear.*   **d.** *Radius.*
   b. *Angular.*   **e.** *All of the above.*
   c. *Diameter.*

2. The associative dimension drawn with the DIMASO variable set to ON has all of its separate parts drawn as separate objects.
   a. *True*   **b.** *False*

3. The Linear Dimensioning command allows you to draw horizontal, vertical, and aligned dimensions.
   a. *True*   **b.** *False*

4. To draw a linear dimension you must (1) specify the first extension line origin, (2) locate the dimension line, and then (3) specify the second extension line.
   a. *True*   **b.** *False*

5. The Angular Dimension command allows you to draw angular dimensions between two parallel lines.
   a. *True*   **b.** *False*

6. By default, the dimension text for a radius dimension is preceded by:
   a. *Radius.*
   b. *Rad.*
   c. *R.*

7. Using the properties (Modify Properties) command will allow you to override the dimensioning system variable settings for a single dimension, without modifying the base dimension style.
   a. *True*   **b.** *False*

8. To turn the associative dimensioning variable on, use the ____ system variable.
   a. *DIMASO*   **c.** *Both a and b*
   b. *DIMCIA*   **d.** *None of the above*

9. You can enter the dimensioning commands:
   a. *From the keyboard.*   **c.** *Both a and b.*
   b. *From the Dimensioning toolbar.*   **d.** *None of the above.*

10. If you pick an associative dimension object, grips will appear at:

    a. *The extension line endpoints.*
    b. *A given distance from the object.*
    c. *The intersection of the dimension and extension lines.*
    d. *Both a and c.*

11. When editing dimension text, you can select:

    a. *Home text.*   **c.** *DDEDIT.*
    b. *Oblique.*   **d.** *All of the above.*

12. The DIMEDIT command allows you to:

    a. *Move existing text to a new location.*   **c.** *Replace dimension text with new text.*
    b. *Rotate existing text.*   **d.** *All of the above.*

13. A leader is used for:

    a. *General annotation.*   **c.** *Pointing from the text to an arc.*
    b. *Pointing from the text to a circle.*   **d.** *All of the above.*

14. Baseline dimensioning is used to dimension multiple points from a single datum baseline.

    a. *True*   **b.** *False*

15. While using the baseline dimensioning variable, you must select your starting point each time you select a new ending point.

    a. *True*   **b.** *False*

creating and modifying layouts

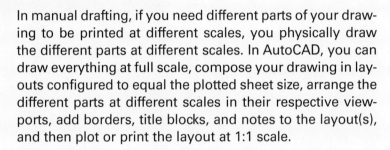

introduction

In manual drafting, if you need different parts of your drawing to be printed at different scales, you physically draw the different parts at different scales. In AutoCAD, you can draw everything at full scale, compose your drawing in layouts configured to equal the plotted sheet size, arrange the different parts at different scales in their respective viewports, add borders, title blocks, and notes to the layout(s), and then plot or print the layout at 1:1 scale.

After completing this chapter, you will be able to do the following:

- Plan the plotted sheet

- Set up a layout

- Create and modify a layout

- Create floating viewports

- Scale viewport contents relative to the layout

- Control the visibility of layers within viewports

OVERVIEW OF A LAYOUT – THE LONE ARRANGER

One of the most powerful and unique features of AutoCAD is the option to work on your drawing in two different environments: model space and paper space. AutoCAD allows you to plot a drawing from model space as well as paper space. In paper space you can create multiple layouts in which you arrange multiple floating viewports for displaying parts of the model space design to be plotted. The purpose of using a paper space layout with viewports is to make it easy to produce plotted sheets with AutoCAD, rearranging and scaling objects that are drawn in model space and adding non-object elements in paper space.

369

---

note Reference to viewports in layouts means floating viewports. The distinction between floating viewports and model space viewports is explained in the section on viewports in this chapter.

The combination of paper space and viewports is a special and powerful application for producing the most commonly used communication tool in the architect/engineer's repertoire, a set of paper drawings, traditionally referred to as "blueprints."

AutoCAD allows you to:

1. Create or import a page setup, applying it to a layout that is the exact width and height of the desired plotted paper sheet.

2. Create (or attach, import or otherwise position) a border, title block, symbols, tables and any other annotation or AutoCAD objects in full size (12" = 1'-0") on the sheet.

3. Open windows of specified sizes and locations (referred to as viewports) at specified scales on the sheet for viewing desired parts of objects drawn in model space.

4. Plot the Layout to 1:1 scale.

When configured, the layout will be like a separate drawing sheet that represents the final printed sheet with a border, title block, and note information, making it easy to arrange objects to their desired scale on the sheet (in their viewports) and a simple task to plot.

Planning to Plot

Figure 15–1 shows a model space drawing of a residential elevation and floor plan without a border, title block, or any notation except dimensions in the plan view. In Figure 15–2, by using the Layout feature, the views are juxtaposed for conventional arrangement and a border, title block , view titles, and text tables are added in paper space. The two figures illustrate the application of model space for creating the design objects and paper space for arranging the design objects on a plot-friendly sheet with border, title block, view titles, and annotation, accomplished in a single drawing file.

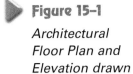

Figure 15-1

Architectural Floor Plan and Elevation drawn in model space

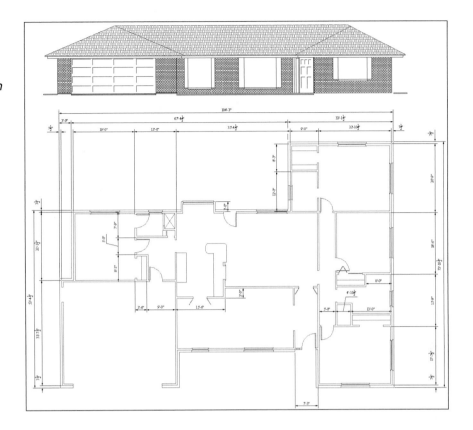

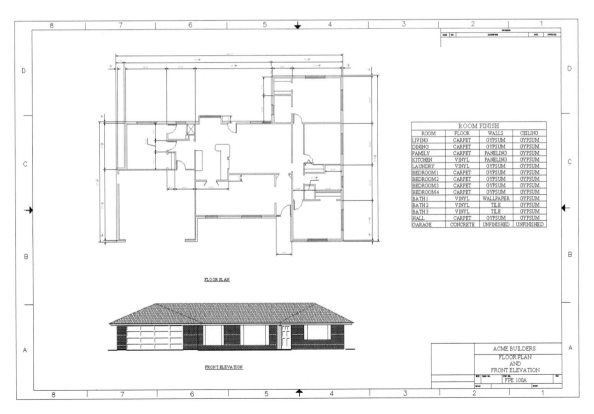

Figure 15-2

Architectural Floor Plan and Elevation in layout format configured for plotting

Chapter 2 gave an example of setting up a drawing for plotting from model space. This chapter describes procedures for setting up a drawing for plotting/printing from layouts. The problem with plotting from model space is that the objects must be located and sized in the coordinate system according to how they will be arranged for plotting. Even with AutoCAD's true-size capability, plotting from model space requires parts that need to be plotted at different scales be drawn at different scales. This is not necessary when the Layout/Paper Space feature is utilized.

Setting up the drawing for plotting from layouts is not as restrictive as it is from model space. Some thought, however, must still be given to the concept of scale, which is the ratio of true size to the size plotted. In other words, before you start drawing, you should have an idea of which parts of the drawing must be plotted at what scale. The example in Chapter 2 showed how to determine the appropriate scale to use so objects will fit on the desired sheet size. That example can be used as a guide for plotting to scale in layouts, except that each view of the objects drawn in model space can be treated as a separate drawing on the sheet, like pictures on the page of a scrapbook.

A Space of Your Own

AutoCAD provides two primary "spaces" in which to work: model space and paper space. When starting out in AutoCAD, it is best to think of these two spaces as "areas" in their own respective planes (model area and paper area or model plane and paper plane) rather than spaces (which suggest mainly three-dimensional use).

In model space, you draw, view, and edit design objects. You draw at 1:1 scale as mentioned in Chapter 2, and you decide whether one unit represents one millimeter, one centimeter, one inch, one foot, or whatever unit is most convenient or customary in your business.

Paper space is a 2D environment used for arranging various views (floating viewports) of what was drawn in model space. It represents the paper on which you arrange the drawing prior to plotting. With AutoCAD, single or multiple paper space layouts can be easily configured and managed in a single drawing file.

By default, every initialized layout has an unnamed page setup associated with it. Once you create a layout, you can change the settings for the layout's page setup, which includes the plot device settings and other settings that affect the appearance and format of the output. The settings you specify in the page setup are stored in the drawing file with the layout. You can modify the settings of a page setup at any time and, if deemed necessary, give the setup a name and save it. You can apply a named page setup saved with one layout to another layout. This creates a new layout with the same configuration as the first one.

Similar to creating tiled viewports in model space (see Chapter 4 for a detailed explanation), you can also create viewports in paper space called floating viewports. In a layout, you can create multiple, overlapping, contiguous, or separated untiled viewports. By default, AutoCAD sets up one floating viewport in each layout. Consider viewports as windows with a view into model space that you can move and resize. AutoCAD treats viewports in a layout like any other object, such as a line, arc, or text object. When you are working in a layout, you can switch from paper space to model space by double-clicking inside one of the floating viewports. If you double-click anywhere outside the viewport, AutoCAD will switch you back to paper space. You must have at least one floating viewport in paper space to view the model.

After arranging the views and scaling appropriately, you can plot the drawing from paper space at 1:1 (full scale). Paper space allows you to plot the drawing in WYSIWYG (What You See Is What You Get) mode. You can switch between model space and paper space by selecting the appropriate tabs provided at the bottom of the drawing window. Model space can be accessed from the Model tab. Selecting one of the available layout tabs will access paper space.

The Default Layouts

By default, a new drawing (using the *acad.dwt* file as a template) starts with two layouts, named Layout1 and Layout2. (A drawing started with another template might have only one layout, for example the *ANSI B -Named Plot Styles.dwt* that is configured with an ANSI B Title Block on an 11 x 17 sheet.) However, each of the two default layouts in the drawing started with *acad.dwt* represents a sheet in landscape mode 11 units by 8.5 units with a dashed rectangle 10.5 units by 8 units outlining the expected printable area.

Figure 15–3 shows how views of a design object drawn in model space will appear with the Model tab selected (on the left) and with the Layout1 tab selected (on the right). Layout1 has the basic elements of a layout: a paper space drawing sheet with specific width and height, and a viewport through which you can view design objects that are drawn in model space. If you erase the viewport, model space objects would not be visible.

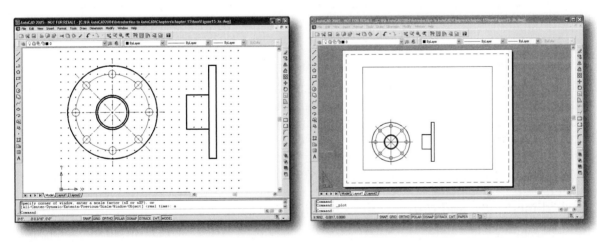

 Figure 15-3

Two views of object in model space (11 x 8.5 limits) and in Layout1

Figure 15–4 shows the basic parts of the default layout with only one viewport and nothing drawn in paper space.

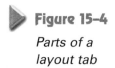

Figure 15–4

Parts of a layout tab

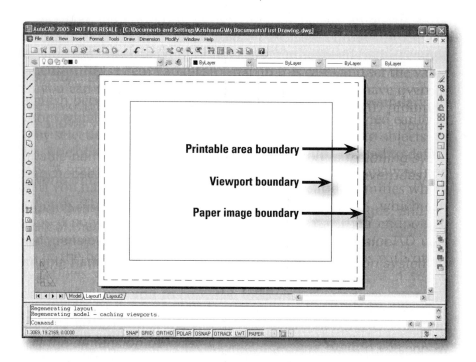

To repeat, by default, every initialized layout has an unnamed page setup associated with it. If necessary, you can change the layout's page setup through the Page Setup Manager (explained in detail in "Reconfiguring the Layout with Page Setup" later in this chapter). You can modify the settings of a page setup at any time.

Layout(s) by way of a Template

The most common procedure for creating a new drawing is to use a template file that has already been configured for the application/discipline in effect. The template will contain one or more layouts, each created at the desired sheet size and normally having a border, title block, revision history table, or other non-object element drawn in paper space on it. Non-object elements are things like borders, title blocks, dimensions, callouts, etc. versus object elements that represent real objects like walls, pipes, switches, streets, etc. A template drawing will also have layers, system variables, and styles for text, dimensions, and other features configured that conform to the standards of the proposed set of drawings.

The model space objects shown in Figure 15–3 will be used as an example of how to use a layout contained in an existing template drawing file (ANSI-A Color Dependent Plot Style template) to produce the desired plot. When the new drawing is created, the starting view will be of the layout named ANSI A Title Block, as shown in Figure 15–5a . After you switch to model space (selecting the Model tab in the Status bar) and draw the objects shown in Figure 15–3, when you switch back to the ANSI A Title Block Layout tab, the objects will appear in the one viewport as shown in Figure 15–5b, and you can plot the drawing from paper space at 1:1 scale. The outline of the single viewport in this layout is not very distinguishable because it coincides with the inside lines of the border/title block. If you double-click inside the viewport, you will be switched to model space in the viewport while still in the layout. (This is different from switching to model space by using the Model tab.) When you do this, the outline of the viewport becomes a heavy line and is more visible, as shown in Figure 15–5b.

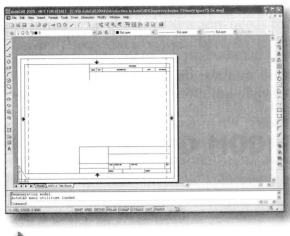

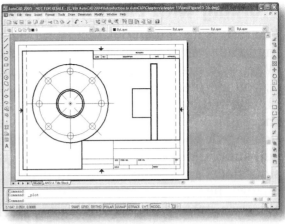

 Figure 15-5a

Newly created ANSI-A A Title Block Layout

 Figure 15-5b

Layout with the objects

VIEWPORTS IN PAPER SPACE

As mentioned earlier, in a layout, you can create multiple, overlapping, contiguous, or separated floating viewports, as shown in Figure 15–6. To repeat, viewports are configurable windows from paper space (a layout) with a view into the model space design that you can move and resize. You can use any of the standard AutoCAD modify commands, such as MOVE, COPY, STRETCH, SCALE, and ERASE, to manipulate the floating viewports. For example, you can use the MOVE command to grab one viewport and move it around the screen without affecting other viewports. A viewport can be of any size and can be located anywhere in the layout. You must have at least one floating viewport to view the objects drawn in model space.

 **Figure 15-6**

Floating viewports in a layout

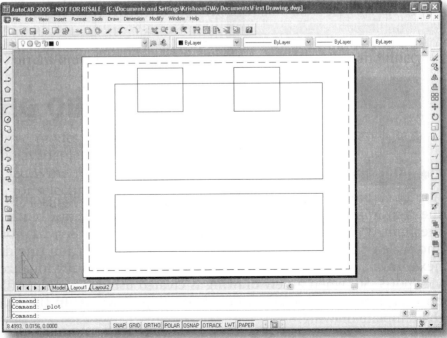

two view
allow for
this, the
border/ti
Window
where th
the bord
it is in F

 Figur

Using
optio
view

Creating

You can
floating
and prop

you are
the boun
been cre
not affec

Creating

 Figur

Choo
View

AutoCA
that can

Speci

Speci

The viewport will be created on the active layer and if necessary, you can move, copy, rotate, scale or stretch the viewport or place it on a different layer just like modifying any other object.

Creating a Floating Polygonal Viewport

 Figure 15-10

Choosing Polygonal Viewport from the Viewports toolbar

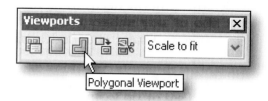

To create an irregularly shaped floating viewport, choose Polygonal Viewport from the Viewports toolbar (see Figure 15–10) and AutoCAD prompts:

Specify start point: *(specify first point to create as irregularly shaped floating viewport)*

Specify next point or ⊥: *(specify next point or choose one of the available options from the Shortcut menu)*

The Arc option adds arc segments to the polygonal viewport.

The Length option draws a line segment of a specified length at the same angle as the previous segment. If the previous segment is an arc, AutoCAD draws the new line segment tangent to that arc segment.

The Undo option removes the most recent line or arc segment added to the polygonal viewport.

The Close option closes the polygon to create the polygonal viewport.

Converting an Object to a Floating Viewport

 Figure 15-11

Choosing Convert Object to Viewport from the Viewports toolbar

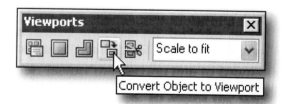

AutoCAD allows you to create a viewport from a closed polyline, ellipse, spline, region, or circle. The polyline you specify must be closed and contain at least three vertices. It can be self-intersecting, and it can contain an arc as well as line segments. AutoCAD prompts:

Select object to clip viewport: *(Select an object)*

Creating Multiple Floating Viewports

 Figure 15–12

Choosing Display Viewports Dialog from the Viewports toolbar

To create multiple floating viewports, open the Viewports dialog box (see Figure 15–13) by choosing the Display Viewports dialog from the Viewports toolbar. AutoCAD lists standard viewport configurations. Choose the name of the configuration you want to use from the **Standard viewports** list. AutoCAD displays the corresponding configuration in the **Preview** window. The Setup menu specifies either a 2D or a 3D setup. When you select 2D, the new viewport configuration is initially created with the current view in all of the viewports. When you select 3D, a set of standard orthogonal 3D views is applied to the viewports in the configuration. The **Preview** section displays a preview of the viewport configuration you select and the default views assigned to each individual viewport in the configuration. Choosing **Change view to** replaces the view in the selected viewport with the view you select from the list. You can choose a named view, or if you have selected 3D setup, you can select from the list of standard views. Use the Preview area to see the choices. After choosing the viewport configuration and setting corresponding values, choose **OK** to close the dialog box; AutoCAD prompts:

Specify first corner or ⊡ *(specify the first corner to define selected viewport configuration or select Fit to create the selected viewport configuration to fit the paper size)*

Specify opposite corner: *(specify opposite corner to define selected viewport configuration)*

 Figure 15–13

Viewports dialog box

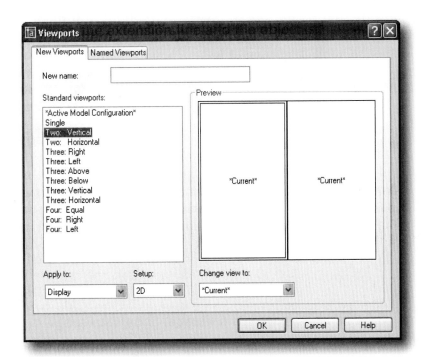

Using the Viewports dialog box for creating new viewports will be the method used in the example of transforming the layout shown in Figure 15–5b to the plot-friendly layout of Figure 15–7. Select the points shown in Figure 15–14a when prompted to specify the corners to determine the rectangle outlining the two vertical viewports. The model space design objects will be displayed the same in both viewports (see Figure 15–14b).

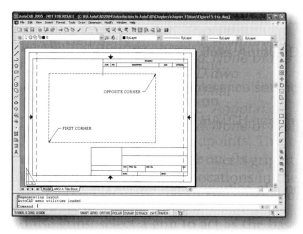

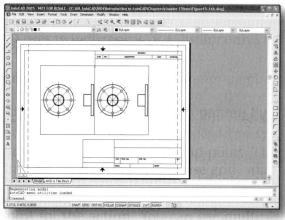

 Figure 15–14a

Selection of points to create two vertical viewports

 Figure 15–14b

Two vertical viewports with the objects

Modifying Floating Viewports

As mentioned above, once you create a viewport, you can change its size and properties, and reposition it as needed. If you want to change the shape or size of a layout viewport, you can use grips to edit the vertices just as you edit any object with grips.

By using grips and/or the STRETCH and MOVE commands, the left viewport is widened and the right viewport is moved to the right and made narrower, as shown in Figure 15–15. The model space design objects are visible and will be arranged properly within each viewport in later steps.

 Figure 15–15

Resizing and rearranging the two viewports

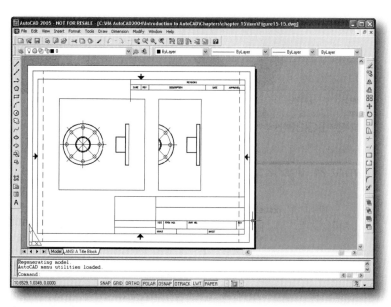

You can also redefine the boundary of a layout viewport by using the VPCLIP command and maximize the viewport by using the VPMAX command. You can control the display of objects in a viewport by changing the Display Viewport Objects setting, and control the setting of the locking feature to prevent the zoom scale factor in the selected viewport from being changed when working in model space.

Clipping an Existing Viewport

 Figure 15–16

Choosing Clip Existing Viewport from the Viewports toolbar

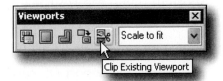

The VPCLIP command (see Figure 15–16) allows you to clip a floating viewports to a user-drawn boundary. AutoCAD reshapes the viewport border to conform to a user-drawn boundary. To clip a viewport, you can select an existing closed object, or specify the points of a new boundary. AutoCAD prompts:

Select viewport to clip: *(select viewport to clip)*

Select clipping object or ⬇: *(select clipping object or select one of the available options from the Shortcut menu)*

If you select an object for clipping, AutoCAD converts the object to a clipping boundary. Objects that are valid as clipping boundaries include closed polylines, circles, ellipses, closed splines, and regions.

The Polygonal option allows you to create a clipping boundary. You can draw line segments or arc segments by specifying points to create a polygonal clipping boundary.

The Delete option deletes the clipping boundary of a selected viewport. This option is available only if the selected viewport has already been clipped. If you clip a viewport that has been previously clipped, the original clipping boundary is deleted and the new clipping boundary is applied.

Maximizing the Floating Viewport

 Figure 15–17

Choosing Maximize Viewport from the Shortcut menu

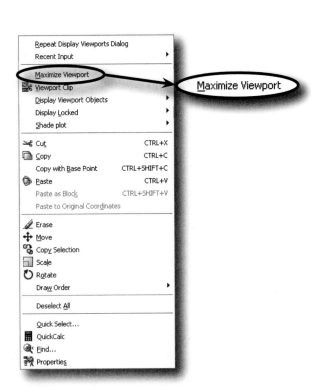

Maximize Viewport (available for selection on the Shortcut menu when an viewport is selected, as shown in Figure 15–17) causes the selected viewport in the current layout to fill the screen drawing area, making the entire drawing area accessible for viewing and editing. The size of the area displayed depends on the zoom factor in effect. When the viewport has been maximized, the Minimize Viewport option available in the Shortcut menu returns the display to the previous layout state. You can also maximize or minimize viewports from the button located on the Status bar.

Controlling the Display of Objects in a Floating Viewport

The Display Viewport Objects selection, available on the Shortcut menu when a viewport is selected, controls the display of objects in the selected viewport. When Off is selected, objects in the selected viewport are not visible, and the Viewport cannot be selected when switching viewports in model space in the current layout. When On (default) is selected, AutoCAD turns on a viewport, making it active and making its objects visible.

Locking the Floating Viewport

The Display Locked selection (on/off), available on the Shortcut menu when a viewport is selected, prevents or enables respectively the zoom scale factor in the selected viewport from being changed when working in model space

Scaling Views Relative to Paper Space

AutoCAD allows you to scale Viewport objects relative to paper space, which establishes a consistent scale for each displayed view. To accurately scale the plotted drawing, you must scale each viewport relative to paper space. Usually the layout is plotted at a 1:1 ratio. The ratio is determined by dividing the paper space units by the model space units. The scale factor of model space design objects in a viewport can be set with the XP option of the ZOOM command while model space is active in that viewport. For example, entering **1/24xp** or **0.04167xp** (1/24 = 0.01467) in response to the ZOOM command prompt will display an image to a scale of 1/2" = 1' 0", which is the same as 1:24 or 1/24. You can also change the plot scale of the viewport using the Viewport Scale Control on the Viewports toolbar (see Figure 15–18).

 Figure 15–18

Viewport Scale Control on the Viewports toolbar

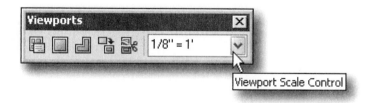

In the case of the viewports in the example drawing shown in Figure 15–15, the views need to be set to half scale. In this case, first double-click in one of the viewports to make it active in model space and then enter a scale factor in the Viewport Scale Control box of the Viewports toolbar of 0.5 or choose 1:2 from the drop-down box. AutoCAD displays 6" = 1', and the objects are rescaled in the selected viewport. Repeat the procedure for the second viewport, and the result will appear as shown in Figure 15–19.

Figure 15-19

Objects in viewports rescaled

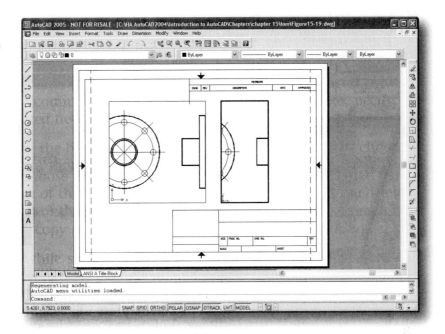

Centering Model Space Objects in a Viewport

Next, in order to center the front view of the object in the left viewport, respond to the ZOOM Center option with the coordinates **3.5,5.0**, which is the center of the circle. Repeat the procedure in the right viewport using the coordinates **9.5,5.0**. Specifying the same Y coordinate for centering the model spaces in both viewports assures that the objects will line up horizontally. Some practice (and some trial and error) is needed to size the viewports and center the model space design objects to assure that the desired object views (and only the desired ones) are visible in the appropriate viewports. The result will appear as shown in Figure 15–20.

Figure 15-20

Objects in viewports centered

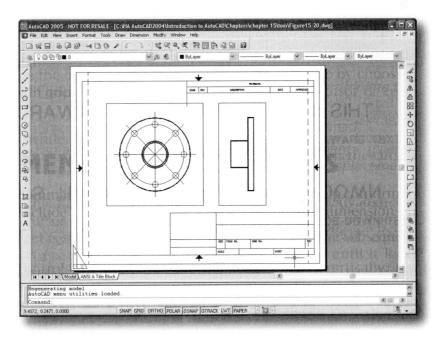

Hiding Viewport Borders

After the viewports are scaled and the objects are centered, double-click outside the viewports to return to paper space. Turn the layer named Viewports off (initially two viewports were created on layer Viewports) and the result will appear as shown in Figure 15–21.

Figure 15-21

Result after the Viewports layer is set to off

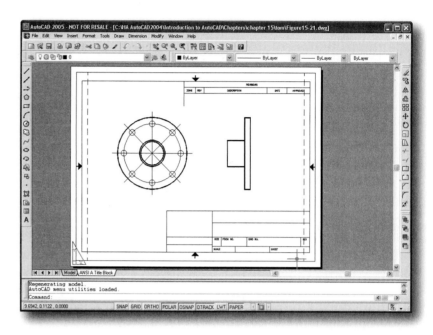

While in paper space, the drawing name and other information can be entered in the title block as appropriate. The views can be named, and the objects can also be dimensioned as shown in Figure 15–22.

Figure 15-22

Completed layout

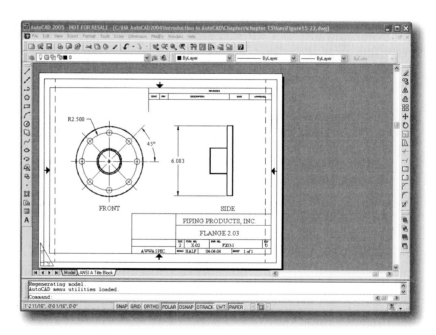

Controlling the Visibility of Layers within Viewports

The Layer Properties Manager dialog box controls the visibility of layers in a single viewport or in a set of viewports. This enables you to select a viewport and freeze a layer in it while still allowing the contents of that layer to appear in another viewport. Figure 15–23 shows two viewports containing the same view of the drawing, but in one viewport the layer containing the dimensioning is set to on, and in the other the dimensioning layer is set to off.

Figure 15–23

One viewport with Dimlayer ON and the other with Dimlayer OFF

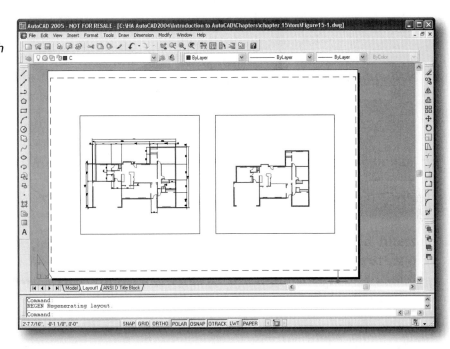

The **Current VP Freeze** column available only from a layout tab (eleventh column from the left as shown in Figure 15–24) freezes selected layers in the current layout viewport. You can freeze or thaw layers in the current viewport without affecting layer visibility in other viewports. **Current VP Freeze** is an override to the Thaw setting in the drawing. In other words, you can freeze a layer in the current viewport if it is thawed in the drawing, but you can't thaw a layer in the current viewport if it is frozen or off in the drawing. A layer is not visible when it is set to Off or Frozen in the drawing.

Figure 15–24

Layer Properties Manager dialog box

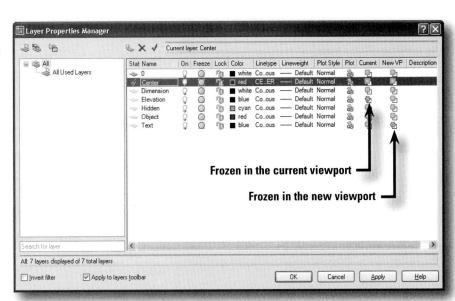

The **New VP Freeze** column available only from a layout tab (twelfth column from the left as shown in Figure 15–24) freezes selected layers in new layout viewports. For example, freezing the TEXT layer in all new viewports restricts the display of text on that layer in any newly created layout viewports but does not affect the TEXT layer in existing viewports. If you later create a viewport that requires text, you can override the default setting by changing the current viewport setting.

In Figure 15–24, the layers Dimension, Elevation, and Hidden are frozen in the current viewport, and Object and Text are frozen in all the new viewports.

CREATING A NEW LAYOUT

 Figure 15–25

Choosing New Layout from the Layouts toolbar

Choosing New Layout from the Layouts toolbar (see Figure 15–25) creates a new layout tab. Up to 255 layouts can be created in a single drawing. AutoCAD prompts:

Enter name of new layout <Layout#>: *(specify a layout name or press ENTER to accept the default name)*

Layout names must be unique. Layout names can be up to 255 characters long and are not case sensitive. Only the first 31 characters are displayed on the tab.

AutoCAD establishes a new layout, adding its tab at the bottom of the screen. By default, every initialized layout has a unnamed page setup associated with it. Once you create a layout, you can change the settings for the layout's page setup with the help of Page Setup Manager dialog box (described later in this chapter), which includes the plot device settings and other settings that affect the appearance and format of the output. The settings you specify in the page setup are stored in the drawing file with the layout.

---

note If you want the Page Setup Manager to be displayed each time you begin a new drawing layout, select the Show Page Setup Manager for New Layouts option on the Display tab in the Options dialog box. If you don't want a viewport to be automatically created for each new layout, clear the Create Viewport in New Layouts option on the Display tab in the Options dialog box.

 Figure 15–26

Choosing Layout from Template from the Layouts toolbar

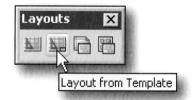

Choosing Layout from Template from the Layouts toolbar (see Figure 15–26) creates a new layout tab based on an existing layout in a template (*.dwt*), drawing (*.dwg*), or drawing interchange (*.dxf*) file. AutoCAD displays a standard file selection dialog box to select a file. Once you select a file, AutoCAD displays the Insert Layouts dialog box as shown in Figure 15–27, which displays the layouts saved in the selected file. After you select a layout, the layout and all objects from the specified template or drawing file are inserted into the current drawing.

Figure 15–27

Insert Layout(s) dialog box

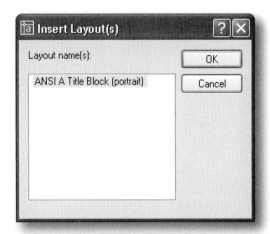

Additional options are available in the LAYOUT command. Copy creates a new layout by copying an existing layout, Delete deletes an existing layout, Rename renames an existing layout, and SaveAs saves a layout as a drawing template (*.dwt*) file without saving any unreferenced symbol table and block definition information. You can access all the available options from the Shortcut menu that appears when you right-click the name of the layout tab.

RECONFIGURING THE LAYOUT WITH PAGE SETUP

Figure 15–28

Choosing Page Setup Manager from the Layouts toolbar

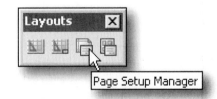

As mentioned earlier, by default, every initialized layout has a unnamed page setup associated with it. You can modify the settings for the layout's page setup with the help of Page Setup Manager dialog box. AutoCAD displays the Page Setup Manager dialog box, as shown in Figure 15–29, when you choose Page Setup Manager from the Layouts toolbar (see Figure 15–28).

Figure 15-29

Page Setup Manager dialog box

AutoCAD displays the current layout name in the **Current layout** box. In the **Page setups** section of the Page Setup Manager dialog box, **Current page setup** displays the name of the page setup that is applied to the current layout. If the name is displayed as <None>, an unnamed page setup is assigned to the current layout. The **Page setups** section lists the page setups that are available to apply to the current layout. If the Page Setup Manager is opened from a layout, the current page setup is selected by default. The list includes the named page setups and layouts that are available in the drawing. Layouts that have a named page setup applied to them are enclosed in asterisks, with the named page setup in parentheses; for example, *ANSI A Title Block (portrait)*. You can double-click a page setup or a layout name that has an unnamed page setup associated in this list to set it as the current page setup for the current layout.

Figure 15-29 lists three layouts: *ANSI A Title Block (portrait)*, *Layout1* and *Layout2* and one page setup: ANSI A Title Block (portrait). **Current layout** is listed as Layout2. Layout1 and Layout2 have unnamed page setups assigned to them.

Changing the Current Page Setup

To change the page setup for the current layout, first select the named page setup or a layout that has an unnamed page setup associated and choose **Set Current** to set the selected page setup as the current page setup.

Modifying the Page Setup

To 3 the page setup assigned to the current layout, choose **Modify**. AutoCAD displays the Page Setup dialog box as shown in Figure 15-30.

Figure 15-30

Page Setup dialog box

The Page Setup dialog box specifies page layout and plotting device settings. The page setup settings that you specify are stored with the layout and can be applied to other layouts or imported into other drawings.

Name of the Page Setup

The **Page setup** section displays the name of the page setup assigned to the selected layout, if any. If the name is listed as <None>, an unnamed page setup is assigned to the selected layout.

Configuration of Printer/Plotter

The **Printer/plotter** section displays the currently configured plotting device. **Name** lists the system printers or PC3 files that are available to select for plotting. **Plotter** lists the plotter selected in the currently selected page setup. **Where** gives the port to which it's connected, or its network location. **Properties** allows access to the Plotter Configuration of the currently configured plotting device. **Description** lists information about the output device specified in the currently selected page setup.

Plotstyle Table (Pen Assignments)

The **Plot style table** section sets the plot style table, edits the plot style table, or creates a new plot style table. Choose **Display plot styles** to have the properties of plot styles assigned to objects displayed on the screen. Choose **Edit** to cause the Plot Style Table Editor to be displayed to view or modify plot styles for the currently assigned plot style table. Refer to Chapter 16 for a detailed explanation about creating and modifying plot style tables.

Shaded Viewport Options

The **Shaded viewport options** section specifies how shaded and rendered viewports are plotted and determines their resolution level with corresponding dpi.

The **Shade Plot** list box displays how views are plotted, and it is specified through the Properties dialog box for the selected viewport.

The **Quality** option menu specifies the resolution at which shaded and rendered viewports are plotted.

The **DPI** text box specifies the dots per inch for shaded and rendered views when Custom is selected in the **Quality** option menu, up to the maximum resolution of the current plotting device.

Paper Size

The **Paper size** section allows you to select the paper size to plot. Select the paper size to plot from the **Paper size** list box for the selected plotting device. Actual paper sizes are indicated by the width (X axis direction) and height (Y axis direction). A default paper size is set for the plotting device when you create a plotter configuration file with the Add-a-Plotter wizard.

Plot Area

The **Plot area** section allows you to select the area to be plotted. You can choose from the **What to plot** list box. Display causes the area displayed on the screen to be plotted. Layout causes everything within the margins of the specified paper size (with the origin calculated from 0,0 in the layout) to be plotted. Extents causes the portion of the current space of the drawing that contains objects to be plotted. Window allows you to specify a window on the screen and plot the objects that are inside the window. The lower left corner of the window becomes the origin of the plot.

Plot Scale

The **Plot scale** section controls the plot area. Select **Fit to paper** to cause the plot to fit within the selected paper size. To plot to a specific scale, select the plot scale from the **Scale** list box. You can create a custom scale by selecting Custom in the **Scale** list box and specify the number of inches (or millimeters) equal to the number of drawing units in the text fields. Select **Scale lineweights** to plot in proportion to the plot scale. Otherwise AutoCAD will plot the objects with assigned lineweights. Lineweights normally specify the linewidth of printed objects and are plotted with the linewidth size regardless of the plot scale.

---

note   If the Layout option is specified in **Plot area**, AutoCAD plots the actual size of the layout and ignores the setting specified in **Scale**.

Plot Offset

The **Plot offset** section specifies an offset of the plotting area from the lower left corner of the paper. In a layout, the lower left corner of a specified plot area is positioned at the lower left margin of the paper. You can offset the origin by entering a positive or negative value. Select **Center the plot** to automatically center the plot on the paper.

Plot Options

The **Plot options** section specifies options for lineweights, plot styles, and the current plot style table. Select **Plot object lineweights** to plot the objects with assigned lineweights. Otherwise AutoCAD will plot with default lineweight. Select **Plot with plot styles** to plot using the object plot styles that are assigned to the geometry, as defined by the plot style table. Select **Plot paperspace last** to plot model space geometry before paper space objects are plotted. Select **Hide paperspace objects** to plot layouts (paper space) with hidden lines removed from objects.

Drawing Orientation

The **Drawing orientation** section specifies the orientation of the drawing on the paper for plotters that support landscape or portrait orientation. Select **Portrait** to orient and plot the drawing so that the short edge of the paper represents the top of the page. Select **Landscape** to orient and plot the drawing so that the long edge of the paper represents the top of the page. Select **Plot upside-down** to orient and plot the drawing upside down.

After making the necessary changes to the Page Setup dialog box, choose **OK** to close the dialog box and save the changes to the selected page setup.

Creating a New Page Setup

To create a new page setup that can be assigned to any of the layouts, choose **New** in the Page Setup Manager dialog box. AutoCAD displays to New Page Setup dialog box as shown in Figure 15–31.

 Figure 15–31

New Page Setup dialog box

Specify the name of the new page setup in the **New page setup name** box. Select one of the available page setups to use as a starting point for the new page setup. <None> specifies that no page setup is used as a starting point. <Default output device> specifies the default output device. Choose **OK** to close the dialog box, and AutoCAD displays the Page Setup dialog box with the settings of the selected page setup, which you can modify as necessary.

Importing a Page Setup

To import a page setup from a drawing template or drawing file, choose **Import**. AutoCAD displays the Select Page Setup From File dialog box (a standard file selection dialog box), in which you can select a drawing format (*.dwg*), or drawing template (*.dwt*) file from which to import one or more page setups. After selecting the appropriate file, choose **Open**; AutoCAD displays the Import Page Setups dialog box. Choose one of the available page setups to import to the current drawing and choose **OK** to close the dialog box.

Selected Page Setup Details

In the **Selected page setup details** section of the Page Setup Manager dialog box (see Figure 15–29), AutoCAD displays information relative to the selected page setup. **Device name** displays the name of the plot device, **Plotter** displays the type of plot device, **Plot size** displays the plot size and orientation, **Where** displays the physical location of the output device, and **Description** displays descriptive text about the output device.

The **Display when creating a new layout** box specifies that the Page Setup Manager dialog box is displayed when a new layout tab is selected or a new layout is created.

After making necessary changes in the Page Setup Manager dialog box, choose **Close** to close the dialog box.

CREATING A LAYOUT BY LAYOUT WIZARD

Once you have mastered the concepts of layouts and viewports, you can capitalize on the time-saving features in the Layout wizard for creating new layouts by invoking Layout wizard from the Tools menu.

MAKING THINGS LOOK RIGHT FOR PLOTTING

Once you have mastered the concepts of layouts and viewports, you can utilize the following features that will make plotting look right.

Setting Paper Space Linetype Scaling

Linetype dash lengths and the space lengths between dots or dashes are based on the drawing units of the model or paper space in which the objects were created. They can be scaled globally by setting the value of the system variable LTSCALE factor, as explained in Chapter 6. If you want to display objects in viewports at different scales in layout, the linetype objects would be scaled to model space rather than paper space by default. However, by setting paper space linetype scaling (system variable PSLTSCALE) to 1 (default), dash and space lengths are based on paper space drawing units, including the linetype objects that are drawn in model space. For example, a single linetype definition with a dash length of 0.30, displayed in several viewports with different zoom factors, would be displayed in paper space with dashes of length 0.30, regardless of the scale of the viewpoint in which it is being displayed (PSLTSCALE set to 1).

---

note   When you change the PSLTSCALE value to 1, the linetype objects in the viewport are not automatically regenerated. Use the REGEN or REGENALL command to update the linetypes in the viewports.

Dimensioning in Model Space and Paper Space

Dimensioning can be done in both model space and paper space. There are no restrictions placed on the dimensioning commands by the current mode. It is advisable to draw associative dimensions in model space, since AutoCAD places the defining points of the dimension in the space where the dimension is drawn. If the model geometry is modified with a command such as STRETCH, EXTEND, or TRIM, the dimensions are updated automatically. In contrast, if the dimensions are drawn in paper space, the paper space dimension does not change if the model geometry is modified.

For dimensioning in model space, the DIMSCALE factor should be set to 0.0. This causes AutoCAD to compute a scale factor based on the scaling between paper space and the current model space viewport.

Figure 15–32 shows dimensions and view labels that have been drawn in paper space of model space objects.

Figure 15–32

Model space objects dimensioned in paper space

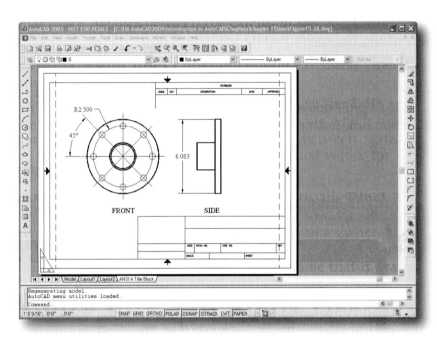

drawing exercises

Open the Exercise Manual PDF file for Chapter 15 on the accompanying CD for discipline-specific exercises.

review questions

1. The drawing created at a scale of 1:1 and plotted to "Scaled to Fit" is plotted:
   a. *At a scale of 1:1.*
   b. *To fit the specified paper size.*
   c. *At the prototype scale.*
   d. *None of the above.*

2. Which of the following determine the relationship between the size of the objects in a drawing and their sizes on a plotted copy?
   a. *Size of the object in the AutoCAD drawing*
   b. *Size of the object on the plot*
   c. *Plot scale*
   d. *All of the above*

3. AutoCAD permits plotting in which of the following environment modes?
   a. *Model space*
   b. *Paper space*
   c. *Layout*
   d. *All of the above*

4. Which TILEMODE system variable setting corresponds to model space?
   a. *0*
   b. *1*
   c. *Either A or B*
   d. *None of the above*

5. When starting a new drawing, how many default plotting layouts does AutoCAD create?
   a. *0*   d. *Unlimited*
   b. *1*   e. *Depends on the selected template*
   c. *2*

6. Within multiple floating viewports, you can establish various scale and layer visibility settings for each individual viewport.
   a. *True*
   b. *False*

7. Paper sizes are indicated by X axis direction (drawing length) and Y axis direction (drawing width).
   a. *True*
   b. *False*

8. Floating viewports, like lines, arcs, and text, can be manipulated using AutoCAD commands such as MOVE, COPY, STRETCH, SCALE, or ERASE.
   a. *True*
   b. *False*

9. While in paper space, both the floating viewports and the 3D model can be modified or edited.
   a. *True*
   b. *False*

10. Which of the following can be converted to a viewport?
   a. *Ellipse*
   b. *Spline*
   c. *Circle*
   d. *All of the above*

11. Which of the following commands allows for the control of layer visibility in a specific viewport?
   a. *VPORTS*
   b. *VPLAYER*
   c. *VIEWLAYER*
   d. *LAYERVIS*

12. Which option under the Additional Parameters section plots the current screen display?
   a. *Current*   **c.** *Extents*
   b. *Display*   **d.** *View*

13. In the Paper Size dialog box, what is the width and height of an A Size sheet of paper?
   a. *10.00 X 8.00*   **c.** *11.00 X 8.50*
   b. *10.50 X 8.00*   **d.** *11.50 X 8.50*

14. In the Paper Size dialog box, what is the width and height of a B Size sheet of paper?
   a. *16.00 X 10.00*   **c.** *17.00 X 10.50*
   b. *16.00 X 10.50*   **d.** *17.00 X 11.00*

15. In the Scale, Rotation, and Origin section, Plotted Inches is equal to:
   a. *Drawing limits.*   **c.** *Plotted feet.*
   b. *Feet.*   **d.** *Units.*

the plot thickens

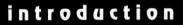

introduction

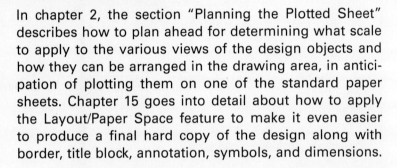

In chapter 2, the section "Planning the Plotted Sheet" describes how to plan ahead for determining what scale to apply to the various views of the design objects and how they can be arranged in the drawing area, in anticipation of plotting them on one of the standard paper sheets. Chapter 15 goes into detail about how to apply the Layout/Paper Space feature to make it even easier to produce a final hard copy of the design along with border, title block, annotation, symbols, and dimensions.

This chapter covers the commands and features used to send the data in the drawing to the plotter in a manner and configuration that will produce the desired results: a working drawing on paper, vellum, mylar or other suitable medium that will communicate the necessary information in a professional format.

After completing this chapter, you will be able to do the following:

• Plot from model space and layout (WYSIWYG)

• Create and modify plot style tables

• Change the Plot Style Property for an object or layer

• Configure plotters

• Apply a plot stamp

• Plot the desired drawing

PLOTTING

 Figure 16–1

Invoking the PLOT command from the Standard toolbar

To plot/print the current drawing from the Model tab, invoke the PLOT command (see Figure 16–1). AutoCAD displays the Plot dialog box, as shown in the Figure 16–2.

 Figure 16–2

Plot dialog box, not expanded

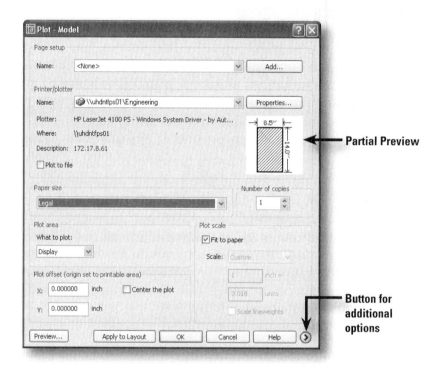

Plot Settings

The Plot dialog box is almost identical to the Page Setup dialog box, except for the title and the **Name** text box window in the **Page setup** section. The Plot dialog box can be expanded or contracted in size by choosing the arrow at the lower right corner of the dialog box. The first seven sections described below are displayed when the dialog box is contracted. The last four described below are displayed only when the dialog box is expanded (see Figure 16–3).

Figure 16-3

Expanded Plot dialog box with additional options

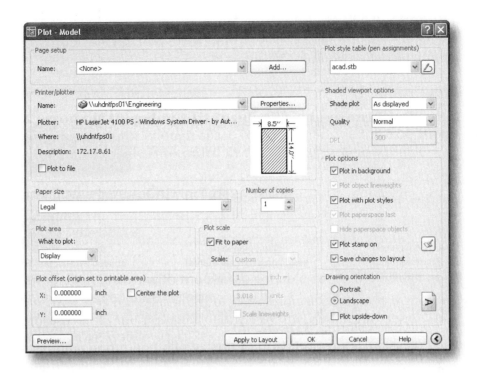

Page Setup

The **Page setup** section displays a list of any named and saved page setups in the drawing. You can base the current setup on a named page setup saved in the drawing, or you can create a new named page setup based on the current settings in the Plot dialog box by choosing **Add**.

Printer/Plotter

The **Printer/plotter** section displays the currently configured plotting device. **Name** lists the system printers/plotters or PC3 files that are available to select for plotting. **Plotter** lists the currently selected plotter or plotter assigned in the currently selected page setup. **Where** gives the port to which the selected plotter is connected, or its network location. **Properties** allows access to the Plotter Configuration of the currently configured plotting device. Refer to the section "Configuring Plotters" later in this chapter for a detailed explanation on plotter configuration. **Description** lists information about the output device currently selected. **Plot to file** plots output to a file rather than to a plotter or printer. If the **Plot to file** option is set to ON, when you choose **OK** in the Plot dialog box, the Plot to File dialog box is displayed. Specify the file name to save the plot file.

Paper Size

The **Paper size** section allows you to select the paper size to plot. Select the paper size to plot from the **Paper size** box for the selected plotting device. Actual paper sizes are indicated by the width (X axis direction) and height (Y axis direction). A default paper size is set for the plotting device when you create a plotter configuration file with the Add-a-Plotter wizard.

Number of Copies

Specify the number of copies to plot in the **Number of copies** box. This option is not available when you plot to file.

Plot Area

The **Plot area** section allows you to select the area to be plotted from the **What to plot** box.

Selecting Display causes the area displayed on the screen to be plotted.

Selecting Limits causes everything within the drawing limits to be plotted to the specified scale when the PLOT command is invoked from the Model tab.

Selecting Layout causes everything within the area of the paper size specified by the layout to be plotted when the PLOT command is invoked from paper space in a layout tab.

Selecting Extents causes the portion of the current space of the drawing that contains objects to be plotted. All visible geometry in the current space is plotted.

Selecting View plots a named view that was previously saved with the VIEW command. You can select a named view from the list. If there are no saved views in the drawing, this option is unavailable.

Selecting Window allows you to specify a window on the screen and plot the area covered by the window. The lower left corner of the window becomes the origin of the plot.

Plot Scale

The **Plot scale** section controls the plot area. Selecting **Fit to paper** causes the plot to fit within the selected paper size. To specify the exact scale for the plot, clear the **Fit to paper** check box and select the plot scale from the **Scale** list box. You can create a custom scale by entering the number of inches (or millimeters) equal to the number of drawing units in the appropriate text fields. When you plot from a layout, the default setting is 1:1 (full scale). Select **Scale lineweights** to plot lineweights in proportion to the plot scale. Otherwise AutoCAD will plot the objects with assigned lineweights. Lineweights normally specify the linewidth of printed objects and are plotted with the linewidth size regardless of the plot scale.

Plot Offset

The **Plot offset** section specifies an offset of the plotting area from the lower left corner of the paper. In a layout, the lower left corner of a specified plot area is positioned at the lower left margin of the plotting area. You can offset the origin by entering a positive or negative value. Select **Center the plot** to automatically center the plot on the paper.

The following sections of the Plot dialog box are displayed only when the dialog box is expanded by choosing the arrow at the lower right corner of the dialog box.

Plot Style Table (Pen Assignments)

The **Plot style table** section sets the plot style table, edits a plot style table, or creates a new plot style table. Plot style tables are settings that give you control over how objects in your drawing are plotted into hard copy plots. By modifying an object's plot style, you override that object's color, linetype, and lineweight. You can also specify end, join, and fill styles, as well as output effects such as dithering, gray scale, pen assignment, and screening. You can use plot styles if you need to plot the same drawing in different ways. This section displays the plot style table that is assigned to the current Model tab or layout tab and provides a list of the currently available plot style tables. If you select **New**, the Add Plot Style Table wizard is displayed, which you can use to create a new plot style table. Choose **Edit** to modify the currently assigned plot style table. Refer to the section "Creating and Modifying Plot Style Tables" later in this chapter for a detailed explanation about creating and modifying a plot style table.

Shaded Viewport Options

The **Shaded viewport options** section specifies how shaded and rendered viewports are plotted and determines their resolution level with corresponding dpi.

The **Shade plot** box displays how views are plotted, and it is specified through the Properties dialog box for the selected viewport.

The **Quality** box specifies the resolution at which shaded and rendered viewports are plotted. The Draft selection sets rendered and shaded model space views to plot as wireframe. The Preview selection sets rendered and shaded model space views to plot at a maximum of 150 dpi. The Normal selection sets rendered and shaded model space views to plot at a maximum of 300 dpi. The Presentation selection sets rendered and shaded model space views to plot at the current device resolution, to a maximum of 600 dpi. DPI values may vary with different computer/plotter configurations. The Maximum selection sets rendered and shaded model space views to plot at the current device resolution with no maximum. The Custom selection sets rendered and shaded model space views to plot at the resolution setting you specify in the **DPI** box, up to the current device resolution.

Plot Options

The **Plot options** section specifies the following options for lineweights, plot styles, and the current plot style table:

Selecting **Plot in background** causes the plot to be processed in the background while you perform other tasks on the computer.

Selecting **Plot object lineweights** plots the objects with assigned lineweights. Otherwise AutoCAD will plot with the default lineweight.

Selecting **Plot with plot styles** plots using the object plot styles that are assigned to the geometry, as defined by the selected plot style table.

Selecting **Plot paperspace last** plots model space geometry before paper space objects are plotted.

Selecting **Hide paperspace objects** plots layouts (paper space) with hidden lines removed from objects.

Selecting **Plot stamp on** includes a plot stamp on the plotted sheet. Refer to the section "Information to Include with Plot Stamp" for a detailed explanation about modifying plot stamp settings.

Select **Save changes to layout** to save the changes that you make in the Plot dialog box to the layout.

Drawing Orientation

The **Drawing orientation** section specifies the orientation of the drawing on the paper for plotters that support landscape or portrait orientation.

Selecting **Portrait** orients and plots the drawing so that the short edge of the paper represents the top of the page.

Selecting **Landscape** orients and plots the drawing so that the long edge of the paper represents the top of the page.

Selecting **Plot upside-down** orients and plots the drawing upside down.

Plot Preview

A preview displays the drawing on the screen as it would appear when plotted. To get a preview of your plot, choose **Preview**. AutoCAD temporarily hides the plotting dialog boxes, draws an outline of the paper size, and displays the drawing as it would appear on the paper when it is plotted (see Figure 16–4).

The cursor changes to a magnifying glass with plus and minus signs. Holding the pick button and dragging the cursor toward the top of the screen enlarges the preview image. Dragging it toward the bottom of the screen reduces the preview image. Right-click, and AutoCAD displays a Shortcut menu offering additional preview options: Pan, Zoom, Zoom window, Zoom original, Plot, and Exit. To end the full preview, choose Exit from the Shortcut menu. AutoCAD returns to the Plot dialog box.

---

note   You can also access a full preview by invoking the Plot Preview command from the File menu.

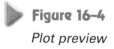

Figure 16–4

Plot preview

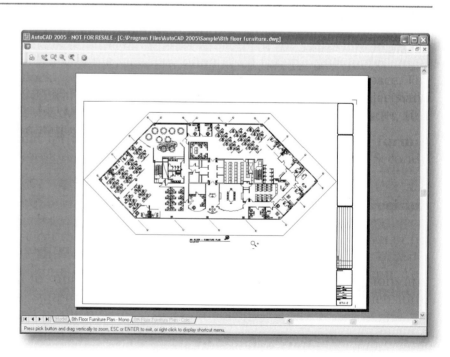

After making the necessary changes in the plot settings, choose **OK**. AutoCAD starts plotting and reports its progress as it converts the drawing to the plotter's graphics language by displaying the number of vectors processed.

If something goes wrong or if you want to stop immediately, choose **Cancel** at any time. AutoCAD cancels the plotting.

Creating and Modifying Plot Style Tables

Plot style tables are settings that control how objects in your drawing are plotted into hard copy plots. By modifying an object's plot style, you can override that object's color, linetype, and lineweight. You can also specify end, join, and fill styles, as well as output effects such as dithering, gray scale, pen assignment, and screening. You can use plot styles if you need to plot the same drawing in different ways.

By default, every object and layer has a plot style property. The actual characteristics of plot styles are defined in plot style tables that you can attach to a Model tab and layouts within drawings. If you assign a plot style to an object, and then detach or delete the plot style table that defines the plot style, the plot style will not have any effect on the object.

AutoCAD provides two plot style modes: color-dependent and named.

> The color-dependent plot styles are based on object color. There are 255 color-dependent plot styles. You cannot add, delete, or rename color-dependent plot styles. You can control the way all objects of the same color plot in color-dependent mode by adjusting the plot style that corresponds to that object color. Color-dependent plot style tables are stored in files with the extension *.ctb*.

> Named plot styles work independently of an object's properties. You can assign any plot style to any object regardless of that object's color. Named plot style tables are stored in files with the extension *.stb*.

By default, all the plot style table files are saved in the path that is listed in the **Files** section of the Options dialog box.

The default plot style mode is set in the Plot Style Table Settings dialog box (see Figure 16–5) which can be opened by choosing **Plot Style Table settings** in the **Plot and Publish** tab of the Options dialog box.

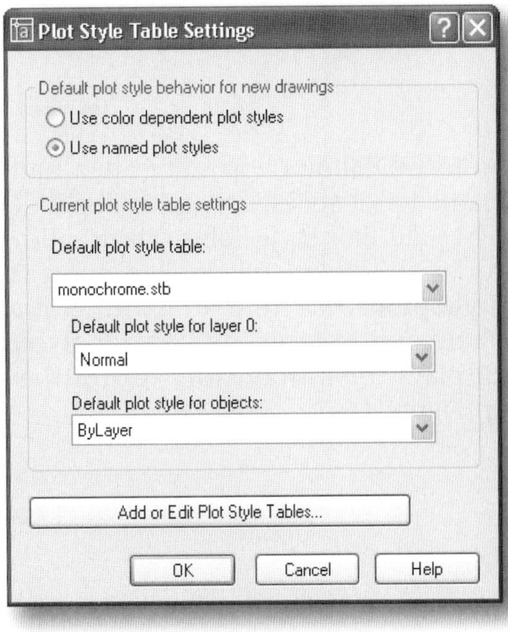

Figure 16–5

Plot Style Table Settings dialog box

Every time you start a new drawing in AutoCAD, the plot style mode that is set in the Options dialog box is applied. Anytime you change the mode, it is applied only for the new drawings or an open drawing that has not yet been saved in AutoCAD.

The CONVERTPSTYLES command converts a currently open drawing from color-dependent plot styles to named plot styles, or from named plot styles to color-dependent plot styles, depending on which plot style method the drawing is currently using.

Creating a New Plot Style Table

Figure 16–6

Choosing Add Plot Style Table from the Wizards flyout of the Tools menu

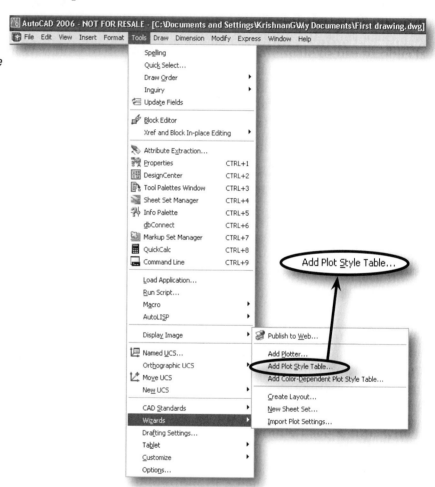

AutoCAD allows you to create a named plot style table to utilize all the flexibility of named plot styles, or a color-dependent plot style table to work in a color-based mode. The Add Plot Style Table wizard allows you to create a plot style from scratch, modify an existing plot style table, import style properties from an *acadr14.cfg* file, or import style properties from an existing *.pcp* or *.pc2* file. After you invoke the Add Plot Style Table option from the Wizards flyout of the Tools menu (see Figure 16–6), AutoCAD displays the introductory text of the Add Plot Style Table wizard, as shown at top left in Figure 16–7.

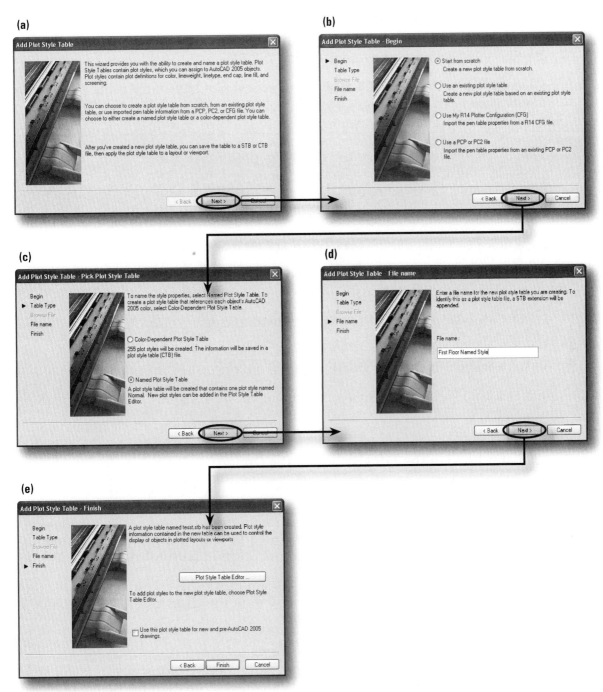

Figure 16–7

Add Plot Style Table wizard pages: (a) Add Plot Style Table – Introductory page, (b) Add Plot Style Table – Begin page, (c) Add Plot Style Table – Pick Plot Style Table page, (d) Add Plot Style Table – File name page, (e) add Plot Style Table1 – Finish Page.

Choose **Next**, and AutoCAD displays the Add Plot Style Table – Begin page, as shown in Figure 16–7(b).

The following four options are available:

Selecting **Start from scratch** allows you to create a new plot style from scratch.

Selecting **Use an existing plot style table** creates a new plot style using an existing plot style table.

Selecting **Use My R14 Plotter Configuration (CFG)** creates a new plot style table using the pen assignments stored in the *acadr14.cfg* file. Select this option if you do not have an equivalent *.pcp* or *.pc2* file.

Selecting **Use a PCP or PC2 file** creates a new plot style table using pen assignments stored in a *.pcp* or *.pc2* file.

---

note   If you selected the **Use an existing plot style table** option to create a new plot style, AutoCAD displays the Add Plot Style Table – Browse File Name page. Specify the plot style table file name from which to create a new plot style name.

To create a new pen table, select **Start from scratch** and choose **Next**. AutoCAD displays the Add Plot Style Table – Pick Plot Style Table page, as shown at in Figure 16–7(c).

Select one of the following options:

Select **Color-Dependent Plot Style Table** to create a plot style table with 255 plot styles.

Select **Named Plot Style Table** to create a named plot style table.

Choose **Next**, and AutoCAD displays the Add Plot Style Table – File name page, as shown in Figure 16–7(d).

Specify the file name in the **File name** box. By default the new style table is saved in the path that is listed in the **Files** section of the Options dialog box.

Choose **Next**, and AutoCAD displays Add Plot Style Table – Finish page, as shown in Figure 16–7(e).

Set the **Use this plot style table for new and pre-AutoCAD 2006 drawings** to ON to attach this plot style table to all new drawings and pre-AutoCAD 2006 drawings by default.

Choose **Finish** to create the plot style table and close the wizard.

Modifying a Plot Style Table

AutoCAD allows you to add, delete, copy, paste, and modify plot styles in a plot style table by using the Plot Style Table Editor. You can open more than one instance of the Plot Style Table Editor at a time and copy and paste plot styles between the tables. Open the Plot Style Table Editor using any of the following methods:

- Choose the **Plot Style Table Editor** button from the Finish screen in the Add Plot Style Table Wizard.
- Open the Plot Style Manager (File menu), right-click a *.ctb* or *.stb* file, and then choose OPEN from the shortcut menu.
- On the **Plot Device** tab of the Plot dialog box or Page Setup dialog box, select the plot style table you want to edit from the Plot Style Table list, and then choose the **Edit** button.
- On the **Plotting** tab of the Options dialog box, choose the **Add or Edit Plot Style Tables** button

Figure 16–8 shows an example of the Plot Style Table Editor for a named plot style table, and Figure 16–9 shows an example of the Plot Style Table Editor for a color-dependent plot style table.

 Figure 16-8

Plot Style Table Editor dialog box for a named Plot Style table: General tab selection, Table View tab selection, Form View tab selection

 Figure 16-9

Plot Style Table Editor dialog box for a Color-Dependent Style table: General tab selection, Table View tab selection, Form View tab selection

Following are the three tabs available in the Plot Style Table Editor:

- **General**—Displays the name of the plot style table, description (if any), location of the file, and version number (see Figure 16–8(a) and Figure 16–9(a)). You can modify the description, and apply scaling to non-ISO lines and to fill patterns.

- **Table View**—Lists entire plot styles in the plot style table and their settings in tabular form (see Figure 16–8(b) and Figure 16–9(b)). The styles are displayed in columns from left to right. The setting names of each row appear at the left of the tab. By default, in the case of a named plot style table, AutoCAD sets up a style named Normal and represents an object's default properties. You cannot modify or delete the Normal style. In the case of a color-dependent plot style table, AutoCAD lists all the 255 color styles in tabular form. In general, this is convenient if you have a small number of plot styles to view them in the tabular form.

- **Form View**—The plot style names are listed under the **Plot styles** list box and the settings for the selected plot style are displayed at the right side of the dialog box (see Figure 16–8(c) and Figure 16–9(c)).

To create a new plot style, choose **Add Style** from the Plot Style Table Editor. AutoCAD adds a new style, and you can change the name to a descriptive name if necessary (cannot exceed 255 characters). You cannot duplicate names within the same plot style table.

> **note** You cannot add or change the name for a plot style in the color-dependent style table.

To delete a pen style, click the gray area above the plot style name in the **Table View** and choose **Delete Style**. In the **Form View**, select the style name from the **Plot styles** list box and choose **Delete Style**.

> **note** You cannot delete a plot style from a color-dependent style table.

The **Description** field allows you to specify a description for plot styles and modify an existing description for a plot style, if necessary. The description cannot exceed 255 characters.

The **Color** list box allows you to assign a plot style color. If you assign a color from one of the available colors, then AutoCAD overrides the object's color at plot time. By default, all of the plot styles are set to use object color.

Set **Dither** to ON for the plotter to approximate colors with dot patterns, giving the impression of plotting with more colors than the ink available in the plotter. If you set **Dither** to OFF, then AutoCAD maps colors to the nearest color, which limits the range of colors used for plotting. The most common reason for turning off dithering is to avoid false line typing from dithering of thin vectors and to make dim colors more visible. If the plotter does not support dithering, the dithering setting is ignored. The default setting is set to ON.

Set **Grayscale** to ON for AutoCAD to convert the object's colors to grayscale, if the plotter supports grayscale. If you set **Grayscale** to OFF, AutoCAD uses the RGB values for the object's colors. The default setting is set to OFF.

The **Pen #** setting in the Plot Style Table Editor specifies which pen to use for each plot style. You can specify a pen to use in the plot style by selecting from a range of pen numbers from 1 to 32. By using the BACKSPACE or DELETE keys, you can set the field to read Automatic. AutoCAD uses the information you provided under Physical Pen Configuration in the Plotter Configuration Editor to select the pen closest in color to the object you are plotting. The default is set to Automatic.

Specify a virtual pen number in the **Virtual pen #** edit field for plotters that do not use pens but can simulate the performance of a pen plotter by using virtual pens. The default is set to Automatic or 0 to specify that AutoCAD should make the virtual pen assignment from the AutoCAD Color Index. You can specify a virtual pen number between 1 and 255. The virtual pen number setting in a plot style is used only by plotters without pens and only if they are configured for virtual pens. If this is the case, all the other style settings are ignored and only the virtual pen is used. The default is set to Automatic.

note   If a plotter without pens is not configured for virtual pens, then both the virtual and the physical pen information in the plot style are ignored and all the other settings are used.

The **Screening** field sets a color intensity setting that determines the amount of ink AutoCAD places on the paper while plotting. The valid range is 0 through 100. Selecting 0 reduces the color to white. Selecting 100 (default) displays the color at its full intensity.

The **Linetype** box allows you to assign a plot style linetype. If you assign a linetype from one of the available linetypes, AutoCAD overrides the object's linetype at plot time. By default, all the plot styles are set to use object linetype.

The **Adaptive adjustment** setting adjusts the scale of the linetype to complete the linetype pattern. Set it to ON to have a complete linetype pattern with no consideration to linetype scaling, and set to OFF if linetype scale is more important than having a complete linetype pattern. The default is set to ON.

The **Lineweight** box allows you to assign a plot style lineweight. If you assign a lineweight from one of the available lineweights, AutoCAD overrides the object's lineweight at plot time. By default, all the plot styles are set to use object lineweight.

The **Line End Style** box allows you to assign a line end style. The line end style options include: Butt, Square, Round, and Diamond. If you assign a line end style from one of the available line end styles, then AutoCAD overrides the object's line end style at plot time. By default, all the plot styles are set to use object end style.

The **Line Join Style** box allows you to assign a line end style. The line join style options include Miter, Bevel, Round, and Diamond. If you assign a line join style from one of the available line join styles, then AutoCAD overrides the object's line join style at plot time. By default, all the plot styles are set to use object line join style.

The **Fill Style** box allows you to assign a fill style. The fill style options include Solid, Checkerboard, Crosshatch, Diamonds, Horizontal Bars, Slant Left, Slant Right, Square Dots, and Vertical Bars. The fill style applies only to solids, plines, donuts, and 3D faces. If you assign a fill style from one of the available fill styles, AutoCAD overrides the object's fill style at plot time. By default, all the plot styles are set to use object fill style.

AutoCAD allows you to edit the available lineweights by choosing **Edit Lineweights**. You cannot add or delete lineweights from the list.

To save the changes and close the Plot Style Table Editor, choose **Save & Close**. To save the changes to another plot style table, choose **Save As**. AutoCAD displays the Save As dialog box. Specify the file name in the **File name** text field and choose **Save** to save and close the Save As dialog box.

In the Add Plot Style Table - Finish page of the wizard, set the **Use this plot style table for new and pre-AutoCAD 2005 drawings** check box to ON to attach this plot style table to all new drawings and pre-AutoCAD 2005 drawings by default (available only to new style tables created based on Named Plot Style table).

Choose **Finish** to create the plot style table and close the wizard.

Changing Plot Style Property for an Object or Layer

As mentioned earlier, every object that is created in AutoCAD has a plot style property in addition to color, linetype, and lineweight. Similarly, every layer has a color, linetype, and lineweight, in addition to a plot style property. The default setting for plot styles for objects and layers are set in the Plot Style Table Settings dialog box (see Figure 16–10) which can be opened from the Options dialog box.

 Figure 16–10

Plot Style Table Settings dialog box

The default plot style for layer 0 can be any of the following when you are using a named plot style table:

- Normal—Uses the object's default properties.
- Named plot style—Uses the properties of the specific named plot style from the currently loaded plot style table.

The default plot style for objects can be any of the following when you are using a named plot style table:

- Normal—Uses the object's default properties.
- BYLAYER—Uses the properties of the layer that contains the object.
- BYBLOCK—Uses the properties of the block that contains the object.
- Named plot style—Uses the properties of the specific named plot style from the currently loaded plot style table.

The default plot setting for an object is BYLAYER, and the initial plot style setting for a layer is Normal. When the object is plotted it retains its original properties.

If you are working in a Named plot style mode, you can change the plot style for an object or layer at any time. If you are working in a color-dependent plot style mode, you cannot change the plot style for objects or layers. By default, they are set to By Color.

To change the plot style for one or more objects, first select the object(s) (system variable PICKFIRST set to ON), and select the plot style from the Plot style control list box on the Properties toolbar, as shown in Figure 16–11. If the plot style does not list the one you want to select, choose Other, and AutoCAD displays the Current Plot Styles dialog box, as shown in Figure 16–12.

 Figure 16–11

Properties toolbar – Plot Style control

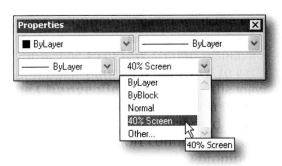

 Figure 16–12

Current Plot Style dialog box

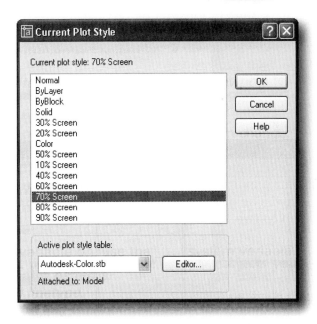

Select the plot style you want to apply to the selected object(s) from the **Current plot style** list box. If you need to select a plot style from a different plot style table, select the plot style table from the **Active plot style table** box. AutoCAD lists all the available plot styles in the **Current plot style** list box; select the one you want to apply to the selected object(s). Choose **OK** to close the dialog box. You can also change the plot style of the selected object(s) from the Properties dialog box.

To change the plot style for a layer, open the Layer Properties Manager dialog box. Select the layer you want to change and select a plot style for the selected layer, similar to changing color or linetype.

Information to Include with Plot Stamp

Figure 16–13

*Choosing Plot Stamp
Settings from the Plot
and Publish tab of the
Options dialog box*

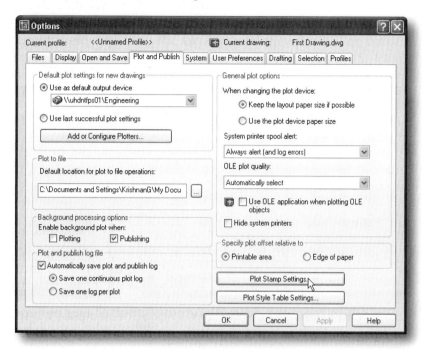

To access plot stamp settings, choose Plot Stamp Settings from the Plot and Publish tab of the Options dialog box (see Figure 16–13). The Plot Stamp dialog box (see Figure 16–14) allows you to specify the information for the plot stamp that can be placed on a specified corner of each drawing and, if necessary, logs the information to a file.

Plot stamp information includes drawing name, layout name, date and time, login name, plot device name, paper size, plot scale and user defined fields, if any. Once you check the **Plot stamp on** check box in the **Plot options** section of the Plot dialog box, it remains active with whatever settings have been most recently entered until you specifically clear the check box. AutoCAD creates a plot stamp at the time the drawing is being plotted, and it is not saved with the drawing. Before you plot the drawing, you can preview the position of the plot stamp (not the contents) in the Plot Stamp dialog box. The plot stamp can be set to plot at one of the four drawing corners and can print up to two lines.

---

note   Plot stamp information is plotted with pen number 7 or the highest numbered available pen if the plotter doesn't hold 7 pens. If you are using a non-pen (raster) device, color 7 is always used for plot stamping.

Figure 16–14

Plot Stamp dialog box

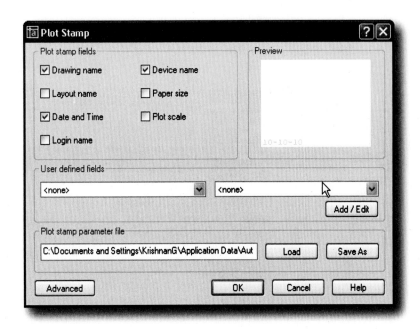

The **Plot stamp fields** section specifies the drawing information you want applied to the plot stamp

> There are seven items that can be included in the plot stamp by selecting them by their field name as follows: **Drawing name**, **Layout name**, **Date and Time**, **Login name**, **Device name**, **Paper size**, and **Plot scale**.

The **Preview** section of the Plot Stamp dialog box provides a visual display of the plot stamp location based on the location and rotation values specified in the Advanced Options dialog box.

---

note The preview displayed in the **Preview** section of the Plot Stamp dialog box is a visual display of the plot stamp location, not a preview of the plot stamp contents.

The **User defined fields** section provides text that can optionally be plotted. You can choose one or both user-defined fields for the plot stamp information. If the user-defined value is set to <none>, no user-defined information is plotted. To add, edit, or delete user-defined fields, choose **Add/Edit**. AutoCAD displays the User Defined Fields dialog box as shown in Figure 16–15.

Figure 16-15

User Defined Fields dialog box

Choose **Add** to add an editable user-defined field to the bottom of the list, choose **Edit** to edit the selected user-defined field, and choose **Delete** to delete the selected user-defined field. Choose **OK** to save the changes and close the User Defined Fields dialog box. Choose **Cancel** to discard the changes and close the User Defined Fields dialog box.

The **Plot stamp parameter file** section displays the name of the file in which the plot stamp settings are stored. If necessary, you can save the current plot stamp settings to a new file by choosing **Save As** and providing an appropriate file name. AutoCAD stores plot stamp information in a file with a *.pss* extension. If you need to load a different parameter file, choose **Load**, AutoCAD displays a standard file selection dialog box, in which you can specify the location of the parameter file you want to use.

To set the location, text properties, and units of the plot stamp, choose **Advanced**. AutoCAD displays Advanced Options dialog box as shown in Figure 16–16.

Figure 16-16

Advanced Options (Plot Stamp) dialog box

Location allows you to select the area where you want to place the plot stamp. The options include Top Left, Bottom Left (default), Bottom Right, and Top Right. The location is relative to the image orientation of the drawing on the page.

Orientation allows you to select the rotation of the plot stamp in relation to the specified page. The options include Horizontal and Vertical for each location.

Stamp upside-down controls whether to rotate the plot stamp upside down.

X offset and **Y offset** determine the offset distance calculated from either the corner of the paper or the corner of the printable area, depending on which setting you specify. Select one of the two options: **Offset relative to printable area** or **Offset relative to paper border** to set the reference point from which to measure the offset distance.

The **Text properties** section determines the font, height, and number of lines you want to apply to the plot stamp text. **Font** specifies the font you want to apply to the text used for the plot stamp information. **Height** specifies the text height you want to apply to the plot stamp information. **Single line plot stamp** controls whether to place the plot stamp information in a single line of text or not. The plot stamp information can consist of up to two lines of text, but the placement and offset values you specify must accommodate text wrapping and text height. If the plot stamp contains text that is longer than the printable area, the plot stamp text will be truncated. If **Single line plot stamp** is set to OFF, plot stamp text is wrapped after the third field.

The **Plot stamp units** section allows you to specify the units used to measure X offset, Y offset, and height. From the **Units** box you can select one of the available units: inches, millimeters, or pixels.

Log file location specifies the name of the file to which the plot stamp information is saved instead of, or in addition to, stamping the current plot. The default log file name is *plot.log*, and it is located in the AutoCAD folder. Choose **Browse** to specify a different file name and path. After the initial plot log file is created, the plot stamp information in each succeeding plotted drawing is added to this file. Each drawing's plot stamp information is a single line of text. If necessary the log file can be placed on a network drive and shared by multiple users.

Choose **OK** to save the changes and close the Advanced Options dialog box. Choose **OK** to save the changes and close the Plot Stamp dialog box.

Configuring Plotters

 Figure 16–17

Choosing Plotter Manager from the File menu

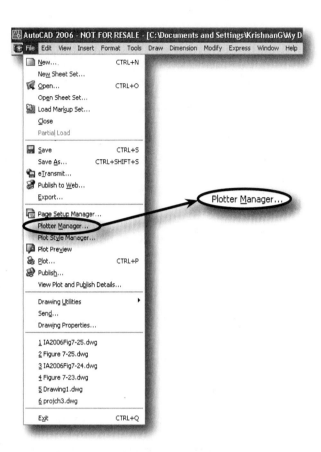

The Plotter Manager allows you to configure a local or network non-system plotter. In addition, you can also configure a Windows system printer with non-default settings. AutoCAD stores information about the media and plotting device in configured plot (*.pc3*) files. The *.pc3* files are stored in the path that is listed in the **Files** section of the Options dialog box. Plot configurations are therefore portable and can be shared in an office or on a project. If you calibrate a plotter, the calibration information is stored in a plot model parameter (*.pmp*) file that you can attach to any *.pc3* files you create for the calibrated plotter.

AutoCAD allows you to configure plotters for many devices and store multiple configurations for a single device. You can create several *.pc3* files with different output options for the same plotter. After you create a *.pc3* file, it's available in the list of plotter configuration names in the **Printer/plotter** section of the Plot dialog box.

Choose Plotter Manager from the File menu (see Figure 16–17); AutoCAD displays the Plotters window explorer, shown in Figure 16–18, listing all the plotters configured.

 Figure 16–18

Plotters window explorer (Windows XP version)

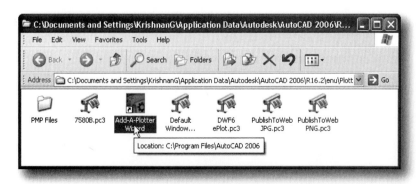

Double-click Add-A-Plotter Wizard, and AutoCAD displays the Add Plotter – Introduction Page, as shown at left in Figure 16–19.

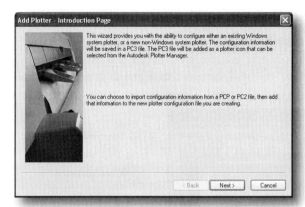 

 Figure 16–19

Add Plotter wizard: Introduction and Begin pages

Choose **Next**, and AutoCAD displays the Add Plotter – Begin page, as shown at right in Figure 16–19.

Choose one of the three available options:

Select **My Computer** to configure a local non-system plotter.

Select **Network Plotter Server** to configure a plotter that is on the network.

Select **System Printer** to configure a Windows system printer. If you want to connect to a printer that is not in the list, you must first add the printer using the Windows Add Printer wizard in the Control Panel.

If you select **My Computer**, the wizard prompts you to select a plotter manufacturer and model number, identify the port to which the plotter is connected, specify a unique plotter name, and choose **Finish** to close the wizard.

If you select **Network Plotter Server**, the wizard prompts you to identify the network server, select the plotter manufacturer and model number, specify a unique plotter name, and choose **Finish** to close the wizard.

If you select **System Printer**, the wizard prompts you to select one of the printers configured in the Windows operating system, specify a unique plotter name, and choose **Finish** to close the wizard.

AutoCAD saves the configuration file in the *.pc3* file format with a unique given name in the path that is listed in the **Files** section of the Options dialog box.

If necessary, you can edit the *.pc3* file using the Plotter Configuration Editor. The Plotter Configuration Editor provides options for modifying a plotter's port connections and output settings including media, graphics, physical pen configuration, custom properties, initialization strings, calibration, and user-defined paper sizes. You can drag these options from one *.pc3* file to another.

You can open the Plotter Configuration Editor using one of the following methods:

- From the File menu, choose Page Setup. Choose **Properties**.
- From the File menu, choose Plot. Choose **Properties**.
- Double-click a *.pc3* file from Windows Explorer and right-click the file and choose Open.
- Choose **Edit Plotter Configuration** on the Add Plotter - Finish page in the Add Plotter wizard.

Figure 16–20 shows the Plotter Configuration Editor for an HP7580B plotter.

 Figure 16–20

Plotter Configuration Editor for an HP7580B plotter (Device and Document Settings tab)

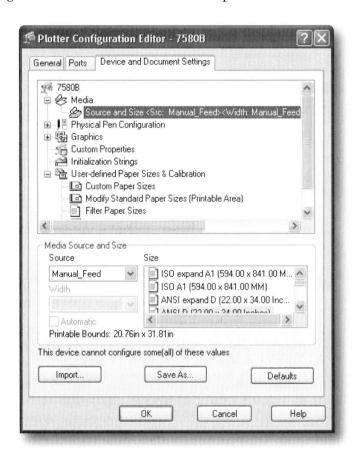

The Plotter Configuration Editor contains three tabs:

- **General** tab—Contains basic information about the configured plotter.
- **Ports** tab—Contains information about the communication between the plotting device and your computer.
- **Device and Document Settings** tab—Contains plotting options.

In the **Device and Document Settings** tab, you can change many of the settings in the configured plot (*.pc3*) file. Following are the six fields in which you make the changes:

- **Media**—Specifies a paper source, size, type, and destination.
- **Physical Pen Configuration**—Specifies settings for pen plotters.

- **Graphics**—Specifies settings for printing vector graphics, raster graphics, and TrueType fonts.
- **Custom Properties**—Displays settings related to the device driver.
- **Initialization Strings**—Sets pre-initialization, post-initialization, and termination printer strings.
- **User-defined Paper Sizes & Calibration**—Attaches a plot model parameter (*.pmp*) file to the *.pc3* file, calibrates the plotter, and adds, deletes, or revises custom or standard paper sizes.

The fields correspond to the categories of settings in the *.pc3* file you're editing. Double-click any of the six categories to view and change the specific settings. When you change a setting, your changes appear in angle brackets (<>) next to the setting name unless there is too much information to display. To save the changes to another *.pc3* file, choose **Save As**. AutoCAD displays the Save As dialog box. Specify the file name in the **File name** box and choose **Save**. To save the changes to the *.pc3* file and close the Plotter Configuration Editor, choose **OK**.

drawing exercises

Open the Exercise Manual PDF file for Chapter 16 on the accompanying CD for discipline-specific exercises.

review questions

1. If you were to plot a drawing at a scale of 1"=60', what should you set LTSCALE to?

   a. *60*
   b. *1/60*
   c. *720*
   d. *1/720*

2. If you want to plot a drawing requiring multiple pens and you are using a single pen plotter, AutoCAD will:

   a. *Not plot the drawing at all.*
   b. *Pause when necessary to allow you to change pens.*
   c. *Invoke an error message.*
   d. *Plot all the drawing using the single pen.*
   e. *None of the above.*

3. The drawing created at a scale of 1:1 and plotted to "Scaled to Fit" is plotted:

   a. *At a scale of 1:1.*
   b. *To fit the specified paper size.*
   c. *At the prototype scale.*
   d. *None of the above.*

4. To plot a full scale drawing at a scale of 1/4"=1', use a plot scale of:

   a. *0.25=12.*   **d.** *12=0.25.*
   b. *0.25=1.*   **e.** *24=1.*
   c. *48=1.*

5. What is the file extension assigned to all files created when plotting to a file?

   a. *DWG*   **d.** *PLO*
   b. *DRW*   **e.** *PLT*
   c. *DRK*

6. When plotting, pen numbers are assigned to:

   a. *Colors.*
   b. *Layers.*
   c. *Thickness.*
   d. *Linetypes.*
   e. *None of the above.*

You need to draw three orthographic views of an airplane whose dimensions are as follows: wingspan of 102 feet, a total length of 118 feet, and a height of 39 feet. The drawing has to be plotted on a standard 12" x 9" sheet of paper. No dimensions will be added, so you will need only 1" between the views. Answer the following five questions using the information from this drawing:

7. What would be a reasonable scale for the paper plot?

    a. *1=5'*
    b. *1=15'*
    c. *1=25'*
    d. *1=40'*

8. What would be a reasonable setting of LTSCALE?

    a. *1*
    b. *5*
    c. *60*
    d. *25*
    e. *300*
    f. *480*

9. If you were plotting from paper space, what ZOOM scale factor would you use?

    a. *1/5X*   **e.** *1/5XP*
    b. *1/25X*   **f.** *1/25XP*
    c. *1/60X*   **g.** *1/60XP*
    d. *1/300X*   **h.** *1/300XP*

10. When inserting your border in paper space, what scale factor should you use?

    a. *1*   **d.** *60*
    b. *5*   **e.** *300*
    c. *25*

11. Which of the following options will the plot preview give you?

    a. *Seeing what portion of your drawing will be plotted*
    b. *Seeing the plotted size of your drawing*
    c. *Seeing rulers around the edge of the plotted page for size comparison*
    d. *None of the above*

12. Which of the following determines the relationship between the size of the objects in a drawing and their sizes on a plotted copy?

    a. *Size of the object in the AutoCAD drawing*
    b. *Size of the object on the plot*
    c. *Maximum available plot area*
    d. *Plot scale*
    e. *All of the above*

13. AutoCAD permits plotting in which of the following environment modes?

    a. *Model space*
    b. *Paper space*
    c. *Layout*
    d. *All of the above*

14. Which TILEMODE system variable setting corresponds to model space?
   a. *0*
   b. *1*
   c. *Either A or B*
   d. *None of the above*

15. Within multiple floating viewports you can establish various scale and layer visibility settings for each individual viewport.
   a. *True*
   b. *False*

16. Paper sizes are indicated by X axis direction (drawing length) and Y axis direction (drawing width).
   a. *True*
   b. *False*

17. Floating viewports, like lines, arcs, and text, can be manipulated using AutoCAD commands such as MOVE, COPY, STRETCH, SCALE, or ERASE.
   a. *True*
   b. *False*

18. Under which menu will you find Plot?
   a. *File*   **c.** *View*
   b. *Edit*   **d.** *Insert*

19. The Plot icon is located on which toolbar?
   a. *Draw*   **c.** *Properties*
   b. *Modify*   **d.** *Standard*

20. Which command plots a drawing to a plotting device or a file?
   a. *PLOT*   **c.** *WINDOW*
   b. *AutoCADPlot*   **d.** *Both a and b*

21. Which option under what to plot plots the drawing limits?
   a. *Display*   **c.** *Limit*
   b. *Extents*   **d.** *Limits*

22. Which option under what to plot allows you to plot the area inside a user specified window?
   a. *Display*   **c.** *View*
   b. *Screen*   **d.** *Window*

Icing on
the Cake

assembled symbols

The AutoCAD BLOCK command feature is a powerful design/drafting tool. It enables a designer to create an object from one or more objects, save it under a user-specified name, and later place it back in the drawing. When block references are inserted in the drawing, they can be scaled up or down in both or either of the X and Y directions. They can also be rotated as they are inserted in the drawing. Block references can best be compared with their manual drafting counterpart, the template. Even though an inserted block reference can be created from more than one object, the block reference acts as a single unit when operated on by certain modify commands such as MOVE, COPY, ERASE, ROTATE, ARRAY, and MIRROR. You can drag and drop blocks from DesignCenter into the current drawing (for details see Chapter 19). You can also export a block reference to become a drawing file outside the current drawing, and it can be inserted into other drawings. Like the plastic template, block references greatly reduce repetitious work.

The BLOCK command can save time because you don't have to draw the same object more than once. Block references save computer storage because the computer needs to store the object descriptions only once.

After completing this chapter, you will be able to do the following:

- Create and insert block references in a drawing

- Convert individual block references to drawing files

- Define attributes, edit attributes, and control the display of attributes

- Use the DIVIDE and MEASURE commands

CREATING BLOCKS

When you invoke the block command to create a block, AutoCAD refers to this as defining the block. The resulting definition is stored in the drawing database. The same block can be inserted as a block reference as many times as needed.

Blocks may comprise one or more objects. The first step in creating blocks is to create a block definition. In order for you to do this, the objects that make up the block must be visible on the screen. That is, the objects that will make up the block definition must have already been drawn so that you can select them when prompted to do so during the BLOCK command.

The layer for the objects comprising the block is very important. Objects that are on layer 0 when the block is created will assume the color, linetype, and lineweight of the layer on which the block reference is inserted. Objects on any layer other than 0, when included in the block definition, will retain the characteristics of that layer, even when the block reference is inserted on a different layer. See Figure 17–1 for an example.

Figure 17–1

Example of inserting block references drawn on different layers

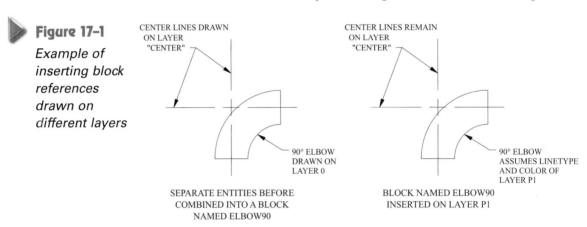

Examples of some common uses of blocks in various disciplines are shown in Figure 17–2.

Figure 17–2

Examples of common uses of blocks in various disciplines

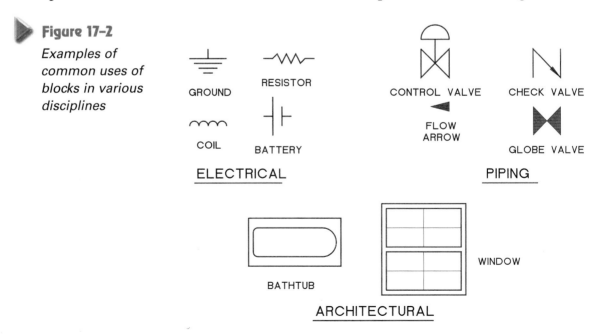

Creating a Block Definition

 Figure 17-3

Invoking the BLOCK command from the Draw toolbar

The BLOCK command creates a block definition for selected objects. Invoke the BLOCK command from the Draw toolbar (see Figure 17–3), AutoCAD displays the Block Definition dialog box, similar to Figure 17–4.

 Figure 17-4

Block Definition dialog box

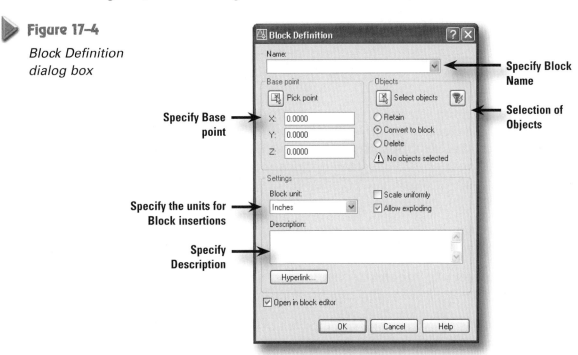

Block Name

Specify the block name in the **Name** box. The block name can be up to 255 characters long and may contain letters, numbers, and blank spaces. To list the block names, if any, in the current drawing, click the down arrow to the right of the **Name** box. AutoCAD lists the blocks in the current drawing.

Base Point

In the **Base point** section, you can specify the insertion point for the block.

The insertion point specified during the creation of the block becomes the base point for future insertions of this block as a block reference. It is also the point about which the block reference can be rotated or scaled during insertion. When determining where to locate the base insertion point it is important to consider what will be on the drawing before you insert the block reference. Therefore, you must anticipate this preinsertion state of the drawing. It is sometimes more advantageous for the insertion point to be somewhere off the object than on it.

You can specify the insertion point on the screen, or you can specify X, Y, and Z coordinates of the insertion point in the **X**, **Y**, and **Z** boxes, respectively, located in the **Base point** section of the Block Definition dialog box. To specify the base point on the screen, choose **Pick point**, located in the **Base point** section of the dialog box. AutoCAD prompts:

Specify insertion point: *(specify the insertion point)*

Once you have specified the insertion point, the Block Definition dialog box reappears.

Selecting Objects

To select objects to include in the block definition, choose **Select Objects** located in the **Objects** section of the Block Definition dialog box. AutoCAD prompts:

Select objects: *(select objects using one of the AutoCAD object selection methods, and press* ENTER *to complete object selection)*

Once the objects are selected, the Block Definition dialog box reappears. You can also use the **Quick Select** option to define the selection of objects by choosing the icon to the right of the **Select Objects** button in the **Objects** section of the Block Definition dialog box. Choose one of the three options located in the **Objects** section to specify whether to retain or delete the selected objects or convert them to a block reference after you create the block.

The **Retain** selection retains the selected objects as distinct objects in the drawing after you create the block.

The **Convert to block** selection converts the selected objects to a block instance in the drawing after you create the block.

The **Delete** selection deletes the selected objects from the drawing after you create the block.

Settings

The **Settings** section controls units, scale, explode options, description, and hyperlink options.

The **Block unit** text box lets you specify the insertion units for the block reference.

Choosing **Scale uniformly** prevents the block from being inserted with different X and Y scale factors.

Choosing **Allow exploding** determines that an inserted reference of the block can be exploded.

The **Description** text box can be used to enter a description of the block if desired.

Choosing **Hyperlink** causes AutoCAD to display the Insert Hyperlink dialog box (see Figure 17–5).

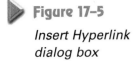

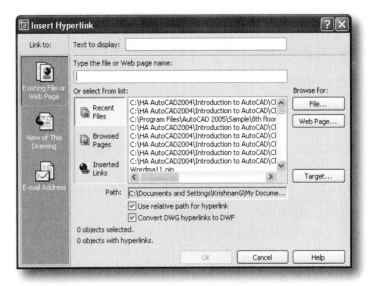

Figure 17–5

*Insert Hyperlink
dialog box*

Hyperlinks are created in AutoCAD drawings as pointers to associated files. Hyperlinks can point to locally stored files, files on a network drive, or files on the Internet. Hyperlinks can launch a word processing program, open a specific file, and even point to a named location in a file. Hyperlinks can open your Web browser and point to a specific Web site. You can specify a view in AutoCAD or a bookmark in a word processing file. Cursor feedback is automatically provided to indicate when the crosshairs are over a graphical object that has an attached hyperlink. You can then select the object and use the Hyperlink Shortcut menu to open the file associated with the hyperlink. This hyperlink cursor and Shortcut menu display can be turned off in the Options dialog box.

Set **Open in block editor** to ON to open the block editor to create a dynamic block.

Choose **OK** to create the block definition with the given name. If the name specified is the same as an existing block in the current drawing, AutoCAD displays a warning like that shown in Figure 17–6.

Figure 17–6

Warning dialog box regarding block definition

To redefine the block, choose **Yes** in the Warning dialog box. The block with that same name is then redefined. Once the drawing is regenerated, any insertion of this block reference already inserted in the drawing is redefined to the new block definition with this name.

Choose **No** in the Warning dialog box to cancel the block definition. Then, to create a new block definition, specify a different block name in the **Name** box of the Block Definition dialog box and choose **OK**.

If you create a block without selecting objects, AutoCAD displays a warning that nothing has been selected and provides an opportunity to select objects before the named block is created.

INSERTING BLOCK REFERENCES

Figure 17–7

Invoking the INSERT *command from the Draw toolbar*

You can insert previously defined blocks into the current drawing by invoking the INSERT command (see Figure 17–7).

---

note   If blocks were created and stored in a template drawing, and you make your new drawing equal to the template, those blocks will be in the new drawing ready to insert. Any drawing inserted into the current drawing will bring with it all of its block definitions, whether they have been inserted or only stored as definitions.

After you invoke the INSERT command from the Draw toolbar, AutoCAD displays the Insert dialog box, similar to Figure 17–8.

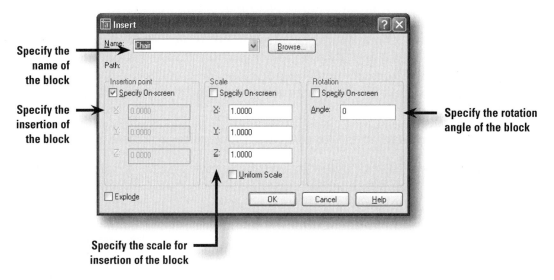

Specify the name of the block

Specify the insertion of the block

Specify the scale for insertion of the block

Specify the rotation angle of the block

Figure 17–8

Insert dialog box

Specify a block name in the **Name** box, or choose the down arrow to display a list of blocks defined in the current drawing and select the block to be inserted. If there is no block definition with the specified name in the current drawing, AutoCAD searches the drives and folders on the path for a drawing of that name and inserts it instead.

The **Insertion point** section of the Insert dialog box allows you to specify the insertion point for inserting a copy of the block definition. You can specify the insertion point in terms of X, Y, and Z coordinates in the **X**, **Y** and **Z** boxes when the **Specify On-screen** is set to OFF. If you prefer to specify the insertion point on the screen, set **Specify On-screen** to ON.

The **Scale** section of the Insert dialog box allows you to specify the scale for the inserted block. The default scale factor is set to 1 (Full scale). You can specify a scale factor between 0 and 1 to insert the block reference smaller than the original size of the block and specify greater than 1 to increase the size from the original size. If necessary, you can specify different X and Y scale factors for insertion of the block reference. If you specify a negative scale factor, AutoCAD inserts a mirror image of the block about the insertion point. As a matter of fact, if –1 were used for both X and Y scale factors, it would "double mirror" the object, the equivalent of rotating it 180 degrees. If you prefer to specify the scale factor on the screen, then set the **Specify On-screen** to ON. If you set **Uniform Scale** to ON, you can enter a value only in the **X** box. The Y and Z scales will be the same as that entered for the X scale.

The **Rotation** section of the Insert dialog box allows you to specify the rotation angle for the inserted block. To rotate the block reference, specify a positive or negative angle, referencing the block in its original position. If you prefer to specify the rotation angle on the screen, then set the **Specify On-screen** to ON.

The **Explode** box allows you to insert the block reference as a set of individual objects rather than as a single unit.

To specify a drawing file to insert as a block definition, specify the drawing file name in the **Name** box, or choose **Browse** to display a standard file dialog box and select the appropriate drawing file.

 The name of the last block reference inserted during the current drawing session is remembered by AutoCAD. The name becomes the default for subsequent use of the INSERT command.

Choose **OK** to insert the selected block.

You can also insert a block by dragging a block from a tool palette (see Figure 17–9).

Figure 17–9

Tool palette with Sample Office Project tab selected

The block name will be added to the **Name** box of the Insert dialog box whenever a block is inserted in the current drawing by dragging from a tool palette.

Tool palettes are tabbed areas within the Tool Palettes window, and blocks that reside on a tool palette are called tools. Several tool properties, including scale, rotation, and layer, can be set for each tool individually. To change the tool properties, choose Properties on the Shortcut menu. Then you can change the tool's properties in the Tool Properties dialog box as shown in Figure 17–10. The Tool Properties dialog box has two categories of properties—the Insert properties category, which controls object-specific properties such as name of the block, name of the original source drawing name, scale, rotation, and angle, and the General properties category, which overrides the current drawing property settings such as layer, color, linetype. lineweight, and plot style.

Figure 17–10

Tool Properties of a selected tool

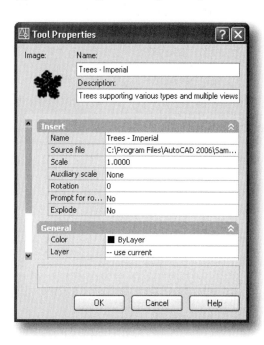

You can place blocks that you use often on a tool palette by dragging blocks from the AutoCAD DesignCenter. See Chapter 19 for details about using the AutoCAD DesignCenter.

NESTED BLOCKS

Blocks can contain other blocks. That is, when you use the block command to combine objects into a single object, one or more of the selected objects can themselves be blocks. And the blocks selected can have blocks nested within them. There is no limitation to the depth of nesting. You may not, however, use the name of any of the nested blocks as the name of the block being defined. This would mean that you were trying to redefine a block, using its old definition in the new.

Any objects within blocks (as nested blocks) that were on layer 0 when made into a block will assume the color, linetype, and lineweight of the layer on which the block reference is inserted. If an object (originally on layer 0 when included in a block definition) is in a block reference that has been inserted on a layer other than layer 0, it will retain the color, linetype, and lineweight of the layer it was on when its block was included in a higher-level block. For example, you draw a circle on layer 0 and include it in a block named Z1. Then, you insert Z1 on layer R, whose color is red. The circle would then assume the color of layer R (in this case it will be red). Create another block called Y3 by including the block Z1. If you insert block reference Y3 on a layer whose color is blue, the block reference Y3 will retain the current color of layer R (in this case it will be red) instead of taking up the color of blue.

SEPARATING INTO INDIVIDUAL OBJECTS

Figure 17–11

Invoking the EXPLODE command from the Modify toolbar

The EXPLODE command causes block references, hatch patterns, and associative dimensioning to be turned into the separate objects from which they were created. It also causes polylines/polyarcs and multilines to separate into individual simple line and arc objects. The EXPLODE command causes 3D polygon meshes to become 3Dfaces, and 3D polyface meshes to become 3Dfaces and simple line and point objects. When an object is exploded, the new, separate objects are created in the space (model or paper) of the exploded objects. Invoke the EXPLODE command from the Modify toolbar (see Figure 17–11), AutoCAD prompts:

Select objects: *(select objects to explode, and press enter to complete object selection)*

You can use one or more object selection methods. The object selected must be eligible for exploding, or an error message will appear. An eligible object may or may not change its appearance when exploded.

Possible Changes Caused by the EXPLODE Command

A polyline segment having width will revert to a zero-width line and/or arc. Tangent information associated with individual segments is lost. If the polyline segments have width or tangent information, the explode command will be followed by the message:

Exploding this polyline has lost (width/tangent) information.

The UNDO command will restore it.

Individual elements within blocks that were on layer 0 when created (and whose color was BYLAYER) but were inserted on a layer with a color different than that of layer 0 will revert to the color of layer 0.

An attribute within a block will revert to the attribute definition when the block reference is exploded and will be represented on the screen by its tag. The value of the attribute specified at the time of insertion is lost. The group will revert to those elements created by the ATTDEF command, prior to combining them into a block via the BLOCK command. The power and usage of attributes are discussed later in this chapter.

Exploding Block References with Nested Objects

Block references containing other blocks and/or polylines are separated for one level only. That is, the highest-level block reference will be exploded, but any nested blocks or polylines will remain block references or polylines. They in turn can be exploded when they come to the highest level.

Block references with equal X, Y, and Z scales explode into their component objects. Block references with unequal X, Y, and Z scales (nonuniformly scaled block references) might explode into unexpected objects.

---

note Block references inserted via the MINSERT command or external references and their dependent blocks cannot be exploded.

MULTIPLE INSERTS OF BLOCK REFERENCES

The minsert (multiple insert) command is used to insert block references in a rectangular array. The total pattern takes the characteristics of a block, except that the group cannot be exploded. This command works similarly to the rectangular array command. Invoke the minsert command at the On-screen prompt, AutoCAD prompts:

> Enter block name or ⬇: (enter name of the block or select one of the available blocks from the Shortcut menu)
>
> Specify insertion point or ⬇: (specify the insertion point or select one of the available options from the Shortcut menu)
>
> Enter X scale factor, specify opposite corner, or [Corner/XYZ] <1>: (specify the X scale factor)
>
> Enter Y scale factor <use X scale factor>: (specify y scale factor or press ENTER)
>
> Specify rotation angle <0>: (specify rotation angle)
>
> Enter number of rows (—) <default>: (specify the number of rows)
>
> Enter number of columns (||||) <default>: (specify the number of columns)
>
> Enter distance between rows or specify unit cell (—): (specify the distance between rows)
>
> Specify distance between columns (||||): (specify the distance between columns)

The row/column spacing can be specified by the delta-X/delta-Y distances between two points picked on the screen. For example, if, in response to the "Distance" prompt, you selected points 2,1 and 6,4 for the first and second points, respectively, the row spacing would be 3 (4 – 1) and the column spacing would be 4 (6 – 2).

CREATING ONE DRAWING FROM ANOTHER

The wblock command permits you to group objects in a manner similar to the block command. But in addition, wblock exports the group to a file, which, in fact, becomes a new and separate drawing. The new drawing (created by the wblock command) might consist of a selected block reference in the current drawing. Or it might be made up of selected objects in the current drawing. You can even export the complete current drawing to the new drawing file. The new drawing assumes the layers, linetypes, styles, and other environmental items, such as system variable settings of the current drawing.

AutoCAD displays the Write Block dialog box when you invoke the WBLOCK command, similar to Figure 17–12, where you provide a drawing name and its path in the **File name and path** box of the **Destination** section of the dialog box as you do when you begin a new drawing. You should enter only the file name and not the extension. AutoCAD appends the *.dwg* extension automatically. The name should comply with the operating system requirements of valid characters and should be unique for drawings on the specified folder/path. Otherwise you will get the message:

Drawingname.dwg already exists. Do you want to replace it?

Choose **Yes** only if you wish to overwrite the existing drawing with the same name. Otherwise choose **No** and enter a different, unique name for the drawing.

Choose one of the three sources from which to create a new drawing in the **Source** section of the Write Block dialog box.

 Figure 17–12

Write Block dialog box

Choose **Block** to create a drawing to be a duplicate of an existing block in the current drawing. Select the name of the desired block from the selection box to the right of the **Block**. By default, the name of the new drawing will be the same as the block chosen. You can enter a different name for the new drawing with its corresponding path in the **File name and path** box.

Choose **Entire drawing** to create a drawing to be a duplicate of the entire current drawing.

Choose **Objects** to create a drawing from objects to be selected in the current drawing. You must select one or more objects to be exported as a new drawing. The object selection process is the same as when you are creating a block with the BLOCK command, described earlier in this chapter.

Specify a base point for the block in the **Base point** section. The default value is 0,0,0. You can specify the insertion point on the screen, or you can specify X, Y, and Z coordinates for the base point in the **X**, **Y**, and **Z** boxes, respectively, located in the **Base point** section of the Write Block dialog box. To specify the base point on the screen, choose **Pick point**, located in the **Base point** section of the dialog box.

Specify the unit value in the **Insert units** box to be used for automatic scaling when the new file is dragged from DesignCenter and inserted as a block in a drawing that uses different units. Select Unitless if you do not want to automatically scale the drawing when you insert it.

Choose **OK** to create the drawing.

One advantage of using the WBLOCK command when the entire drawing is written to a file is that all of the unused blocks, layers, linetypes, and other unused named objects are not written. That is, the drawing is automatically purged. This means that unused items will not be written to the new drawing file, (for example, unused items include block definitions that have not been inserted, noncurrent layers that have no objects drawn on them, and styles that are not being used). This can be useful if you just wish to clean up a cluttered drawing, especially one that has had other drawing files inserted into it, each bringing with it various unused named objects.

REDEFINING THE INSERTION POINT

Figure 17-13

Invoking the BASE command from the Draw menu

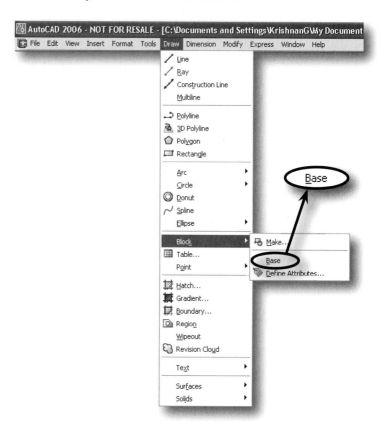

The BASE command (see Figure 17–13), allows you to establish a base insertion point for the whole drawing in the same manner that you specify a base insertion point when using the BLOCK command to combine elements into a block. The purpose of establishing this base point is primarily so that the drawing can be inserted into another drawing by way of the INSERT command and have the specified base point coincide with the specified insertion point. The default base point is the origin (0,0,0). You can specify a 2D point, and AutoCAD will use the current elevation as the base *Z* coordinate. Or you can specify the full 3D point.

ATTRIBUTES

Attributes can be used for automatic annotation during insertion of a block reference. Attributes are special text objects that can be included in a block definition and must be defined beforehand and then selected when you are creating a block definition.

Attributes have two primary purposes:

> The first use of attributes is to permit annotation during insertion of the block reference to which the attributes are attached. Depending on how you define an attribute, it either appears automatically with a preset (constant) text string or it prompts the user for a string to be written as the block reference is inserted. This feature permits you to insert each block reference with a string of preset text or with its own unique string.

> The second (perhaps the more important) purpose of attaching attributes to a block reference is to have extractable data about each block reference stored in the drawing database file. Then, when the drawing is complete (or even before), you can invoke the ATTEXT (short for "attribute extract") command to have attribute data extracted from the drawing and written to a file in a form that database-handling programs can use. You can have as many attributes attached to a block reference as you wish. As just mentioned, the text string that makes up an attribute can be either constant or user specified at the time of insertion.

A Definition Within a Definition

When creating a block, you select objects to be included. Objects such as lines, circles, and arcs are drawn by means of their respective commands. Normal text is drawn with the text or mtext command.

As with drawing objects, attributes must be drawn before they can be included in the block. AutoCAD calls this procedure defining the attribute. Therefore, an attribute definition is simply the result of defining an attribute by means of the ATTDEF command. The attribute definition is the object that is selected during the BLOCK command. Later, when the block reference is inserted, the attributes that are attached to it and the manner in which they become a part of the drawing are a result of how you created (defined) the attribute definition.

Visibility and Plotting

If an attribute is to be used only to store information, you can, as part of the definition of the attribute, specify whether or not it will be visible. If you plan to use an attribute with a block as a note, label, or callout, you should consider the effect of scaling (whether equal or unequal X/Y factors) on the text that will be displayed. The scaling factor(s) on the attribute will be the same as on the block reference. Therefore, be sure that it will result in the size and proportions desired. You should also be aware of the effect of rotation on visible attribute text. Attribute text that is defined as horizontal in a block will be displayed vertically when that block reference is inserted with a 90-degree angle of rotation.

Creating an Attribute Definition

▶ **Figure 17-14**

Invoking the ATTDEF command from the Draw menu

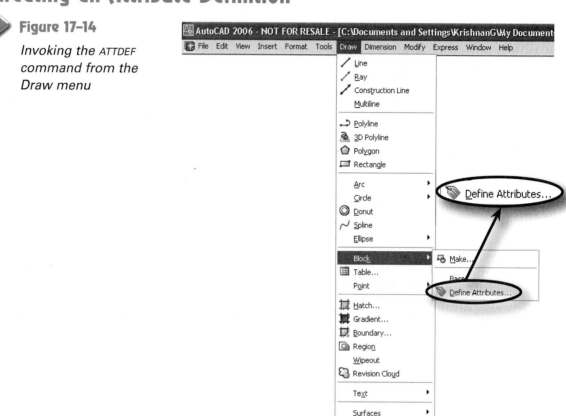

The ATTDEF command (see Figure 17–14) defines the mode; attribute tag, prompt, and value; insertion point; and text options for an attribute. After you invoke the ATTDEF command, AutoCAD displays the Attribute Definition dialog box, as shown in Figure 17–15.

▶ **Figure 17-15**

Attribute Definition dialog box

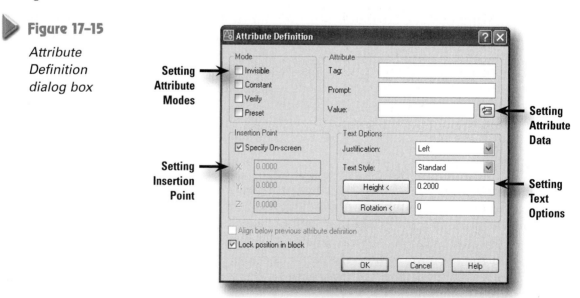

The **Mode** section sets options for attribute values associated with a block when you insert the block in a drawing. Set one or more of the available modes:

The **Invisible** mode specifies that attribute values are not displayed or printed when you insert the block. Setting **Invisible** to ON causes the attribute value *not* to be displayed when the block reference is completed. Even if visible, the value will not appear until the insertion is completed. Attributes needed only for data extraction should be invisible, to quicken regeneration and to avoid cluttering your drawing. You can use the ATTDISP command to override the Invisible mode setting.

The **Constant** mode provides attributes a fixed value for block insertions. If **Constant** is set to ON, you must enter the value of the attribute while defining it. That value will be used for that attribute every time the block reference to which it is attached is inserted. There will be no prompt for the value during insertion, and you cannot change the value.

The **Verify** mode prompts you to verify that the attribute value is correct when you insert the block. If **Verify** is set to ON, you will be able to verify its value when the block reference is inserted. For example, if a block reference with three (nonconstant value) attributes is inserted, once you have completed all prompt/ value sequences that have displayed the original defaults, you will be prompted again, with the latest values as new defaults, giving you a second chance to be sure the values are correct before the INSERT command is completed. Even if you press ENTER to accept an original default value, it also appears as the second-chance default. If, however, you make a change during the verify sequence, you will *not* get a third chance, that is, a second verify sequence.

The **Preset** mode sets the attribute to its default value when you insert a block containing a preset attribute. If **Preset** is set to ON, the attribute automatically takes the value of the default that was specified at the time of defining the attribute. During a normal insertion of the block reference, you will not be prompted for the value. You must be careful to specify a default during the ATTDEF command, or the attribute value will be blank. A block consisting of only attributes whose defaults were blank when **Preset** modes were set to ON could be inserted, but it would not display anything and cannot be purged from the drawing. The only adverse effect would be that of adding to the space taken in memory. One way to get rid of a non-displayable block reference like this is to use a visible entity to create a block with the same name, thereby redefining it to something that can be edited, that is, erased and subsequently purged.

You can duplicate an attribute definition with the COPY command, and use it for more than one block. Or you can explode a block reference and retain one or more of its attribute definitions for use in subsequent blocks.

The **Attribute** section of the Attribute Definition dialog box allows you to set attribute data. Enter the attribute's tag, prompt, and default value in the text boxes.

Specify the attribute's tag in the **Tag** box, which identifies each occurrence of an attribute in the drawing. The tag can contain any characters except spaces. AutoCAD changes lowercase letters to uppercase. The tag is the identifier of the attribute definition and is displayed where this attribute definition is located, depicting text size, style, and angle of rotation. Two attribute definitions with the same tag should not be included in the same block. Tags appear in the block definition only, not after the block reference is inserted. However, if you explode a block reference, the attribute value (described herein) changes back into the tag. If multiple attributes are used in one block, each must have a unique tag in that block.

Specify the prompt in the **Prompt** box that is displayed when you insert a block reference containing the attribute definition. If you do not specify the prompt, AutoCAD uses the attribute tag as the prompt. If you turn on the Constant mode, the **Prompt** field is disabled. The prompt is what you see when inserting a block reference with an attribute whose value is not constant or preset.

Specify the default attribute value in the **Value** box. This is optional, except if you turn on the Constant mode, for which the default value needs to be specified. The value of an attribute is the actual string of text that appears (if the visibility mode is set to ON) when the block reference (of which it is a part) is inserted. Whether visible or not, the value is tied directly to the attribute, which, in turn, associates it with the block reference. It is this value that is written to the database file. It might be a door or window size or, in a piping drawing, the flange rating, weight, or cost of a valve or fitting. In an architectural drawing, the value might represent the manufacturer, size, color, cost or other pertinent information attached to a block representing a desk.

---

note   When an extraction of attribute data is performed, it is the value of an attribute that is written to a file, but it is the tag that directs the extraction operation to that value. This will be described in detail in the later section, "Extracting Attributes."

The **Insertion Point** section of the dialog box allows you to specify a coordinate location for the attribute in the drawing, either by choosing **Specify On-screen** to specify the location on the screen or by entering coordinates in the text boxes provided.

The **Text Options** section of the dialog box allows you to set the justification, text style, height, and rotation of the attribute text.

Choose **Align below previous attribute definition** to place the attribute tag directly below the previously defined attribute. If you haven't previously defined an attribute definition, this option is unavailable.

Selecting **Lock position in block** locks the position of the attribute within the block reference.

Choose **OK** to define the attribute definition.

After you close the Attribute Definition dialog box, the attribute tag appears in the drawing. Repeat the procedure to create additional attribute definitions.

Inserting a Block Reference with Attributes

Blocks with attributes may be inserted in a manner similar to that for inserting regular block references. If there are any nonconstant attributes, you will be prompted to enter the value for each. You may set the system variable called attreq to 0 (zero), thereby suppressing the prompts for attribute values. In this case the values will either be blank or be set to the default values if they exist. You can later use the eattedit command to establish or change values.

Controlling the Display of Attributes

The attdisp command controls the visibility of attributes. Attributes will normally be visible if the Invisible mode is set to N (normal) when they are defined. Invoke the attdisp command options by selecting one of the available options from the Attribute Display flyout menu of Display on the View menu.

Choosing On makes all attributes visible;

Choosing Off makes all attributes invisible.

Choosing Normal displays the attributes as you created them.

If REGENAUTO is set to ON, changing the ATTDISP setting causes drawing regeneration.

Editing Attribute Values

Figure 17–16

Invoking the EATTEDIT command from the Modify II toolbar

Unlike other objects in an inserted block reference, attributes can be edited independently of the block reference. The EATTEDIT command allows you to change the value of attributes in blocks that have been inserted. This permits you to insert a block reference with generic attributes; that is, the default values can be used in anticipation of changing them to the desired values later. Or you can copy an existing block reference that may need only one or two attributes changed to make it correct for its new location. And, of course, there is always the chance that either an error was made in entering the value or design changes necessitate subsequent changes.

Invoke the EATTEDIT command (see Figure 17–16), AutoCAD prompts:

Select block reference: *(select the block reference)*

AutoCAD displays the Enhanced Attribute Editor dialog box, similar to the one shown in Figure 17–17. Selecting objects that are not block references or block references that contain no attributes will cause an error message to appear.

 Figure 17–17

Enhanced Attribute Editor dialog box

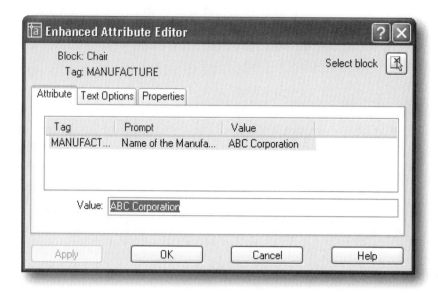

The dialog box lists all the attributes defined with values for the selected block reference. Make the necessary changes to the selected attribute values. In addition, you can also change the text attributes and object properties of the selected attribute tag. Accept the changes by choosing **OK**. Choosing the **Cancel** button terminates the command, returning all values to their original state.

Block Attribute Manager

 Figure 17–18

Invoking the BATTMAN command from the Modify II toolbar

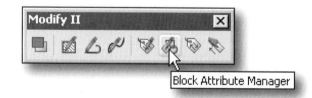

The BATTMAN command (see Figure 17–18) provides a means of managing blocks that contain attributes. AutoCAD displays the Block Attribute Manager dialog box, similar to Figure 17–19.

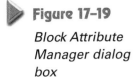

Figure 17-19

Block Attribute Manager dialog box

The BATTMAN command lets you edit the attribute definitions in blocks, change the order in which you are prompted for attribute values when inserting a block, and remove attributes from blocks. AutoCAD displays attributes of the selected block in the attribute list of the Block Attribute Manager dialog box. By default, the Tag, Prompt, Default, and Modes attribute properties are shown in the attribute list. You can specify which attribute properties you want displayed in the list by choosing **Settings**. The number of instances of the selected block is shown in a description below the attribute list.

The **Block** box lists all block definitions in the current drawing that have attributes from which you can select the block whose attributes you want to modify.

Choose **Select block** to select a block from the drawing area with your pointing device.

Choose **Sync** to update all instances of the selected block with the attribute properties currently defined.

Choose **Move Up** to move the selected attribute tag earlier in the prompt sequence. **Move Up** is not available when a constant attribute is selected.

Choose **Move Down** to move the selected attribute tag later in the prompt sequence. **Move Down** is not available when a constant attribute is selected.

Choose **Edit** to display the Edit Attribute dialog box (see "Editing Attribute Values" earlier in this chapter) where you can modify attribute properties.

Choose **Remove** to remove the selected attribute from the block definition. **Remove** is not available for blocks with only one attribute.

Choose **Settings** to open the Settings dialog box, where you can customize how attribute information is listed in the Block Attribute Manager.

Choose **Apply** to update the drawing with the attribute changes you have made and leave the Block Attribute Manager open.

Choose **OK** to close the Block Attribute Manager dialog box and save the changes.

Extracting Attributes

Extracting data from a drawing is one of the most innovative features in CAD. Paper copies of drawings have long been used to communicate more than just how objects look. In addition to dimensions, drawings tell builders or fabricators what materials to use, quantities of objects to make, manufacturers' names and models of parts in an assembly, coordinate locations of objects in a general area, and what types of finishes to apply to surfaces. But, until computers came into the picture (or pictures came into the computer), extracting data from manual drawings involved making lists (usually by hand) while studying the drawing, often checking off the data with a marker. The AutoCAD attribute feature and the attext command combine to allow complete, fast, and accurate extraction of, first, data consciously put in for the purpose of extraction, second, data used during the drawing process, and third, data that AutoCAD maintains about all objects (block references in this case).

 Figure 17–20

Invoking the EATTEXT command from the Modify II toolbar

Invoke the EATTEXT command (see Figure 17–20) to extract attribute objects through the Attributes Extraction wizard. AutoCAD displays the Attribute Extraction – Select Drawing page of the wizard, as shown at top left in Figure 17–21.

Figure 17–21

Attribute Extraction wizard pages

You can select a drawing file from which to extract information from block attributes or select blocks that have attributes in the current drawing. Three options are available:

Select Objects lets you select specific blocks and Xrefs in the current drawing for extracting attribute data.

Current Drawing selects all blocks that have attributes from the current drawing for extracting attribute data.

Select Drawings lets you select multiple drawing files for extracting attribute data.

After selecting the desired objects or drawing(s), choose **Next>**, and AutoCAD displays the Attribute Extraction – Settings page, as shown at top right in Figure 17–21, which allows you to specify whether to extract block attribute information from external reference files and nested blocks.

Select **Include xrefs** to extract block attribute information from external references (xrefs).

Select **Include nested blocks** to extract from blocks nested in other blocks.

Choose **Next>**, and AutoCAD displays the Attribute Extraction – Use Template page, as shown at second row left in Figure 17–21, which allows you to use the block attribute settings from templates previously saved in a template file. The template is a file saved in ASCII format; it lists the fields that specify the tags and determine which block references will have their attribute data extracted.

Choose **Next>**, and AutoCAD displays the Attribute Extraction – Select Attributes page, as shown at second row right in Figure 17–21, which allows you to select blocks and attributes. Blocks in the selected drawing are listed in the **Blocks** section. In the Block Alias column, you can assign an alias to the block. The Number column displays the number of instances of the block that are present in the selected drawings. Select the box next to a block name to display that block's attributes in the **Attributes for block** section. The Value column displays the value of the block attribute. Select the check box next to an attribute name to extract the information for that attribute. Choose **Check All** to select all blocks or block attributes for extraction, and choose **Uncheck All** to clear the selection of all blocks or block attributes.

Choose **Next>**, and AutoCAD displays the Attribute Extraction – View Output page, as shown at third row left in Figure 17–21, which allows you to preview the block attributes to be extracted. The list displays the attributes currently selected for extraction.

Choose **Next>**, and AutoCAD displays the Attribute Extraction – Save Template page, as shown at third row right in Figure 17–21, which allows you to save the attribute extraction settings you have made to a template file.

Choose **Next>**, and AutoCAD displays the Attribute Extraction – Export page, as shown at bottom left in Figure 17–21, which allows you to specify the attribute extraction file name and format and export the attribute information to the specified file.

Choose **Finish** to extract the block attribute information, export it to the file specified, and close the wizard.

Changing Objects in a Block Reference Without Losing Attribute Values

Sometimes it might be desirable to make changes to the objects within block references that have been inserted with attributes and have had values assigned to the attributes. Remember that the values of the attributes can be edited with the eattedit command. But, in order to change the geometry of a block reference, it is normally required that you explode the block reference, make the necessary changes, and then redefine the block. As long as the attribute definitions keep the same tags in the new definition, the redefined blocks that have already been inserted in the drawing will retain those attributes with the original definitions.

Another method of having a new block definition applied to existing block references is to create a new drawing with the new block definition. This can be done by using the WBLOCK command to create a drawing that conforms to the old definition and then to edit that drawing. Or you can start a new drawing with the name of the block that you wish to change.

In order to apply the new definition to the existing block reference, you can call up the drawing with the OPEN command, and then use the INSERT command with the **Enter block name** option. For example, if you wish to change the geometry of a block named Part_1, and the changes have been made and stored in a separate drawing with the same name (Part_1), the sequence of prompts is as follows:

-insert

Enter block name or ⬇: **part_1=**

Block "part_1" already exists. Redefine it? [Yes/No] <N>: **y**

Block "Part_1" redefined

Regenerating model.

Specify insertion point or ⬇: (ENTER)

It is not necessary to specify an insertion point or to respond to scale factor and rotation angles. Simply inserting the block reference with the = (equals) behind the name will cause the definition of the drawing to become the new definition of the block with the same name residing in the current drawing.

Redefining a Block with Attributes

The attredef command allows you to redefine a block reference and updates associated attributes. After you invoke the attredef command, AutoCAD prompts:

Enter name of the block you wish to redefine: *(specify the block name to redefine)*

Select objects for new Block...

Select objects: *(select objects for the block to redefine and press ENTER)*

Specify insertion base point of new Block: *(specify the insertion base point of the new block)*

New attributes assigned to existing block references are given their default values. Old attributes in the new block definition retain their old values. Old attributes not included in the new block definition are deleted from the old block references.

DIVIDING OBJECTS

Figure 17–22

Invoking the DIVIDE command from the Point flyout of the Draw menu

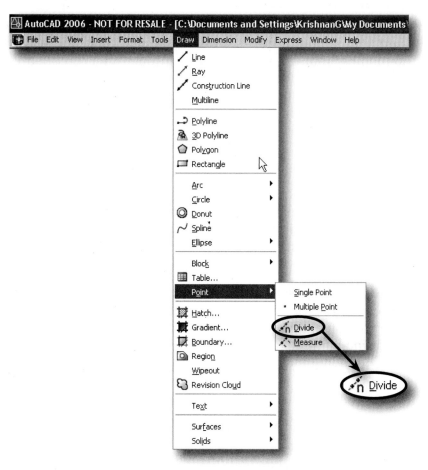

The DIVIDE command causes AutoCAD to divide an object into equal-length segments, placing markers at the dividing points. Objects eligible for application of the DIVIDE command are the line, arc, circle, ellipse, spline, and polyline. Selecting an object other than one of these will cause an error message to appear, and you will be returned to the Command: prompt. Invoke the DIVIDE command (see Figure 17–22), AutoCAD prompts:

Select the object to divide: *(select a line, arc, circle, ellipse, spline, or polyline)*

Enter the number of segments or ⬇: *(specify the number of segments)*

You can respond with an integer from 2 to 32767, causing points to be placed along the selected object at equal distances but not actually separating the object. You can use the Node Object Snap mode to snap to the divided points. Logically, there will be one less point placed than the number entered, except in the case of a circle. The circle will have the first point placed at the angle from the center of the current snap rotation angle. A closed polyline will have the first point placed at the first point drawn in the polyline. The total length of the polyline will be divided into the number of segments entered without regard to the length of the individual segments that make up the polyline. An example of a closed polyline is shown in Figure 17–23.

Figure 17-23

The DIVIDE command as used with a closed polyline

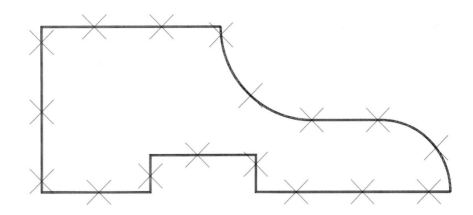

---

note It is advisable to set the PDSIZE and PDMODE system variables to values that will cause the points to be visible.

The Block option allows a named block reference to be placed at the dividing points instead of a point. The sequence of prompts is as follows:

divide (enter)

Select object to divide: *(select a line, arc, circle, ellipse, spline, or polyline)*

Enter the number of segments or ⊕: *(choose Block from the Shortcut menu)*

Enter name of block to insert: *(enter the name of the block)*

Align block with object? ⊕ <Y>: *(press ENTER to align the block reference with the object, or enter **n**, for not to align with the block reference)*

Enter the number of segments: *(specify the number of segments)*

If you respond with **No** or **N** to the "Align block with object?" prompt, all of the block references inserted will have a zero angle of rotation. If you respond with **Yes**, the angle of rotation of each inserted block reference will correspond to the direction of the linear part of the object at its point of insertion or to the direction of a line tangent to a circular part of an object at the point of insertion.

MEASURING OBJECTS

 Figure 17-24

Invoking the MEASURE command from the Point flyout of the Draw menu

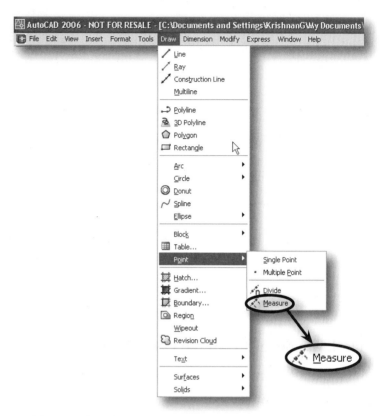

The MEASURE command causes AutoCAD to divide an object into specified-length segments, placing markers at the measured points. Objects eligible for application of the MEASURE command are the line, arc, circle, ellipse, spline, and polyline. Selecting an object other than one of these will cause an error message to appear, and you will be returned to the Command: prompt. Invoke the MEASURE command (see Figure 17–24), AutoCAD prompts:

Select object to measure: *(select a line, arc, circle, ellipse, spline, or polyline)*

Specify length of segment or ⬇: *(specify the length of the segment, or enter* **b** *for the Block option)*

If you reply with a distance, or show AutoCAD a distance by specifying two points, the object is measured into segments of the specified length, beginning with the closest endpoint from the selected point on the object. The Block option allows a named block reference to be placed at the measured point instead of a point. The sequence of prompts is as follows:

measure (ENTER)

Select object to measure: *(select a line, arc, circle, ellipse, spline, or polyline)*

Specify length of segment or ⬇: *(choose Block from the Shortcut menu)*

Enter name of block to insert: *(enter the name of the block)*

Align block with object? ⬇: *(press* ENTER *to align the block reference with the object, or enter* **n,** *for not to align with the block reference)*

Specify length of segment: *(specify the length of segments)*

INTRODUCTION TO DYNAMIC BLOCKS

The following is a brief explanation of the Dynamic Block feature, introduced in AutoCAD 2006. It is an extremely powerful new addition to the already formidable design/engineering program provided by Autodesk.

A dynamic block has flexibility and intelligence. Individual objects or groups of objects within a dynamic block reference can easily be changed in a drawing while you work. You can manipulate the geometry in a dynamic block reference through custom grips or custom properties. This allows you to adjust the block in-place, as necessary, rather than searching for another block to insert or having to redefine the existing one.

For example, after inserting a block in a drawing representing a door, you might need to change the size of the door while you're editing the drawing. If the block is dynamic and defined to have an adjustable size, you can change the size of the door simply by dragging the custom grip or by specifying a different size in the Properties palette. You might also need to change the open angle of the door. The door block might also contain an alignment grip, which allows you to align the door block reference easily to other geometry in the drawing.

Figure 17–25 shows three insertions of a block representing a simplified instrument panel. The block on the left is shown with the geometry as it was drawn when the block was defined with the ON/OFF switch in the center. The middle and right insertions (called block references) are references of the same block with the ON/OFF switch moved to the left and right respectively. All three references are insertions of the same block definition. The dynamic capability of the block allows the user to change the size, shape, location or orientation of pre-selected geometry in the reference (after it has been inserted). In Figure 17–25, all three references are drawn from a single block definition.

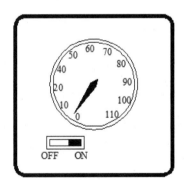
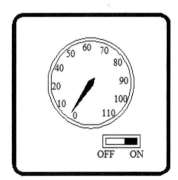

 Figure 17-25

Three Block References of an Instrument Panel with the ON/OFF switch in the Center (as defined), on the left, and on the right

As with almost every complicated feature, learning Dynamic Blocks requires advanced skills and an investment in time and effort. Full coverage is beyond the scope of this text. Extensive coverage of Dynamic Blocks is available in the Stellman/Krishnan text *Harnessing AutoCAD 2006*, published by Thomson/Delmar Learning.

drawing exercises

Open the Exercise Manual PDF file for Chapter 17 on the accompanying CD for discipline-specific exercises.

review questions

1. The maximum number of characters for a block name is:

    a. *8.* **d.** *31.*

    b. *16.* **e.** *255.*

    c. *23.*

2. A block is:

    a. *A rectangular-shaped figure available for insertion into a drawing.*

    b. *A single element found in a block formation of a building drawn with AutoCAD.*

    c. *One or more objects stored as a single object for later retrieval and insertion.*

    d. *None of the above.*

3. A block reference cannot be exploded if it:

    a. *Consists of other blocks (nested).*

    b. *Has a negative scale factor.*

    c. *Has been moved.*

    d. *Has different X and Y scale factors.*

    e. *None of the above.*

4. All of the following can be exploded, except:

    a. *Block references.*

    b. *Associative dimensions.*

    c. *Polylines.*

    d. *Block references inserted with MINSERT command.*

    e. *None of the above.*

5. To insert a block reference called "TABLE" and have the block reference converted to an individual object as it is inserted, what should you use for the block name?

    a. */TABLE* **d.** *TABLE*

    b. **TABLE* **e.** *None of the above*

    c. *?TABLE*

6. To return a block reference back to its original objects, use:

    a. *EXPLODE.* **d.** *UNDO.*

    b. *BREAK.* **e.** *STRETCH.*

    c. *CHANGE.*

7. To identify a new insertion point for a drawing file which will be inserted into another drawing, invoke the:

    a. *BASE command.* **d.** *WBLOCK command.*

    b. *INSERT command.* **e.** *DEFINE command.*

    c. *BLOCK command.*

8. MINSERT places multiple copies of an existing block similar to the command:

   a. *ARRAY.*   **d.** *COPY.*

   b. *MOVE.*   **e.** *MIRROR.*

   c. *INSERT.*

9. If one drawing is to be inserted into another drawing and editing operations are to be performed on the inserted drawing, you must first:

   a. *Use the PEDIT command.*

   b. *EXPLODE the inserted drawing.*

   c. *UNDO the inserted drawing.*

   d. *Nothing, it can be edited directly.*

   e. *None of the above.*

10. The BASE command:

   a. *Can be used to move a block reference.*

   b. *Is a subcommand of PEDIT.*

   c. *Will accept 3D coordinates.*

   d. *Allows one to move a dimension baseline.*

   e. *None of the above.*

11. Attributes are associated with:

   a. *Objects.*   **d.** *Layers.*

   b. *Block references.*   **e.** *Shapes.*

   c. *Text.*

12. To merge two drawings, use:

   a. *INSERT.*   **d.** *BLOCK.*

   b. *MERGE.*   **e.** *IGESIN.*

   c. *BIND.*

13. The DIVIDE command causes AutoCAD to:

   a. *Divide an object into equal length segments.*

   b. *Divide an object into two equal parts.*

   c. *Break an object into two objects.*

   d. *All of the above.*

14. One cannot explode:

   a. *Polylines containing arcs.*

   b. *Blocks containing polylines.*

   c. *Dimensions incorporating leaders.*

   d. *Block references inserted with different X, Y, and Z scale factors.*

   e. *None of the above.*

15. The DIVIDE command will:

   a. *Place points along a line, arc, polyline, or circle.*

   b. *Accept 1.5 as segment input.*

   c. *Place markers on the selected object and separate it into different segments.*

   d. *Divide any object into the equal number of segments.*

16. The MEASURE command causes AutoCAD to divide an object:

    a. *Into specified length segments.*
    b. *Into equal length segments.*
    c. *Into two equal parts.*
    d. *All of the above.*

17. A command used to edit attributes is:

    a. *EDDATTE.*   **d.** *ATTFILE.*
    b. *EDIT.*   **e.** *None of the above.*
    c. *EDITATT.*

18. Attributes are defined as the:

    a. *Database information displayed as a result of entering the LIST command.*
    b. *X and Y values that can be entered when inserting a block reference.*
    c. *Coordinate information of each vertex found along a SPLINE object.*
    d. *None of the above.*

drawings in drawings

introduction

One of the most powerful time-saving features of AutoCAD is the ability to have one drawing (referred to as an external reference) become part of a second drawing while maintaining the integrity and independence of the first one. And if the external reference is changed, those changes will be reflected in the drawing in which it is referenced. AutoCAD lets you display or view the contents of other drawing files while working in your current drawing file. This feature is provided by the XREF command, short for external reference.

After completing this chapter, you will be able to do the following:

- Attach and detach reference files

- Change the path for reference files

- Load and unload reference files

- Decide whether to attach or overlay an external reference file

- Clip external reference files

- Control dependent symbols

- Edit external references

- Manage external references

- Use the BIND command to add dependent symbols to the current drawing

- Attach and detach image files

ALL ABOUT EXTERNAL REFERENCES

As mentioned in Chapter 17, existing AutoCAD drawings can be combined by means of the INSERT command, inserting one drawing into another. When one drawing is inserted into another, it is actually a duplicate of the inserted drawing that becomes a part of the drawing into which it is inserted. The data from the inserted drawing is added to the data of the current drawing. Once the duplicate of a drawing is inserted, no link or association remains between the original drawing from which the inserted duplicate came and the drawing it has been inserted into.

The external reference feature gives users another method for combining existing drawing files. The XREF command (external reference) does not obsolete the INSERT feature; users can decide which method is more appropriate for the current application.

When a drawing is externally referenced (instead of inserted as a block), the user can view and object snap to the external reference from the current drawing, but each drawing's data is still stored and maintained in a separate drawing file. The only information in the reference drawing that becomes a permanent part of the current drawing is the name of the reference drawing and its folder path. If necessary, externally referenced files can be scaled, moved, copied, mirrored, or rotated by using the AutoCAD modify commands. You can control the visibility, color, and linetype of the layers belonging to an external drawing file. This lets you control which portions of the external drawing file are displayed, and how. No matter how complex an external reference drawing may be, it is treated as a single object by AutoCAD. If you invoke the MOVE command and select one line, for example, the entire object is highlighted and moves, not just the line you selected. You cannot explode the externally referenced drawing. A manipulation performed on an external reference will not affect the original drawing file, because an external reference is only an image, however scaled or rotated.

Accuracy and efficient use of drawing time are other important design benefits that are enhanced through external reference files. When an addition or change is made to a drawing file that is being used as an external reference file, all drawings that use the file as an external reference will reflect the modifications. Figure 18–1 shows a drawing that externally references four other drawings.

When you attach a drawing file as an external reference file, it is permanently attached until it is detached or bound to the current drawing. When you load the drawing with external references, AutoCAD automatically reloads each external reference drawing file; thus, each external drawing file reflects the latest state of the referenced drawing file.

External reference files will save time and ensure the drawing accuracy required to produce a professional product. Figure 18–1E is the foundation plan and Figure 18–1F is the roof plan and front elevation of a residence. Figure 18–1A is the floor plan of the residence and Figure 18–1B is a typical border/title block used in the layout of the drawings for a set of drawing for the residence. shows a drawing that externally references four other drawings to illustrate the doorbell detail. Figures 18–1C and 18–1D show how the foundation and roof plans used the floor plan drawing as an external reference to create the geometry based on the dimensions and details in the floor plan. When these drawings are configured for plotting, the floor plan reference is hidden but not removed from the drawing. In case any changes are made to the floor plan, they will be reflected automatically so corresponding changes can be made in the foundation and roof plans. Figure 18–1F also shows an external reference of the drawing of the front elevation, which is included in the final plot. Figures 18–1E and 18–1F can now be completed by adding dimensions, filling in the title block and adding notes, call-outs and labeling the views.

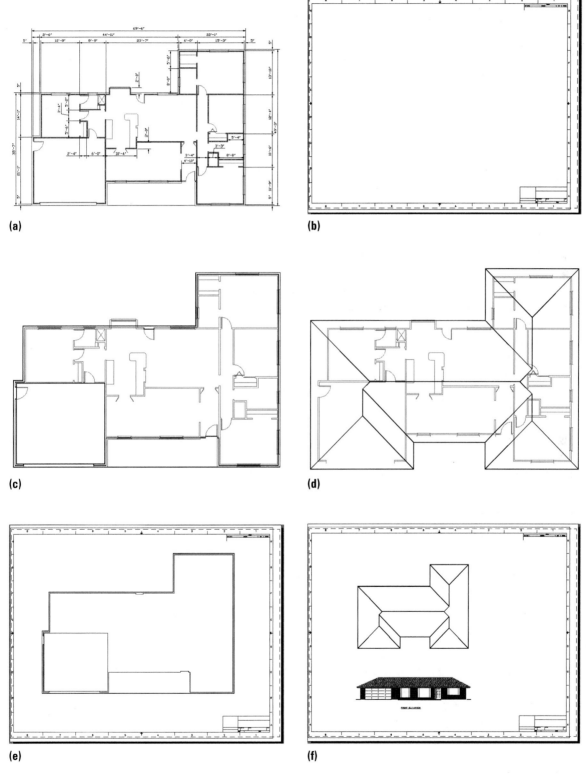

(a) (b)

(c) (d)

(e) (f)

Figure 18–1A-1F

Drawings 18-1E and 18-1F appear to be single drawings but they each reference one or more other drawings

The XREF command, when combined with the networking capability of AutoCAD, gives the project manager powerful tools for coping with the problems of file management. The project manager instantaneously sees the work of the departments and designers working on aspects of the contract. If necessary, you can overlay a drawing where appropriate, track the progress, and maintain document integrity. At the same time, departments need not lose control over individual designs and details.

If you need to make changes to an attached external reference file while you are in the host drawing, you can do so by using the REFEDIT command. AutoCAD also allows you to open an attached external reference in a separate window. With the XOPEN command, the external reference opens immediately in a new window. You can make necessary changes and save the changes. Immediately, the changes will be reflected in the host drawing.

In AutoCAD you can control the display of the external reference file by means of clipping, so you can display only a specific section of the reference file.

External References and Dependent Symbols

The symbols that are carried into a drawing by an external reference are called dependent symbols, because they depend on the external file, not on the current drawing, for their characteristics. The symbols have arbitrary names and include blocks, layers, linetypes, text styles, and dimension styles.

When you attach an external reference drawing, AutoCAD automatically renames the xref's dependent symbols. AutoCAD forms a temporary name for each symbol by combining its original name with the name of the xref itself. The two names are separated by the vertical bar (|) character. Renaming the symbols prevents the xref's objects from taking on the characteristics of existing symbols in the drawing.

For example, you created a drawing called PLAN1 with layers 0, First-fl, Dim, and Text, in addition to blocks Arrow and Monument. If you attach the PLAN1 drawing as an external reference file, the layer First-fl will be renamed as PLAN1|First-fl, Dim as PLAN1|Dim, and Text as PLAN1|Text, as shown in Figure 18–2. Blocks Arrow and Monument will be renamed as PLAN1|Arrow and PLAN1|Monument. The only exceptions to renaming are unambiguous defaults like layer 0 and linetype continuous. The information on layer 0 from the reference file will be placed on the active layer of the current drawing when the drawing is attached as an external reference of the current drawing. It takes on the characteristics of the current drawing.

Figure 18-2

*Layer
Properties
Manager*

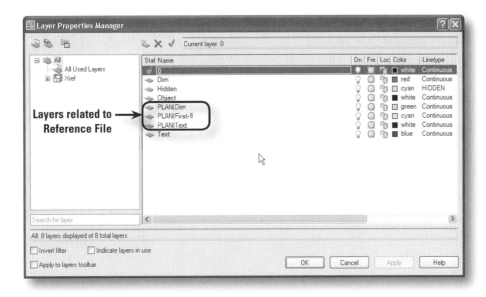

This prefixing is carried to nested xrefs. For example, if the external file PLAN1 included an xref named "Title" that has a layer Legend, it would get the symbol name PLAN1|Title|Legend if PLAN1 were attached to another drawing.

This automatic renaming of an xref's dependent symbols has two benefits:

- It allows you to see at a glance which named objects belong to which external reference file.
- It allows dependent symbols to have the same base name in both the current drawing and an external reference, and coexist without any conflict.

The AutoCAD commands and dialog boxes for manipulating named objects do not let you select an xref's dependent symbols. Usually, dialog boxes display these entries in lighter text.

For example, you cannot insert a block that belongs to an external reference drawing in your current drawing, nor can you make a dependent layer the current layer and begin creating new objects. These types of tasks are made easier by using AutoCAD's DesignCenter, explained in Chapter 19

You can control the visibility of the layers (ON/OFF, Freeze/Thaw) of an external reference drawing and, if necessary, you can change the color and linetype. When the VISRETAIN system variable is set to 0 (default), any changes you make to these settings apply only to the current drawing session. They are discarded when you end the drawing. If VISRETAIN is set to 1, the current drawing visibility, color, and linetype for xref dependent layers take precedence. They are saved with the drawing and are preserved during xref reload operations.

There may be times when you want to make your xref data a permanent part of your current drawing. To make an xref drawing a permanent part of the current drawing, use the Bind option of the XREF command. With the Bind option, all layers and other symbols, including the data, become part of the current drawing. This is similar to inserting a drawing via the INSERT command.

If necessary, you can make dependent symbols such as layers, linetypes, text styles, and dimension styles part of the current drawing by using the XBIND command instead of binding the whole drawing. This allows you to work with the symbol just as if you had defined it in the current drawing.

ATTACHING EXTERNAL REFERENCES

 Figure 18–3

Invoking the XATTACH command from the Reference toolbar

The XATTACH command (see Figure 18–3) attaches an external reference to the current drawing. After you invoke the XATTACH command, AutoCAD displays the Select Reference File dialog box. Select the drawing file from the appropriate directory to attach to the current drawing. AutoCAD displays the External Reference dialog box, as shown in Figure 18–4.

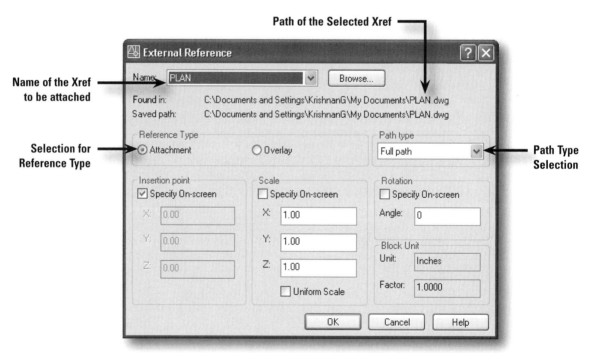

Figure 18–4

External Reference dialog box

The **Name** box displays the external reference drawing name attached to the current drawing.

Choose **Browse** to display the Select a Reference dialog box, in which you can select additional drawing files to attach to the current drawing.

Found in displays the path where the external reference was found, and **Saved path** displays the saved path, if any, that is used to locate the external reference.

The **Reference Type** section specifies whether the external reference is an attachment or an overlay. If you attach an external reference in the **Attachment** mode, the external reference will be included in the drawing when the drawing itself is attached as an external reference to another drawing. If you attach an external reference in the **Overlay** mode, however, it is not included in a drawing when the drawing itself is attached as an external reference or overlaid external reference to another drawing. The only behavioral difference between overlays and attachments is how nested references are handled. Overlaid external references are designed for data sharing. If necessary, you can change the status from Attachment to Overlay, or vice versa.

The **Path type** box specifies whether the saved path to the xref is set to No path, Full path or Relative path. When you set the path type to No path, AutoCAD first looks for the xref in the folder of the host drawing. This option is useful when the xref files are in the same folder as the host drawing. Instead, if you set the path type to Full path, AutoCAD saves the xref's precise location to the host drawing. This option is the most precise but the least flexible. If you move a project folder, AutoCAD cannot resolve any xrefs that are attached with a full path. And if you set the path type to Relative path, AutoCAD saves the xref's location relative to the host drawing. If you move a project folder, AutoCAD can resolve xref's attached with a relative path, as long as the xref's location relative the host drawing has not changed. You must save the current drawing before you can set the path type to Relative path.

Insertion point specifies the insertion point of the external reference.

Scale specifies the X scale factor, Y scale factor, and Z scale factor of the external reference. Selecting **Uniform Scale** causes the Y and Z scale factors to be the same as the specified X scale factor.

Rotation specifies the rotation angle of the external reference.

The **Insertion point**, **Scale**, and **Rotation** features are similar to those in the insertion of a block, explained in Chapter 17.

The **Block Unit** section displays the units (inches, millimeters, etc.) that were used when the selected reference drawing was created and the unit scale factor, calculated based on the reference drawing units value and the current drawing units.

Choose **OK** to attach the selected external reference drawing to the current drawing.

---

note   Once a drawing is attached as a reference file, the Manage Xref's icon is displayed in the Status bar, allowing you to open the Reference Manager.

MANAGING EXTERNAL REFERENCES

Figure 18-5

Open Xref Manager from the Reference toolbar

Open the Xref Manager by choosing External Reference on the Reference toolbar (see Figure 18–5). The Xref Manager (see Figure 18–6) allows you to manage various tasks related to reference files attached to the current drawing.

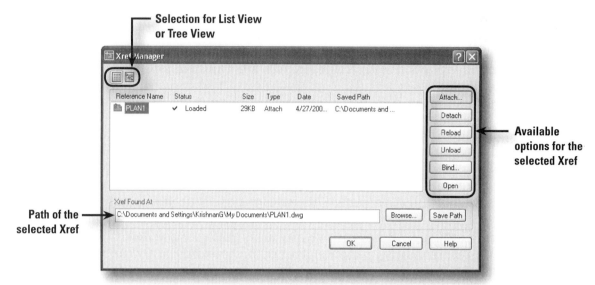

 Figure 18–6

Xref Manager dialog box

The Xref Manager lists the xrefs in the drawing in a tree view or a list view. You can use the F3 and F4 keys to switch between list view and tree view. By default, the List View option (see Figure 18–6) displays a list of the attached reference files and their associated data. To sort a column alphabetically, select the column heading. A second click sorts it in reverse order. To resize a column's width, select the separator between columns and drag the pointing device to the right or left.

In a Tree View listing, AutoCAD displays a hierarchical representation of the external reference in alphabetical order, as shown in Figure 18–7. The Tree View shows the level of nesting relationship of the attached external references, whether they are attached or overlaid, whether they are loaded, unloaded, marked for reload or unload, or not found, unresolved, or unreferenced.

 **Figure 18–7**

Xref Manager dialog box with Tree View listing

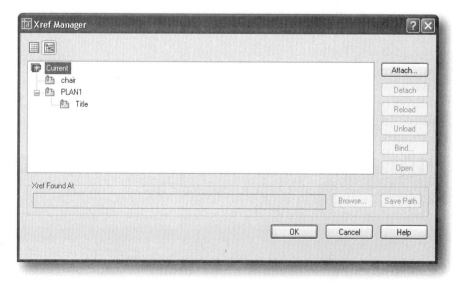

The **Reference Name** column (see Figure 18–6) lists the drawings that are attached to the current drawing. The external reference file's Reference Name does not have to be the same as its original file name. To rename the external file, click the name twice or press F2. The new name can contain up to 255 characters, including embedded spaces and punctuation.

The **Status** column displays the state of the external reference file, which can be Loaded, Unloaded, Unreferenced, Unresolved, Orphaned, or Not found. Loaded indicates that the external drawing is attached and displayed. Unloaded indicates that the external reference is attached but not displayed. To unload, first select the drawing name(s) in the list, and choose **Unload**. To display the drawing again, choose **Reload**. Unreferenced indicates that the drawing is attached but erased. Unresolved indicates that AutoCAD cannot read. Orphaned indicates that the drawing is attached to another xref that is unreferenced, unresolved, or not found. Not found indicates that the drawing no longer exists in the valid search paths.

The **Size** column shows the file size of the corresponding external reference drawing. The size is not displayed if the external reference is unloaded, not found, or unresolved.

The **Type** column indicates whether the external reference is an attachment or an overlay.

The **Date** column displays the last date the associated drawing was modified. The date is not displayed if the external reference is unloaded, not found, or unresolved.

The **Saved Path** column shows the saved path of the associated external reference file. To change to a different path or file name for the currently selected external reference file, choose **Browse** in the Xref Manager dialog box. AutoCAD displays the Select New Path dialog box, in which you can specify a different path or file name. To save the path as it appears in the Saved Path field of the currently selected external reference file (and displayed in the **Xref Found At** box), select **Save Path**. AutoCAD saves the path of the currently selected external reference file.

Detaching External Referencs

The **Detach** option detaches one or more external reference drawings from the current drawing. To detach, first select the drawing name(s) from the displayed list in the Xref Manager dialog box, and then choose **Detach**. AutoCAD detaches the selected external reference drawing(s) from the current drawing, erasing all instances of the specified xref(s) and marking the xref definition(s) for deletion from the definition table. Only the xrefs attached or overlaid directly to the current drawing can be detached; nested xrefs cannot be detached. AutoCAD cannot detach an xref referenced by another xref or block.

Reloading External References

The **Reload** option allows you to display the attached external reference drawings as most recently saved. To reload external reference drawing(s), first select the drawing name(s) from the displayed list in the Xref Manager dialog box, and then choose **Reload**. AutoCAD reloads the selected external reference drawing(s). When you open a drawing, it automatically reloads any external references attached. It is helpful in a network environment to get the latest version of the reference drawing while you are in an AutoCAD session.

Unloading External References

The **Unload** option unloads one or more xrefs. Unloaded xrefs can be easily reloaded. Unlike detaching, unloading merely suppresses the display and regeneration of the external reference definition, to help current session editing and improve performance. Unloading can also be useful when a series of external reference drawings needs to be viewed during a project on an as-needed basis. Rather than have the referenced files displayed at all times, you can reload the drawing when you require the information.

Binding External References

The Bind option makes the selected xref and its dependent named objects (such as blocks, text styles, dimension styles, layers, and linetypes) a part of the current drawing. In other words, the Bind option allows you to make your external reference drawing data a permanent part of the current drawing. To bind one or more external reference drawings, first select the drawing name(s) from the displayed list in the Xref Manager dialog box, and then choose **Bind**. AutoCAD displays the Bind Xrefs dialog box shown in Figure 18–8. Select one of the two available bind types: **Bind** or **Insert**.

Figure 18–8

Bind Xrefs dialog box

The **Bind** selection binds the selected xref definition(s) to the current drawing, and it becomes an ordinary block in your current drawing. It also adds the dependent symbols to your drawing, letting you use them as you would any other named objects. In the process of binding, AutoCAD renames the dependent symbols. The vertical bar symbol (|) is replaced with three new characters: a $, a number, and another $. The number is assigned by AutoCAD to ensure that the named object will have a unique name. If you do not want to bind the entire external reference drawing, but only specific dependent symbols, such as a layer, linetype, block, dimension style, or text style, then you can use the XBIND command, explained later in this chapter, in "Adding Dependent Symbols to the Current Drawing."

The **Insert** selection binds the xref to the current drawing in a way similar to detaching and inserting the reference drawing. It is just like inserting a drawing with the INSERT command. AutoCAD adds the dependent symbols to the current drawing by stripping off the external reference drawing name.

Opening the External Reference

The **Open** option opens the selected xref for editing in a new window. The new window is displayed after the Xref Manager is closed. To open one or more xrefs, first select the drawing name(s) from the displayed list in the Xref Manager dialog box, and then choose **Open**. You can also open an external reference for editing by invoking the XOPEN command.

After making the necessary changes in the Xref Manager dialog box, choose **OK** to keep the changes and close the dialog box.

ADDING DEPENDENT SYMBOLS TO THE CURRENT DRAWING

Figure 18-9

Invoking the XBIND command from the Reference toolbar

The XBIND command (see Figure 18–9) lets you permanently add a selected subset of external reference-dependent symbols to your current drawing. The dependent symbols include block, layer, linetype, dimension style, and text style. Once the dependent symbol is added to the current drawing, it behaves as if it were created in the current drawing and saved with the drawing when you close the drawing session. While adding the dependent symbol to the current drawing, AutoCAD removes the vertical bar symbol (|) from each dependent symbol's name, replacing it with three new characters: a $, a number, and another $ symbol.

For example, you might want to use a block that is defined in an external reference. Instead of binding the entire external reference, it is advisable to use the XBIND command. With the XBIND command, the block and the layers associated with the block will be added to the current drawing. If the block's definition contains a reference to an external reference, AutoCAD binds that xref and all its dependent symbols as well. After binding the necessary dependent symbols, you can detach the external reference file. After you invoke the XBIND command, AutoCAD displays the Xbind dialog box, similar to the Figure 18–10.

Figure 18-10

Xbind dialog box

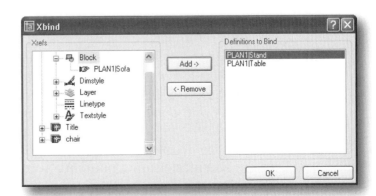

On the left side of the Xbind dialog box, AutoCAD lists the external reference files currently attached to the current drawing. Double-click on the name of the external reference file, and AutoCAD expands the list by listing the dependent symbols. Select the dependent symbol from the list and choose **Add**. AutoCAD moves the selected dependent symbol into the **Definitions to Bind** list. If necessary, return it to the external reference dependent list from the **Definitions to Bind** list by choosing **Remove** after selecting the appropriate dependent symbol.

Choose **OK** to bind the selected definitions to the current drawing.

CONTROLLING THE DISPLAY OF EXTERNAL REFERENCES

 Figure 18–11

Invoking the XCLIP command from the Reference toolbar

The XCLIP command (see Figure 18–11) allows you to control the display of unwanted information by clipping the external reference drawings and blocks. Clipping does not edit or change the external reference or block, it just prevents part of the object from being displayed. The defined clipping boundary can be visible or hidden. You can also define the front and back clipping planes.

You can define a rectangular or polygonal clipping boundary or generate a polygonal clipping boundary from an existing closed polyline. Valid boundaries are 2D polylines with straight or spline-curved segments. Polylines with arc segments, or fit-curved polylines, can be used as the definition of the clip boundary, but the clip boundary will be created as a straight segment representation of that polyline. If the polyline has arcs, the clip boundary is created as if it had been decurved prior to being used as a clip boundary.

The XCLIP command can be applied to one or more external references or blocks. If you set the clip boundary to OFF, the entire external reference or block is displayed. If you subsequently set the clip boundary to ON, the clipped drawing is displayed again. If necessary, you can delete the clipping boundary; AutoCAD redisplays the entire external reference or block. You can define only one clipping boundary at any time. In addition, AutoCAD also allows you to generate a polyline from the clipping boundary. After you invoke the XCLIP command, AutoCAD prompts:

Select objects: *(select one or more external references and/or blocks to be included in the clipping and press* ENTER *to complete the selection)*

Enter clipping option [ON/OFF/Clipdepth/Delete/generate Polyline/New boundary] <New> *(select one of the available options from the Shortcut menu)*

Choose New (default) to define a rectangular or polygonal clip boundary or generate a polygonal clipping boundary from a polyline. AutoCAD prompts:

Specify clipping boundary:

[Select polyline/Polygonal/Rectangular] <Rectangular>: *(select one of the available options from the Shortcut menu)*

The Rectangular (default) option allows you to define a rectangular boundary by specifying the opposite corners of a window. The clipping boundary is applied in the current UCS and is independent of the current view.

The Select polyline option defines the boundary by using a selected closed polyline.

The Polygonal option allows you to define a polygonal boundary by specifying points for the vertices of a polygon.

Once the clipping boundary is defined, AutoCAD displays only the portion of the drawing that is within the clipping boundary and then exits the command. If you already have a clipping boundary of the selected external reference drawing, and you invoke the New boundary option, then AutoCAD prompts:

Delete old boundary(s)? [Yes/No] <Yes>: *(select one of the two available options)*

If you choose Yes, the entire reference file is redrawn and the command continues; if you choose No, the command sequence is terminated.

note The display of the boundary border is controlled by the XCLIPFRAME system variable. If it is set to 1 (ON), AutoCAD displays the boundary border; if it is set to 0 (OFF), the default setting, AutoCAD does not display the boundary border.

The ON/OFF option controls the display of the clipped boundary. OFF displays all of the geometry of the external reference or block, ignoring the clipping boundary. ON displays the clipped portion of the external reference or block only.

The Clipdepth option sets the front and back clipping planes on an external reference or block. Objects outside the volume defined by the boundary and the specified depth are not displayed.

The Delete option removes the clipping boundary for the selected external reference or block. To turn off the clipping boundary temporarily, use OFF, explained earlier. Delete erases the clipping boundary and the clipdepth, and displays the entire reference file.

note The ERASE command cannot be used to delete clipping boundaries.

AutoCAD draws a polyline coincident with the clipping boundary. The polyline assumes the current layer, linetype, and color settings. When you delete the clipping boundary, AutoCAD deletes the polyline. If you need to keep a copy of the polyline, invoke the generate Polyline option. AutoCAD makes a copy of the clipping boundary. You can use the PEDIT command to modify the generated polyline, and then redefine the clipping boundary with the new polyline. To see the entire external reference while redefining the boundary, use OFF to turn off the clipping boundary.

EDITING EXTERNAL REFERENCES

 Figure 18–12

Invoking the REFEDIT *command (Edit Reference In-Place) from the Refedit toolbar*

You can edit block references and external references while working in a drawing session by means of the REFEDIT command (see Figure 18–12). This is referred to as in-place reference editing. If you select a reference for editing and it has attached xrefs or block definitions, the nested references and the reference are displayed and available for selection in the Reference Edit dialog box.

note You can edit only one reference at a time. Block references inserted with the MINSERT command cannot be edited. You cannot edit a reference file if it is in use by someone else.

You can also display the attribute definitions for editing if the block reference contains attributes. The attributes become visible, and their definitions can be edited along with the reference geometry. Attributes of the original reference remain unchanged when the changes are saved back to the block reference. Only subsequent insertions of the block will be affected by the changes.

As mentioned earlier, you can also edit an external reference by opening it in a separate window with the XOPEN command. After you invoke the REFEDIT command, AutoCAD prompts:

> Select reference: *(select a xref or block in the current drawing)*

AutoCAD displays the Reference Edit dialog box as shown in Figure 18–13.

 Figure 18–13

Reference Edit dialog box (Identify Reference tab)

The **Identify Reference** tab provides visual aids, as shown in Figure 18–13, for identifying the reference to edit and controls how the reference is selected. If you select an object that is part of one or more nested references, the nested references are displayed in the dialog box. Objects selected that belong to any nested references cause all the references to become candidates for editing. Select the specific reference you want to edit by choosing the name of the reference in the **Reference name** list box of the Reference Edit dialog box. This will lock the reference file to prevent other users from opening the file. Only one reference can be edited in place at a time. The path of the selected reference is displayed at the bottom of the dialog box. If the selected reference is a block, no path is displayed.

The **Preview** section of the dialog box displays a preview image of the currently selected reference. The preview image displays the reference as it was last saved in the drawing. The reference preview image is not updated when changes are saved back to the reference.

Choosing **Automatically select all nested objects** controls whether nested objects are included automatically in the reference editing session. If this option is chosen, all the objects in the selected reference will be automatically included in the reference editing session (becoming part of the working set). If **Prompt to select nested objects** is chosen instead, then nested objects must be selected individually in the reference editing session. If this option is chosen, after you close the Reference Edit dialog box and enter the reference edit state, AutoCAD prompts you to select the specific objects in the reference that you want to edit.

The **Settings** tab, as shown in Figure 18–14, provides options for editing references. **Create unique layer, style, and block names** controls whether layers and other named objects extracted from the reference are uniquely altered. If this option is set to ON, named objects in xrefs are altered (names are prefixed with $#$), similar to the way they are altered when you bind xrefs. If it is set to OFF, the names of layers and other named objects remain the same as in the reference drawing. Named objects that are not altered to make them unique assume the properties of those in the current host drawing that share the same name.

Figure 18–14

Reference Edit dialog box (Settings tab)

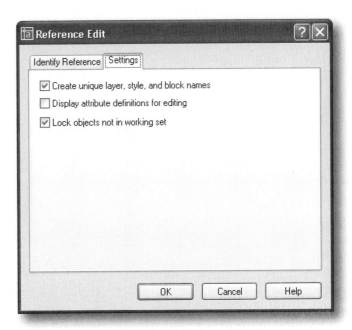

Display attribute definitions for editing controls whether all variable attribute definitions in block references are extracted and displayed during reference editing. If this option is set to ON, the attributes (except constant attributes) are made visible, and the attribute definitions are available for editing along with the selected reference geometry. When changes are saved back to the block reference, the attributes of the original reference remain unchanged. The new or altered attribute definitions affect only subsequent insertions of the block; the attributes in existing block instances are not affected. Xrefs and block references without definitions are not affected by this option.

Lock objects not in working set locks all objects not in the working set. If this option is set to ON, it will prevent you from accidentally selecting and editing objects in the host drawing while in a reference editing state. The behavior of locked objects is similar to objects on a locked layer. If you try to edit locked objects, they are filtered from the selection set.

Choose **OK** to close the Reference Edit dialog box. AutoCAD prompts to select objects if **Prompt to select nested objects** is selected.

The objects you choose are temporarily extracted for modification in the current drawing and become the *working set*. The working set objects stand out so they can be distinguished from other objects. All other objects not selected appear faded. You can now perform modifications on the working set objects.

> **note** Make sure **Reference Edit fading intensity** is set to the appropriate setting in the Options dialog box (**Display** tab).

If a new object is created while editing a reference, it is usually added to the working set automatically. However, if making changes to objects outside the working set causes a new object to be created, it will not be added to the working set.

Objects removed from the working set are added to the host drawing and removed from the reference when the changes are saved back. Objects created or removed are automatically added to or deleted from the working set. You can tell whether an object is in the working set or not by the way it is displayed on the screen; a faded object is not in the working set. When a reference is being edited, the Refedit toolbar is displayed, as shown in Figure 18–15.

Figure 18–15

The Refedit toolbar

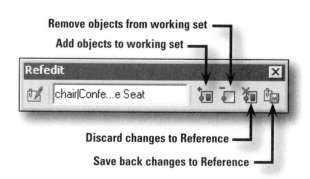

To add objects to the working set, choose Add to Working set from the Refedit toolbar and select objects to be added. You can only select items when the type of space (model or paper) is in effect that was in effect when the REFEDIT command was initiated. To remove objects from the working set, choose Remove from Working set from the Refedit toolbar and select objects to be removed. As mentioned earlier, if you remove objects from the working set and save changes, the objects are removed from the reference and added to the current drawing. Any changes you make to objects in the current drawing (not in the xref or block) are not discarded. Once the modifications are complete, choose SaveReference Edits to save the changes to the reference file. To discard the changes, choose Close Reference from the Refedit toolbar.

IMAGES

AutoCAD allows you to attach and detach a raster or bit-mapped bi-tonal, 8-bit gray, 8 bit color, or 24-bit color image file. The image formats that can be inserted into AutoCAD include *.BMP*, *.TIFF*, *.RLE*, *.DIB*, *.JPG*, *.PCX*, *.FLIC*, *.GIF*, *GEOSPOT*, *.IG4*, *.IGS*, *.RLC*, *.PCT*, *.CALSI*, *.PNG*, and *.TGA*. More than one image can be displayed in any viewport, and the number and size of images is not limited.

Attaching an Image

Figure 18–16

Invoking the IMAGEATTACH command from the Reference toolbar

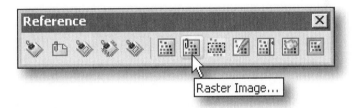

The IMAGEATTACH command (see Figure 18–16) attaches an image object to the current drawing. After you invoke the IMAGEATTACH command, AutoCAD displays the Select Image dialog box. Select the image file from the appropriate directory and choose **Open**. AutoCAD displays the Image dialog box, similar to Figure 18–17.

Figure 18–17

Image dialog box

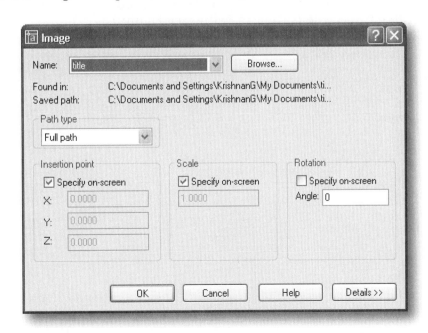

The **Name** box displays the image name attached to the current drawing.

Choose **Browse** to display the Select an Image File dialog box, in which you can select additional image files to attach to the current drawing.

Found in displays the path where the image file was found, and **Saved path** displays the saved path, if any, that is used to locate the image file.

The **Path type** box specifies whether the saved path to the image is set to No path, Full path or Relative path. The functionality of the options are similar to xref's explained earlier.

The **Insertion point**, **Scale**, and **Rotation** features are similar to those in the insertion of a block, explained in Chapter 17.

Choose **Details** to display the image information for the selected image file. This information includes image resolution in horizontal and vertical units, image size by width and height in pixels, and image size by width and height in the current selected units.

Choose **OK** to attach the selected image file to the current drawing.

MANAGING IMAGES

Figure 18–18

Open the Image Manager from the Reference toolbar

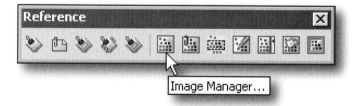

Open the Image Manager by choosing Image on the Reference toolbar (see Figure 18–18). The Image Manager allows you to manage various tasks related to images files attached to the current drawing (see Figure 18–19).

Figure 18–19

Image Manager dialog box

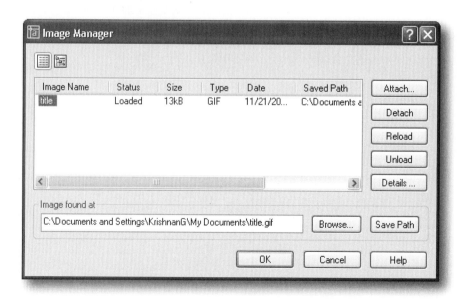

The Image Manager dialog box is similar to the Xref Manager dialog box. In the Image Manager dialog box, AutoCAD lists the images attached to the current drawing. The information provided in the list box of the Image Manager dialog box, such as **Image Name**, **Status**, **Size**, **Type**, **Date**, and **Saved Path**, is similar to the information provided in the Xref Manager dialog box. The **Saved Path** column shows the saved path of the associated external reference file. To change to a different path or file name for the currently selected external reference file, choose **Browse** in the Image Manager dialog box. AutoCAD displays the Select Image File dialog box, in which you can specify a different path or file name. To save the path as it appears in the **Saved Path** field of the currently selected image file, select **Save Path**. AutoCAD saves the path of the currently selected image file. You can switch between List View and Tree View, as in the Xref Manager dialog box, by clicking the two buttons at the top left of the Image Manager dialog box.

Detaching an Image

To detach one or more images from the current drawing, first select the image name(s) from the displayed list in the Image Manager dialog box, and then choose **Detach**. AutoCAD detaches the selected image(s) from the current drawing, erasing all instances of a specified image(s).

Reloading an Image

The **Reload** option allows you to display the most recent version of an image. To reload the image file to the current drawing, first select the image file name from the displayed list in the Image Manager dialog box, and then choose **Reload**. AutoCAD reloads the selected image file into the current drawing.

Unloading an Image

The **Unload** option unloads the image data from working memory without erasing the image objects from the drawing. It is highly recommended that you unload images that are no longer needed for editing. By unloading the images, you can improve performance by reducing the memory requirement for AutoCAD. To unload the image file from the current drawing, first select the image file name from the displayed list in the Image Manager dialog box, and then select **Unload**. AutoCAD unloads the selected image file from the current drawing.

Details of an Image

The **Details** option provides detailed information about the selected image, including the image name, saved path, active path, file creation date and time, file size and type, color, color depth, width and height in pixels, resolution, default size in units, and a preview image. To display the detailed information for the selected image file, first select the image file name from the displayed list in the Image Manager dialog box, and then select **Details**. AutoCAD displays the detailed information for the selected image file in the Image File Details dialog box, similar to the Figure 18–20.

Figure 18–20

Image File Details
dialog box

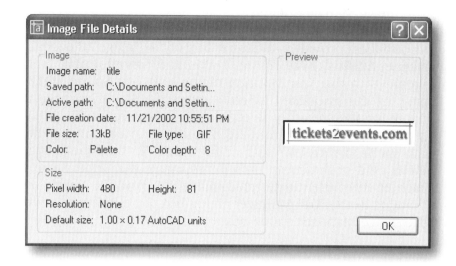

After making the necessary changes in the Image Manager dialog box, choose **OK** to keep the changes and close the dialog box.

Controlling the Display of the Image Objects

 Figure 18–21

Invoking the IMAGECLIP *command from the Reference toolbar*

The IMAGECLIP command (see Figure 18–21) allows you to control the display of unwanted information by clipping the image object; this is similar to the use of the XCLIP command for external references and blocks.

Adjusting the Image Settings

 Figure 18–22

Invoking the IMAGEADJUST *command from the Reference toolbar*

The IMAGEADJUST command (see Figure 18–22) controls the brightness, contrast, and fade values of the selected image. After you invoke the IMAGEADJUST command, AutoCAD displays the Image Adjust dialog box, similar to Figure 18–23.

Figure 18-23

Image Adjust dialog box

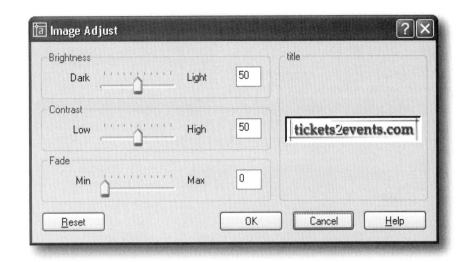

You can adjust the **Brightness**, **Contrast**, and **Fade** within the range of 0 to 100.

Select **Reset** to reset values for the brightness, contrast, and fade parameters to the default settings of 50, 50, and 0, respectively.

After making the necessary changes in the Image Adjust dialog box, choose **OK** to keep the changes and close the dialog box.

Adjusting the Display Quality of Images

Figure 18-24

Invoking the IMAGEQUALITY command from the Reference toolbar

The IMAGEQUALITY command controls the display quality of images. The quality setting affects display performance. A high-quality image takes longer to display. Changing the setting updates the display immediately without causing a regeneration. Images are always plotted using a high-quality display.

Invoke the IMAGEQUALITY command (see Figure 18–24), AutoCAD prompts:

Enter image quality setting [High/Draft]: *(select one of the available options from the Shortcut menu)*

High produces a high-quality image on screen, and Draft produces a lower-quality image on screen.

Controlling the Transparency of an Image

 Figure 18-25

Invoking the TRANSPARENCY command from the Reference toolbar

The TRANSPARENCY command controls whether the background pixels in an image are transparent or opaque. Invoke the TRANSPARENCY command (see Figure 18–25), AutoCAD prompts:

Select image: *(select the images, and press ENTER to complete the selection)*

Enter transparency mode [ON/OFF] <current>: *(select one of the two available options)*

ON turns transparency on, so that objects beneath the image are visible. OFF turns transparency off, so that objects beneath the image are not visible.

Controlling the Frame Display of an Image

 Figure 18-26

Invoking the IMAGEFRAME command from the Reference toolbar

The IMAGEFRAME command controls whether image frames are displayed or hidden from view. Invoke the IMAGEFRAME command (see Figure 18-26), AutoCAD prompts:

Select image: *(select the images, and press ENTER to complete the selection)*

Enter image frame setting [ON/OFF]: *(select one of the available options from the Shortcut menu)*

The ON selection displays the image frame around images, and OFF selection turns off image frames around images.

drawing exercises

Open the Exercise Manual PDF file for Chapter 18 on the accompanying CD for discipline-specific exercises.

review questions

1. If an externally referenced drawing called "*FLOOR.DWG*" contains a block called "TABLE" and is permanently bound to the current drawing, the new name of the block is:

   a. *FLOOR0TABLE.*
   b. *FLOOR|TABLE.*
   c. *FLOOR$0TABLE.*
   d. *FLOOR_TABLE.*
   e. *FLOOR$|$TABLE.*

2. If an externally referenced drawing called "*FLOOR.DWG*" contains a block called "TABLE," the name of the block is listed as:

   a. *FLOOR0TABLE.*
   b. *FLOOR|TABLE.*
   c. *FLOOR$0TABLE.*
   d. *FLOOR_TABLE.*
   e. *FLOOR$|$TABLE.*

3. The maximum number of files that can be externally referenced into a drawing is:

   a. *32.*   d. *32,000.*
   b. *1,024.*   e. *Only limited by memory.*
   c. *8,000.*

4. If you want to retain, from one drawing session to another, any changes you make to the color or visibility of layers in an externally referenced file, the system variable that controls this is:

   a. *XREFRET.*   d. *VISRETAIN.*
   b. *RETXREF.*   e. *These changes cannot be saved from one session to another.*
   c. *XREFLAYER.*

5. XREFs are converted to blocks if you detach them.

   a. *True*   b. *False*

6. When detaching XREFs from your drawing, it is acceptable to use wild cards to specify which XREF should be detached.

   a. *True*   b. *False*

7. Overlaying Xrefs rather than attaching them causes AutoCAD to display the file as a bit map image, rather than a vector-based image.

   a. *True*   b. *False*

8. To make a reference file a permanent part of the current drawing database, use the XREF command with the:

    a. Attach option.   **d.** Reload option.
    b. Bind option.   **e.** Path option.
    c. ? option.

9. The XREF command is invoked from which toolbar?

    a. Draw
    b. Modify
    c. External Reference
    d. Any of the above
    e. None of the above

10. The Attach option of the XREF command is used to:

    a. Bind the external drawing to the current drawing.
    b. Attach a new external reference file to the current drawing.
    c. Reload an external reference drawing.
    d. All of the above.

11. The following are the dependent symbols that can be made a permanent part of your current drawing, except:

    a. Blocks.
    b. Dimstyles.
    c. Text Styles.
    d. Linetypes.
    e. Grid and Snap.

12. Including an image file in a drawing will incorporate the image similar to the way a drawing file is merged by using:

    a. INSERT.
    b. XREF.
    c. WBLOCK.

13. Which of the following is not a valid file type to use with the IMAGE command?

    a. .BMP   **d.** .JPG
    b. .TIF   **e.** .GIF
    c. .WMF

14. Which of the following parameters can be adjusted on a bitmapped image?

    a. Brightness   **d.** All of the above
    b. Contrast   **e.** None of the above
    c. Fade

mission control

introduction

The AutoCAD DesignCenter makes it much easier to manage content within your drawing. Content includes blocks, external references, layers, raster images, linetypes, hatch and gradient fills, layouts, text styles, dimension styles, and custom content created by third-party applications. You can manage content between your drawing and other sources such as other drawings, whether currently open, stored on any drive, or even elsewhere on a network or somewhere on the Internet and support files containing linetypes, text styles and patterns.

After completing this chapter, you will be able to do the following:

- Open, undock, move, resize, dock, and close the AutoCAD DesignCenter

- Locate drawings, files, and their content in a manner similar to Windows Explorer

- Use the DesignCenter content area

- Preview images, drawings, content, and their written descriptions

- Manage blocks, layers, xrefs, layouts, dimstyles, textstyles, and raster images

- Manage Web-based content and custom content from third-party applications

- Browse sources for content and drag and drop (copy) into drawings

- Access symbols and information directly over the Internet

DESIGNCENTER

The DesignCenter provides a program window with a specialized drawing file–handling section. It allows you to drag and drop content and images into your current drawing or attach a drawing as an external reference. AutoCAD has a page in the DesignCenter called DC Online that gives immediate and direct access to thousands of symbols, manufacturers' product information, and content aggregators' sites. Content in the AutoCAD DesignCenter can be dragged into a tool palette for use in the current drawing.

The DesignCenter displays two panes (see Figure 19–1) from which you can manage drawing content, and in addition, two additional panes that can be turned ON and OFF when you click Preview and Description display panes.

Figure 19-1

DesignCenter window

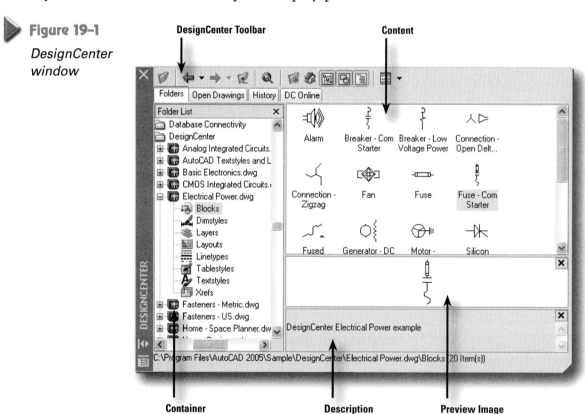

Use the Tree view to browse sources of Containers and to display content in the content area. Use the content area to add items to a drawing or to a Tool Palette.

Tree View

The Tree view pane displays the hierarchy of files and folders on your computer and network drives, a list of open drawings, custom content, and a history of the last locations you accessed. Select an item in the Tree view to display its contents in the content area.

In AutoCAD, the primary container is the drawing. It contains the blocks, images, linetypes, and other definitions that are most commonly sought to add to your current drawing. A folder can be considered a container because it contains files, and a drawing is a file. An image can also be a file in a folder. An image is considered as content rather than a container like a drawing. Drawings and images can be dragged and dropped into your current drawing from the content area. So, a drawing can be both content and a container.

Content

The content area displays the content of the "container" currently selected in the Tree view. If you select a folder, all of the files in the folder will be listed in the content area. If you select a drawing file, the content area displays icons for the different content types. If you select a content type such as Blocks under a specific drawing name, all the blocks that are in the selected drawing will be listed in the content area. Instead if you select Layers, all the layers in the selected drawing will be listed in the content area. Content includes block definitions, external references, layer names and compositions, raster images, linetypes, layouts, text styles, dimension styles, and custom content created by third-party applications. For example, a layer of one name may have the same properties as a layer of another name. But, because its name is part of its composition, it can be identified as a unique layer. Also, two items of content can have the same name if they are not the same type of content. For example, you can name both a text style and a dimension style "architectural."

note You can only view the name of an item of content in the content area along with a raster image if it is a block or image. Similarly, you can view a description only if one has been written. The item itself still resides in its container. Through the DesignCenter, you can drag and drop copies of the item's definition into your current drawing, but you cannot edit the item itself from the DesignCenter.

Preview

The Preview pane displays a preview of the selected item in a pane below the content area. If there is no preview image saved with the selected item, the Preview pane is empty. Preview on the DesignCenter toolbar toggles the Preview pane.

Description

The Description pane displays a text description of the selected item in a pane below the content area. If a preview image is also displayed, the description is displayed below it. If there is no description saved with the selected item, the Description area is empty. Description on the DesignCenter toolbar toggles the Description pane.

Figure 19–2 shows an example of a content (block named Conference Seat), content type (Block), and Container (drawing named 8th floor furniture).

Figure 19–2

An example showing content, content type, and container in the DesignCenter window

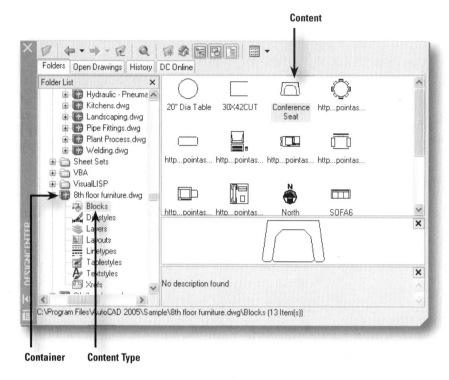

Working with the DesignCenter

 Figure 19–3

Invoking the AutoCAD DesignCenter from the Standard toolbar

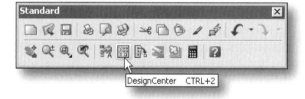

As mentioned earlier, AutoCAD DesignCenter is a window rather than a dialog box. It is like calling up a special program that runs along with AutoCAD and expedites file managing and drawing content–handling tasks. After you invoke the AutoCAD ADCENTER command, AutoCAD displays the AutoCAD DesignCenter window, similar to Figure 19–4.

 Figure 19–4

AutoCAD DesignCenter window

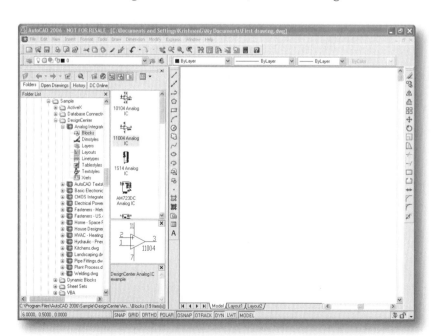

Positioning the DesignCenter Window

The default position of the DesignCenter is docked at the left side of the drawing area. This is where it will be located when you open the DesignCenter window after AutoCAD is started for the first time, as shown in Figure 19–4. However, once the DesignCenter has been repositioned and the drawing session is closed with the DesignCenter window open, the next time you open AutoCAD it will be at its relocated position.

You can undock the DesignCenter by double-clicking on its title bar, and then it can be repositioned anywhere on the screen, as shown in Figure 19–5.

Figure 19–5

DesignCenter floating (undocked) in the drawing area

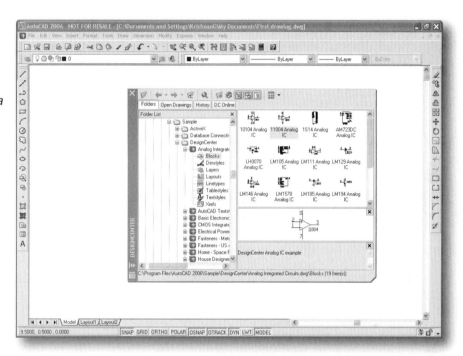

To redock the DesignCenter, double-click the border or drag the DesignCenter to the left or right side of the drawing area. Dragging the DesignCenter window while holding CTRL prevents docking. It cannot be docked at the top or bottom of the screen. When undocked, the DesignCenter can be resized by holding down the pick button over an edge or corner of the window.

You can also auto-hide the DesignCenter window similar to the tool palettes. If you right-click on the DesignCenter window title bar and select Auto-hide from the Shortcut menu, only the DesignCenter title bar is displayed when the cursor moves outside the DesignCenter window. The DesignCenter window will appear again when cursor is moved again on the top of the title bar.

Accessing Content in the DesignCenter

The Tree view in the left portion of the DesignCenter window together with the four DesignCenter tabs (**Folders**, **Open Drawings**, **History**, and **DC Online**) help you find and load content into the content area. The names of content type, such as blocks, linetypes, text styles, and so on, and containers such as drawings, image files, folders, drives, networks, and Internet locations, can be viewed in the Tree view. Content itself can be viewed in the content area, but not in the Tree view pane. Either pane can be used to move up and down through the path from the drive to the item of content. It is usually quicker to navigate the path in the Tree view because of its ability to display multiple levels of hierarchy. When the container (a folder for drawings/images and a drawing/content type for blocks, images, and other items) appears in the Tree view, select it and then view the name of the items of content (if any) in the content area.

The **Folders** tab displays a hierarchy of navigational icons, including networks and computers, Web addresses (URLs), computer drives, folders, drawings and related support files, xrefs, layouts, hatch styles, and named objects, including blocks, layers, linetypes, text styles, dimension styles, and plot styles within a drawing.

The **Open Drawings** tab displays a list of the drawings that are currently open. Click a drawing file and then click one of the content types from the list to load the content into the content area.

The **History** tab displays the last 20 items opened in DesignCenter. Double-click a drawing file from the list to navigate to the drawing file in the Tree view of the Folders tab and to load the content into the content area.

The **DC Online** tab provides content from the DesignCenter Online Web page, including blocks, symbol libraries, manufacturers' content, and online catalogs.

Using the Tree View

As mentioned earlier, in the Tree view you can view the content type and container, but not the actual content. You can use the Tree view to display the icon for an item of content in the content area. Just doing this will not, however, cause a raster image to be displayed in the Preview pane, as discussed earlier in this section.

The folder named *Sample* that comes with the AutoCAD program contains numerous drawings that will be used in this section as examples. First, select the folder named AutoCAD2006 (or the folder in which AutoCAD has been installed) and then the folder named Sample. This, again, might require going all the way back to a particular drive. Then select the plus sign or double-click the folder name or drive name to display the folders within a folder. Figure 19–6 shows the folder named Sample highlighted in the Tree view. Note that the list of content folders is displayed in the content area.

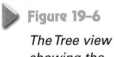

Figure 19–6

The Tree view showing the folder named Sample selected

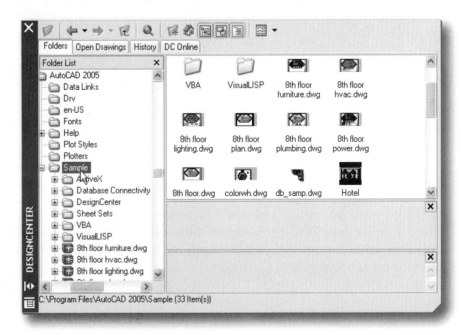

Select the Taisei Detail Plan in the Tree view. Figure 19–7 shows the drawing named Taisei Detail Plan highlighted in the Tree view. Note that the list of content types is displayed in the content area.

Figure 19-7

The Tree view showing Taisei Detail Plan drawing selected

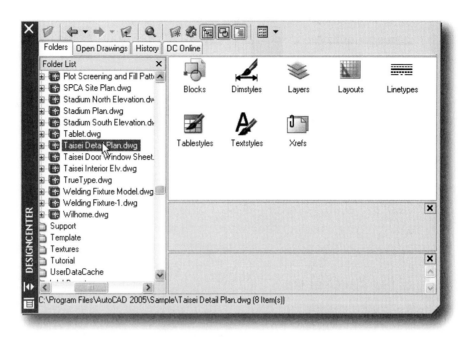

Next, double-click on the drawing named Taisei Detail Plan in the Tree view or select the box with the plus in it to the left of the drawing name. The content types will be listed below the drawing name, as well as in the content area. To display the names of individual items of content, blocks for example, select the content type Blocks in the Tree view, as shown in Figure 19–8.

When one of the block icons displayed in the content area is chosen, AutoCAD displays the corresponding preview image and description, if any, in the Preview pane and Description pane, respectively. Figure 19–9 shows the selection of a block named C720_P, and its corresponding preview image and description are shown in the Preview pane and in the Description pane.

Figure 19-8

The Tree view showing Taisei Detail Plan drawing with content type Blocks selected

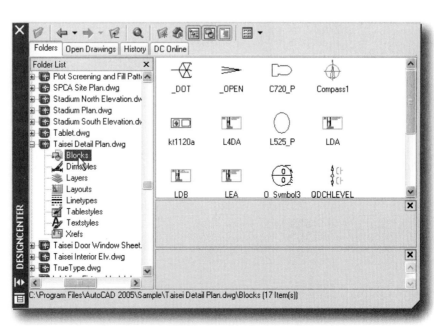

Figure 19-9

Selection of a block named C720_P with the corresponding preview image and description shown in the Preview pane and in the Description pane

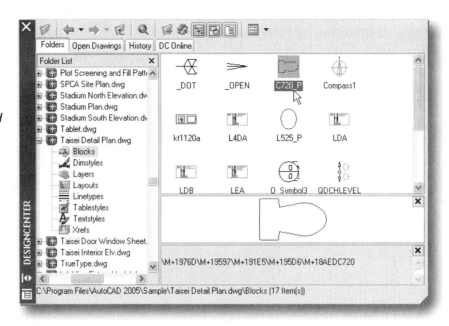

Using the Content Area

As mentioned earlier, the content area is used for displaying content. It can also display containers and content type. You can navigate up through the hierarchy of drives, folders, drawings, content types, and content in the content area by selecting Up from the Shortcut menu that is displayed when you right-click the content area background. By double-clicking a container in the content area, you can navigate down through the hierarchy until you reach the content itself. It is not as easy as in the Tree view, however, because only one level is displayed at a time. If the content is a drawing, block, or image, you can view a raster image of it in the Preview pane. Simply select the item in the content area. AutoCAD also displays the path to the selected drawing at the bottom of the DesignCenter window, as shown in Figure 19–9.

You can also load a drawing into the content area by invoking the LOAD command from the DesignCenter toolbar. LOAD causes the Load dialog box to be displayed, from which you can select a drawing file whose contents will be loaded onto the content area.

note The content area displays only items of the same level that are members of a single container one level above them. The content area does not display more than one level at a time. For example, content types such as blocks, xrefs, and layers are members of one drawing. If there are any blocks in the drawing, then they are listed when you have selected that drawing's content type called Blocks. Several drawings may be members of a particular folder. Folders are members of one drive, or folders may actually be subfolders of a folder one level up. Remember that in the content area, only one level of the hierarchy is displayed, and all items shown are members of the same component/container one level above. For displaying more than one level at a time, use the Tree view.

Searching for Content

The Search feature provides a means to locate containers, content type, and content in a manner similar to the Windows Find feature. Choosing SEARCH from the DesignCenter toolbar causes the Search dialog box to be displayed as shown in Figure 19–10.

Figure 19–10

Search dialog box with the Drawings tab displayed

From the **Look for** box in the Search dialog box, select the type of content you wish to find. Available options include Blocks, Dimstyles, Drawings, Drawings and Blocks, Attach Pattern Files, Hatch Patterns, Layers, Layouts, Linetypes, Textstyles, Tablestyles and Xrefs.

From the **In** box, select the location for searching. Available options include My Computer, local hard drives (*C:*) (and any others), 3½ Floppy (*A:*), and network drives.

Choosing **Browse** causes the Browse For Folder dialog box to be displayed, which has a file manager window as shown in Figure 19–11. Here you can specify a path by stepping through the levels to the location that you wish to search. When the final location is highlighted, choose **OK**, and the path to this location is displayed in the **In** box.

Figure 19–11

Browse For Folder dialog box

The number of tabs that will be displayed in the Search dialog box depends on the type of content selected in the **Look for** box. Each of the content types will have a tab that corresponds to the type of content selected. If Drawings is selected, there are two additional tabs: **Date Modified** and **Advanced**.

On the **Drawings** tab, the **Search for the word(s)** box lets you enter the word(s), such as a drawing name or author, to determine what to search for. The **In the field(s)** box lets you select a field type to search. Fields include File Name, Title, Subject, Author, and Keywords.

On the **Date Modified** tab (available only when Drawings is selected in the **Look for** box of the Search dialog box) you can specify a search of drawing files by the date they were modified, as shown in Figure 19–12. The search will include all files that comply with other filters if **All files** is selected. If **Find all files created or modified** is selected, you can limit the search to the range of dates specified by selecting one of the following secondary options: The **between** and **and** boxes let you limit the search to drawings modified between the dates entered. **During the previous months** lets you limit the search to drawings modified during the number of previous months entered. **During the previous days** lets you limit the search to drawings modified during the number of previous days entered.

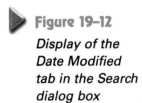

Figure 19–12

Display of the Date Modified tab in the Search dialog box

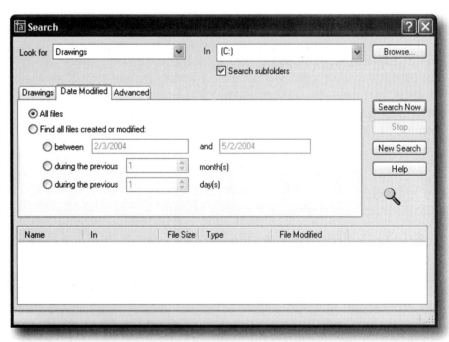

On the **Advanced** tab (available only when Drawings is selected in the **Look For** box of the Search dialog box) you can specify a search of files by additional parameters as shown in Figure 19–13. In the **Containing** box you can specify a file by one of four options: Block Name, Block and Drawing Description, Attribute Tag, and Attribute Value. The search will then be limited to items containing the text entered in the **Containing text** box. The **Size is** box lets you limit the search to drawings that are At Least or At Most as many KBs (size of the drawing file) as the number entered into the second text box.

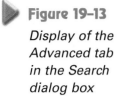

Figure 19-13

Display of the Advanced tab in the Search dialog box

All of the other tabs (Blocks, Dimstyles, Drawings and Blocks, Layers, Layouts, Linetypes, Textstyles, and Xrefs) have a **Search for the name** box in which you can enter the name of the item you wish to find. This can be a block name, dimstyle, or one of the other content types.

Once the parameters have been specified, such as the path, content type, and date modified, selecting **Search Now** initiates the search. Select **Stop** to terminate the search. Selecting **New Search** lets you specify new parameters for another search. Selecting **Help** causes the AutoCAD Command Reference (Help) dialog box to be displayed.

Each time a search is performed, the text entered in the **Search for the name** text box or the **Search for the word(s)** box is saved. If you wish to repeat the same search again, select the down arrow next to the text box and select the text associated with the search you wish to repeat.

Content from DesignCenter Online

DesignCenter Online provides access to predrawn content such as blocks, symbol libraries, manufacturers' content, and online catalogs. This content can be used in common design applications to assist you in creating your drawings. To access DesignCenter Online, click the **DC Online** tab in DesignCenter. Once the DesignCenter Online window is open (see Figure 19-14), you can browse, search, and download content to use in your drawing.

Figure 19-14

The DesignCenter window with the DC Online tab displayed and Category Listing selected.

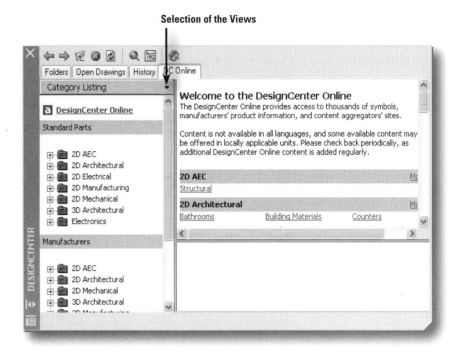

In the DesignCenter Online window, two panes are displayed—a right pane and a left pane. The right pane is the content area from which you can drag and drop the items in the current drawing. The content area displays the items or folders that you selected in the left pane. The left pane can display one of the four views: **Category Listing**, **Search**, **Settings**, and **Collections**. You choose the view by clicking the heading at the top of the left pane.

Category Listing

The **Category Listing** view displays folders containing libraries of standard parts, manufacturer-specific content, and content aggregator Web sites (see Figure 19–14).

The **Standard Parts** category includes groups of drawings and images that can be used in architectural and engineering design disciplines such as architecture, landscaping, mechanical, and GIS. For example, you can select the box to the left of the 2D Architectural group and then select the box to the left of the Landscaping subgroup when expanded. This expands to the list of types of content. If you select Tables from this list, the right pane will display thumbnail sketches of drawings or images that are available for download (see Figure 19–15). When you select an image, additional links are displayed in the lower half of the right pane along with a larger sketch of the content.

Figure 19-15

*Category
Listing with 2D
Architectural/
Landscaping/
Tables group
displayed*

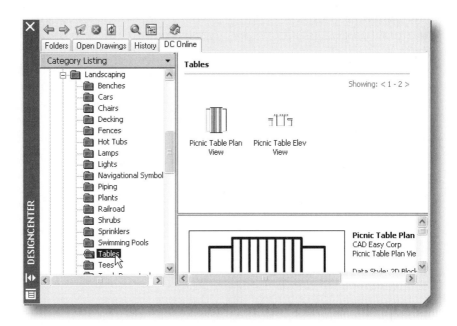

Another group of content related to tables can be found under the 2D Architectural group by selecting the Furniture subgroup and then selecting Tables (see Figure 19–16).

Figure 19-16

*Category
Listing with 2D
Architectural/
Furniture/Tables
group displayed*

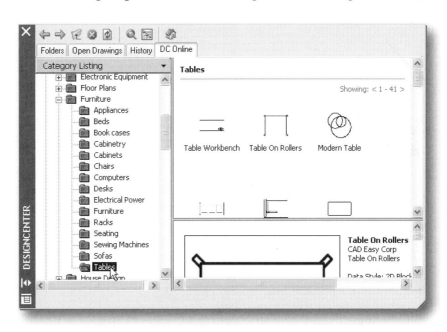

note If you are looking for drawings, images, or links to drawings and images of a particular type of content (such as tables in the example above), there is usually more than one group or path to a group that includes such drawings, images or links. See the explanation for using the **Search** view in this section.

The **Manufacturers** category includes groups of Web sites or Internet addresses of manufacturers of products used in architectural and engineering construction. Through these Web sites, drawings and images can be downloaded where available. For example, in the Manufacturers group (similar to the selection in the Standard Parts group), you can select the box to the left of the 2D Architectural group and then select the box to the left of the Landscaping subgroup when expanded. This expands to the list of types of content. If you select Outdoor Furnishing from this list, the right pane will display one or more Internet addresses of Web sites from which you can access drawings, images, or other content data that are available for download (see Figure 19–17).

 Figure 19–17

Manufacturers with 2D Architectural/ Landscaping/ Outdoor Furnishings group displayed

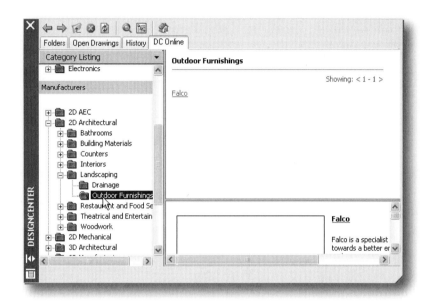

The **Aggregators** category contains lists of libraries compiled by commercial catalog providers. Just as with the items in the **Manufacturers** category, you can access Web sites that contain or lead to drawings and blocks for use in architectural and engineering design and drafting. For example, in the Aggregators group (similar to the selection in the Manufacturers group), you can select the AEC Aggregators from this list, and the right pane will display one or more Internet addresses of Web sites from which you can access drawings, images, or other content data that are available for download (see Figure 19–18).

 Figure 19–18

Aggregators groups displayed

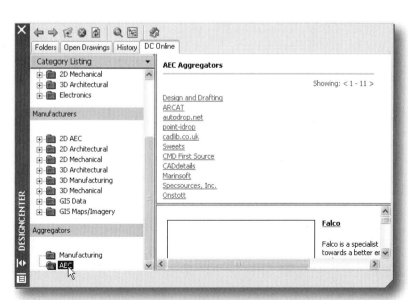

Search

The **Search** view allows you to search for online content. You can query items with Boolean and multiple-word search strings. In the **Search** view there is a text box in which you can enter words or combinations of characters to tell AutoCAD what type of content to search for. You can display details for how to use Boolean and multiple-word search strings by selecting **Need Help?** Figure 19–19 shows an example of typing **table** in the text box and selecting **Search**. The right pane shows the result of the search. Selecting one of the content in the right pane causes additional links to be displayed in the lower half of the right pane along with a larger sketch of the content.

 Figure 19–19

The DesignCenter window with the DC Online tab Search view selected

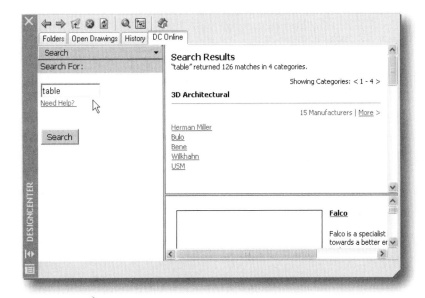

Settings

The **Settings** view controls how many categories and items are displayed on each page in the content area as a result of a search or folder navigation. You can select number of categories from 5, 10, 15 in the **Number of Categories per page** box and number of items from 50, 100, and 200 in the **Number of Items per page** box (see Figure 19–20). Choose **Update Settings** to update the changes.

 Figure 19–20

The DesignCenter window with the DC Online tab Settings view selected

Collections

The **Collections** view specifies the discipline-specific content types that are displayed in DesignCenter Online. AutoCAD lists collections for each of the three categories with a check box beside each collection (see Figure 19–21). Place a check in a particular collection's check box that you wish to be displayed in the **Category Listing** view. Once you have selected/deselected the desired collections, choose **Update Collections**, and AutoCAD will return to the **Category Listing** view displaying the list of collections whose check boxes have checks in them.

 Figure 19–21

The DesignCenter window with the DC Online tab Collections view selected

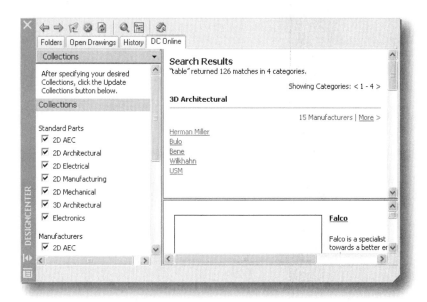

Adding Content to Drawings

As mentioned earlier, blocks, external references, layers, raster images, linetypes, layouts, text styles, dimension styles, and custom content created by third-party applications are the content types that can be added to the current drawing session by using the AutoCAD DesignCenter. Content in the AutoCAD DesignCenter can also be dragged into a tool palette for use in the current drawing.

You can add content from the content area into your current drawing using several methods:

- Drag an item to the graphics area of a drawing to add it using default settings, if any.
- Right-click an item in the content area to display a Shortcut menu with several options, and choose one of the available options.
- Double-click a block to display the Insert dialog box; double-click a hatch to display the Boundary Hatch and Fill dialog box.

Layers, Linetypes, Text Styles, and Dimension Styles

A definition of a layer or linetype or a style created for text or dimensions can be dragged into the current drawing from the content area. It will become part of the current drawing as if it had been created in that drawing.

Blocks

A block definition can be inserted into the current drawing by dragging its icon from the content area into the drawing area. You cannot, however, do this while a command is active. It must be done at the Command prompt. Only one block definition at a time can be inserted from the content area.

There are two methods of inserting blocks from the AutoCAD DesignCenter. One method uses *Autoscaling*, which scales the block reference as needed based on the comparison of the units in the source drawing to the units in the target drawing. Another method is to use the Insert dialog box to specify the insertion point, scale, and rotation.

note When dragging and dropping using the automatic scaling method, the dimension values inside the blocks will not be true.

When you drag a block definition from the content area into your drawing, you can release the button on the pointing device (drop) when the block is at the desired location. This is useful when the desired location can be specified with a running object snap mode in effect. The block will be inserted with the default scale and rotation.

To invoke the Insert dialog box, double-click the block icon or right-click the block definition in the content area, and then select Insert Block from the Shortcut menu. In the Insert dialog box, specify the **Insertion point**, **Scale**, and **Rotation**, or select **Specify On-screen**. You can select **Explode** to have the block definition exploded on insertion.

Unlike xrefs, when the source file of a block definition is changed, block definitions in the drawings that contain that block are not automatically updated. With DesignCenter, you decide whether a block definition should be updated in the current drawing. The source file of a block definition can be a drawing file or a nested block in a symbol library drawing. From the Shortcut menu displayed when you right-click a block or drawing file in the content area, choose Redefine Only or Insert and Redefine to update the selected block.

Raster Images

A raster image such as a digital photo, print screen capture saved in a paint program as a bitmap, or a company logo can be copied into the current drawing by dragging its icon from the content area into the drawing area. Then specify the **Insertion point**, **Scale**, and **Rotation**. Or you can right-click the image icon and choose Attach Image from the Shortcut menu.

External References

To attach an xref to the current drawing, right-click the drawing file and then select Attach as Xref from the Shortcut menu. The Attach Xref dialog box is displayed, from which you can choose between **Attachment** or **Overlay** as the **Reference Type** option. Set additional parameters, such as the **Insertion point**, **Scale**, **Rotation**, and attach the selected reference file to the current drawing.

When you copy, insert, or attach content into a drawing that already has an item of the content type with the same name, AutoCAD will display a warning. The item is not added to the drawing. If the item is a block, AutoCAD checks to determine if the name is already listed in the database. The warning "Duplicate definition of [object][name] ignored" is displayed. If the xref exists in the drawing, the warning "Xref [name] has already been defined. Using existing definition" is displayed. If the item with the duplicate name is a layer, linetype, or other item that is not a block or xref, the warning "Add [object] operation performed. Duplicate definitions will be ignored" is displayed.

Open Drawings with DesignCenter

With DesignCenter, you can open a drawing from the content area using the Shortcut menu, pressing CTRL while dragging a drawing, or dragging a drawing icon to any location outside the graphics area of a drawing area. The drawing name is added to the DesignCenter history list for quick access in future sessions.

Adding Items from DesignCenter to a Tool Palette

AutoCAD allows you to add drawings, blocks, and hatches from DesignCenter to the Tool Palettes. Select one or more items in the content area of the DesignCenter window and drag and drop to one of the tabbed pages of the Tool Palettes as shown in Figure 19–22. You can also create a new tool palette and add the selected items. To create a new tool palette, first select one or more items, right-click, and from the Shortcut menu select Create Tool Palette. AutoCAD prompts for a name and creates a tool palette, and adds the selected items to the newly created palette.

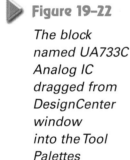

Figure 19–22

The block named UA733C Analog IC dragged from DesignCenter window into the Tool Palettes

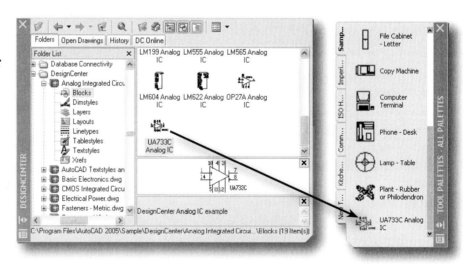

Exiting Autocad DesignCenter

You can exit AutoCAD DesignCenter by either selecting the **X** in the DesignCenter window or entering **adcclose** at the AutoCAD Command: prompt.

review questions

1. What command is used to invoke the AutoCAD DesignCenter?
   a. *DSNCEN*
   b. *DGNCEN*
   c. *ADCENTER*
   d. *DGNCTR*

2. Which of the following are considered to be drawing content types?
   a. *Blocks*
   b. *Layers*
   c. *Linetypes*
   d. *All the above*

3. AutoCAD allows items to be directly edited from the DesignCenter.
   a. *True*
   b. *False*

4. Which of the following are not drawing content types?
   a. *Lines*
   b. *Circles*
   c. *Arcs*
   d. *All of the above*

5. Within the DesignCenter a drawing is considered to be a _____.
   a. *Content*
   b. *Container*
   c. *Folder*
   d. *a & b*
   e. *a & c*

6. Invoking the AutoCAD DESIGNCENTER command opens the DesignCenter dialog box.
   a. *True*
   b. *False*

7. The default position for the DesignCenter is in the lower right corner.
   a. *True*
   b. *False*

8. Which of the following areas of the DesignCenter displays the names and icons representing content?
   a. *Toolbar*
   b. *Content area*
   c. *Tree*
   d. *Preview pane*
   e. *Description pane*

9. Large Icons, Small Icons, List, and Details are four optional modes of displaying content using which of the following buttons?

  a. *LOAD*
  b. *FIND*
  c. *UP*
  d. *VIEWS*

10. Within the Tree view which can display the previous twenty items and their paths accessed through the AutoCAD DesignCenter?

  a. *Open Drawings*
  b. *Desktop*
  c. *History*
  d. *None of the above.*

deliverables

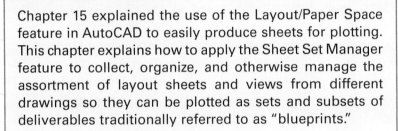

introduction

Chapter 15 explained the use of the Layout/Paper Space feature in AutoCAD to easily produce sheets for plotting. This chapter explains how to apply the Sheet Set Manager feature to collect, organize, and otherwise manage the assortment of layout sheets and views from different drawings so they can be plotted as sets and subsets of deliverables traditionally referred to as "blueprints."

After completing this chapter, you will be able to do the following:

- Collect, sort, create, and manage drawing sheets

- Create layout views automatically

- Automate the numbering of sheets

- Archive a set of drawings

- Publish sets and subsets of drawing sheets

DRAWING SETS

A drawing set is just that, a set of drawings. A small design project might require only two or three sheets, while in a major construction project, a drawing set might consist of hundreds of sheets. And there are often subsets for various disciplines: civil/survey, structural, mechanical, architectural, and electrical. The mechanical subset might have its own subsets, plumbing and heating-ventilating-air conditioning, and the electrical its sub-subsets, power and communications. When all of these sheets are scattered throughout one or more offices in one or more locations, and some sheets are just one layout or view in one drawing and other sheets are layouts from many different drawings, it becomes a difficult and almost impossible task to organize, manage, and continually update the sheets individually and as a set and subsets.

With the Sheet Set Manager, you can manage drawings as sheet sets. A sheet set is an organized and named collection of sheets from several drawing files. A sheet is a selected layout from a drawing file. You can import a layout from any drawing into a sheet set as a numbered sheet. You can manage, transmit, publish, and archive sheet sets as a unit.

CREATING A NEW SHEET SET

The Create Sheet Set wizard contains a series of pages that step you through the process of creating a new sheet set. You can choose to create a new sheet set from existing drawings, or use an existing sheet set as a template on which to base your new sheet set.

The following steps should be performed before creating a sheet set:

- Move the drawing files to be used in the sheet set into a minimum number of folders. This will make sheet set management easier.
- Have only one layout tab (in addition to Model tab) in each drawing in the sheet set. This affects access to sheets by multiple users. Only one sheet in each drawing can be open at a time.
- Specify or create a drawing template (*.dwt*) file for use by the sheet set for creating new sheets. This template file is called the sheet creation template and can be specified in the Sheet Set Properties dialog box or the Subset Properties dialog box.
- Create a page setup overrides file. Specify or create another *.dwt* file for storing page setups for plotting and publishing. This file is called the page setup overrides file and can be used to apply a single page setup to each sheet in a sheet set. This will override the individual page setups stored in each drawing.

 Figure 20-1

Choosing New Sheet Set from the File menu

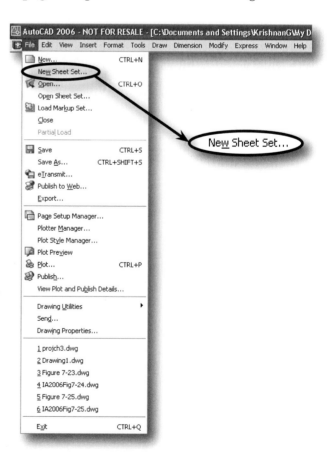

Invoke the NEWSHEETSET command (see Figure 20–1) to open the Create Sheet Set – Begin wizard as shown in Figure 20–2.

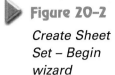

Figure 20-2

Create Sheet Set – Begin wizard

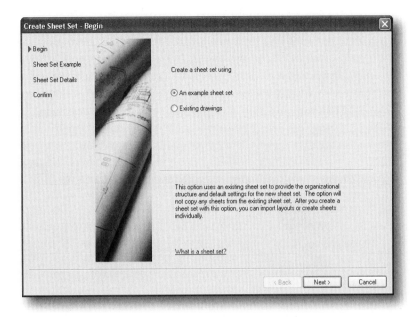

On the first page of the Create Sheet Set wizard, select a method for creating a sheet set. Two methods are provided: **An example sheet set** and **Existing drawings** (see Figure 20-2).

Choosing **An example sheet set** allows you to use a sample sheet set format, structure, and default settings without copying any of the sheets in the sample sheet set to the new set. After you create a sheet set using this option, you can import layouts or create sheets individually.

Choosing **Existing drawings** allows you to import layouts from existing drawings in the specified folder(s). The layouts from these drawings are imported into the newly created sheet set automatically.

Creating Sheet Sets from Examples

To create a sheet set from a sample sheet set, select **An example sheet set** in the Create Sheet Set – Begin wizard page and choose **Next**. AutoCAD displays the Create Sheet Set – Sheet Set Example page as shown in Figure 20-3.

Figure 20-3

Create Sheet Set – Sheet Set Example page

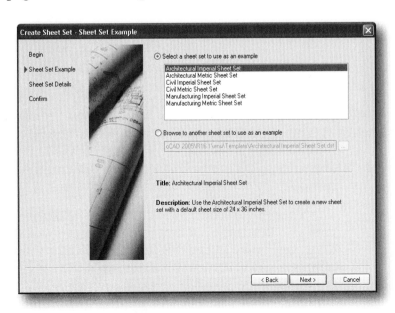

AutoCAD displays a list of sample sheet sets in the box from which you can choose one to use as the basis of the new sheet set. If you select **Browse to another sheet set to use as an example**, you can use the standard Windows-type browsing mechanism to search folders for a sheet set to use as the basis of a new sheet set.

After selecting an existing sheet set from one of the optional methods, choose **Next**. AutoCAD displays the Create Sheet Set – Sheet Set Details page as shown at left in Figure 20–4.

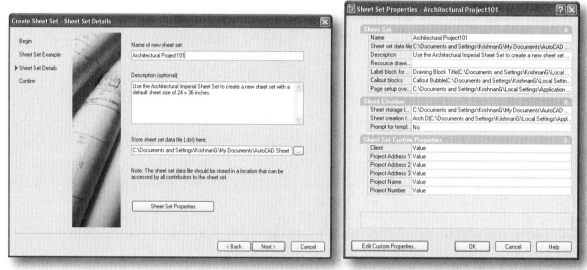

Figure 20-4

Create Sheet Set – Sheet Set Details page and Sheet Set properties

Specify the name and description of the newly created sheet set in the **Name of new sheet set** box and **Description (optional)** box respectively. Specify the folder in which to store the newly created sheet set in the **Store sheet set data file (.dst) here**.

> **note**   The Sheet set data file should be stored in a location that can be accessed by all contributors to the sheet set.

Choose **Sheet Set Properties** to view or edit the sheet set properties. AutoCAD displays the Sheet Set Properties dialog box as shown at right in Figure 20–4.

The Sheet Set Properties dialog box displays information specific to the sheet set selected. The information can be modified in the boxes to the right of individual data descriptions. This includes information such as the path and file name of the sheet set data (*.dst*) file, folder paths that contain drawing files included in the sheet set, and custom properties associated with the sheet set.

The **Sheet Set** section includes the name of the sheet set, location of the sheet set data file, description if any for the newly created sheet set, path for the resource drawings, name of the drawing that contains label block for views, callout block names associated with the sheet set, and name of the page setup override file.

The **Sheet Creation** section includes the paths of the folders that contain the drawing files associated with the sheet set, name of the template for creation of new sheet, and setting for whether to prompt for new template.

The **Sheet Set Custom Properties** section contains user-defined properties. Choose **Edit Custom Properties** to add or remove custom properties associated with a sheet set. Custom properties can be used to store information such as a contract number, the name of the designer, and the release date.

Choose **OK** to return to the Create Sheet Set – Sheet Set Details wizard page and then choose **Next**. AutoCAD displays the Create Sheet Set – Confirm page as shown in Figure 20–5.

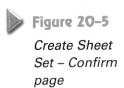

Figure 20–5

Create Sheet Set – Confirm page

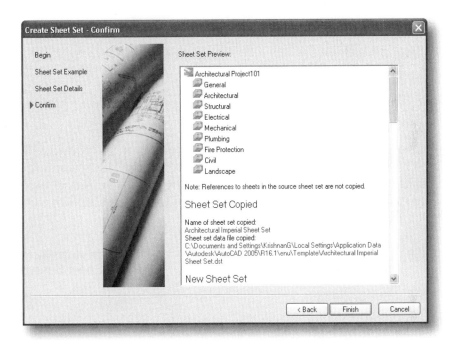

In the Create Sheet Set – Confirm wizard page, the **Sheet Set Preview** box lists the information about the sheet set for review before you accept it. If the information is acceptable, choose **Finish** to complete the creation of the new sheet set. If any of the information needs to be modified, choose **Back** and make the necessary changes. AutoCAD creates a new sheet set in the form of a sheet set data file with the extension of *.dst* in the specified location with the specified properties.

Creating Sheet Sets from Existing Drawings

To create a sheet set from existing drawings, select **Existing drawings** on the Create Sheet Set – Begin wizard page and choose **Next**. AutoCAD displays the Create Sheet Set – Sheet Set Details page similar to creating a sheet from example sheet set (at left in Figure 20–4). Specify the name and description of the newly created sheet set in the **Name of new sheet set** box and **Description (optional)** box respectively. Specify the folder in which to store the newly created sheet set in the **Store sheet set data file (.dst) here**. Choose **Sheet Set Properties** to view or edit the sheet set properties. Choose **Next**, and AutoCAD displays the Create Sheet Set – Choose Layouts page as shown in Figure 20–6.

Figure 20-6

*Create Sheet
Set – Choose
Layouts page*

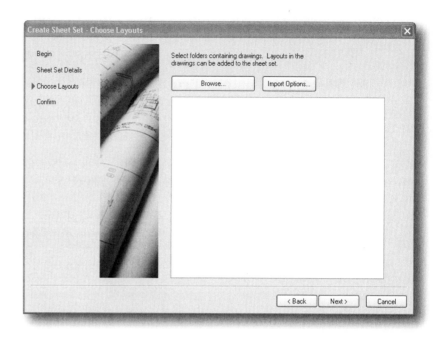

Choose **Browse** to select a folder or folders to list drawing files from which AutoCAD will import layouts in drawing files to create the sheet set. You can easily add more folders containing drawings by choosing **Browse** for each additional folder. Choosing **Import Options** causes the Import Options dialog box to be displayed as shown in Figure 20–7.

Figure 20-7

Import Options dialog box

When **Prefix sheet titles with file name** is checked, AutoCAD automatically adds the drawing file name to the beginning of the sheet title. When **Create subsets based on folder structure** is checked, AutoCAD automatically creates subsets in the newly created sheet sets based on folder structure. Choose **OK** to close the dialog box and accept the check box settings.

On the Create Sheet Set – Choose Layouts page, choose **Next**, and AutoCAD displays the Create Sheet Set – Confirm wizard page similar to creating a sheet from the example sheet set (Figure 20–5). The **Sheet Set Preview** box lists the information about the sheet set for review before you accept it. If the information is acceptable, choose **Finish** to complete the creation of the new sheet set. If any of the information needs to be modified, choose **Back** and make the necessary changes. AutoCAD creates a new sheet set in the form of a sheet set data file with the extension of *.dst* in the specified location with the specified properties.

SHEET SET MANAGER

Figure 20-8

Invoking the Sheet Set Manager from the Standard toolbar

The Sheet Set Manager provides the tools to organize, manage, and update a set of drawings. The Sheet Set Manager not only accesses the drawings associated with a project, but also lets you access the layouts and views that become the plotted sheets making up the final set of plotted drawings. As mentioned earlier, sheet sets are stored in the form of a sheet set data file with the extension of *.dst*. Each sheet in a sheet set is a layout in a drawing (*.dwg*) file. Open the Sheet Set Manager from the Standard toolbar (see Figure 20–8). AutoCAD displays the Sheet Set Manager palette with the **Sheet List** tab displayed as shown on the left in Figure 20–9.

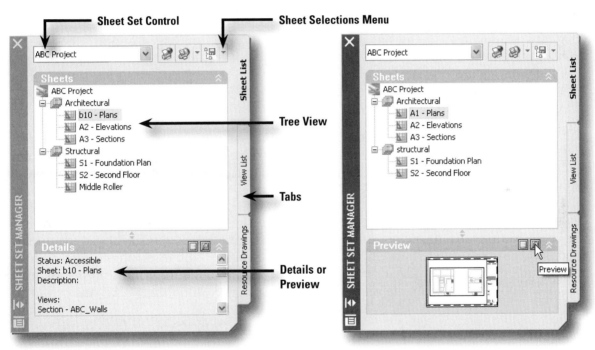

Figure 20-9

Sheet Set Manager palette with the Sheet List tab displayed on the left, Sheet Set Manager palette with the preview image of the selected sheet on the right

The **Sheet Set Control** lists menu options to create a new sheet set, open an existing sheet set, or switch between open sheet sets.

The **Sheet List** tab displays an organized list of all sheets in the sheet set. Each sheet in a sheet set is a specified layout in a drawing file.

The **View List** tab displays an organized list of all sheet views in the sheet set. Only sheet views created with AutoCAD 2005 and later are listed.

The **Resource Drawings** tab displays a list of folders, drawing files, and model space views available for the current sheet set. You can add and remove folder locations to control which drawing files are associated with the current sheet set.

The **Details** and **Preview** buttons display either descriptive information or a thumbnail preview of the currently selected item in the tree view.

Viewing and Modifying a Sheet Set

The **Sheet List** box displays the name of the sheet set that is open (see Figure 20–9 on the left), or if no sheet set is open, then you can choose **Open** and open an existing sheet set. You can also create a new sheet set by choosing **New**, which in turn will start the wizard to create a new sheet set.

The Sheet Set Manager shown on the left in Figure 20–9 contains two subsets called Architectural and Structural in the sheet set called ABC Project. The Architectural subset contains three numbered sheets: Plans, Elevations, and Sections. The Structural subset contains three numbered sheets: Foundation Plan, Second Floor, and Middle Roller. Each of these sheets is a layout in a *.dwg* file. When you select a sheet, information about it is displayed under the Details section of the palette. Figure 20–9 on the left shows details of the Plans sheet. To view a thumbnail preview of the selected sheet, choose the **Preview** button. Figure 20–9 on the right shows preview image of the Plans sheet.

You can also open the Sheet Set Properties as shown on the right in Figure 20–4 from the Shortcut menu that is displayed when the name of the sheet set is selected.

Instead of using the OPEN command to open a drawing, you can open the drawing using the Sheet Set Manager. Double-click the name of the sheet, and it will open in a new window. When the sheet is open, it will be locked automatically and no other user can open it at the same time. The lock status is indicated in the details section of the selected item (see Figure 20–10). When you close the drawing, the status will be changed to Accessible.

 Figure 20–10

Sheet Set Manager palette with the Details section indicating the status of the selected item

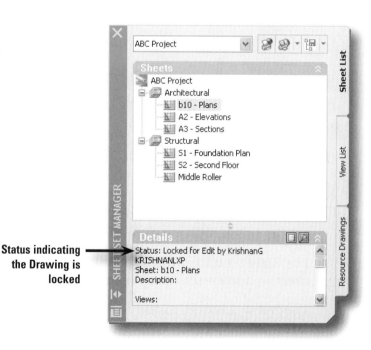

Status indicating the Drawing is locked

In a similar manner, you can create a new sheet. Choose New Sheet from the Shortcut menu, and AutoCAD displays the New Sheet dialog box, in which you can specify the name and number of the new sheet based on the default template set in the sheet set properties. When you create a new sheet, you create a new layout in a new drawing file, which is stored in the location specified in the sheet set properties. Instead of manually creating new drawing files, use the Sheet Set Manager.

To import a layout from an existing drawing, choose Import layout as a sheet from the Shortcut menu. AutoCAD displays the Import Layouts as Sheets dialog box. In the **Select a drawing file containing layouts** box, enter the drawing file name or browse to find a drawing from which to import layouts. In the **Select layouts to import as sheets** box, select layouts to import. You can choose to have AutoCAD prefix sheet titles with the file name. Choose **OK** to import the layout(s).

With a large sheet set, you will find it necessary to organize sheets and views in the tree view. On the **Sheet List** tab, sheets can be arranged into collections called subsets. To create a new subset, first select the name of the sheet set and choose New subset from the Shortcut menu. AutoCAD displays the Subset Properties dialog box, where you can create a new sheet subset for organizing the sheets. Figure 20–10 shows two subsets (Architectural and Structural) in the ABC Project sheet set.

If necessary, you can rename and renumber the sheet. To rename and renumber the selected sheet, choose Rename & Renumber from the Shortcut menu. AutoCAD displays the Rename & Renumber Sheet dialog box as shown in Figure 20–11, where you can specify the sheet number and title for the selected sheet. The **Number** box specifies the sheet number of the selected sheet. The **Sheet title** box specifies the sheet title of the selected sheet. The **Next** option loads the next sheet into this dialog box.

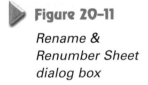

Figure 20–11

Rename & Renumber Sheet dialog box

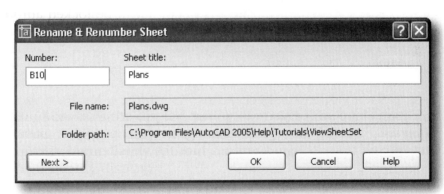

Sheet Selections (see Figure 20–9) causes a menu to be displayed where you can save, manage, and restore sheet selections by name. This makes it easy to specify a group of sheets for publish, transmit, or archive operations. To create a sheet selection, first select several sheets from the sheet list and choose Create from the Sheet Selections menu (see Figure 20–12). AutoCAD displays the New Sheet Selection dialog box, where you can specify a sheet selection name and choose **OK** to create the new sheet selection. To restore the selection, choose the name of the sheet selection from the Sheet Selections menu. Choosing Manage causes the Sheet Selection dialog box to be displayed, where you can rename or delete the selected Sheet Selection.

To remove the selected sheet from the sheet set, choose Remove sheet from the Shortcut menu. AutoCAD removes the currently selected sheet from the sheet set.

Figure 20-12

Sheet Set Manager with Create selected from the Sheet Selections menu

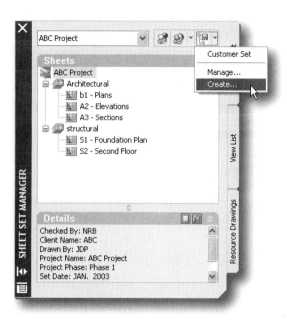

Placing a View on a Sheet

The Sheet Set Manager automates and enhances the process for placing a view of a drawing on a sheet. First open the sheet where you want to place the view. To find the view to add to this sheet, click the **Resource Drawings** tab. On this tab, you can browse for drawings that contain the views you want to add to your sheet. Any drawings you want to use must be listed at this location. To add a folder that contains drawings to the list, choose **Add New Location** and select the folder. When you select a view, information about it is displayed in the **Details** section of the palette. To view a thumbnail preview of the selected view, choose **Preview**.

To place a view, first select the view and from the Shortcut menu, select Place on Sheet (see Figure 20-13). Before you place the view, right-click to view or change the scale of the view. Click anywhere on the sheet to insert it. A block label is also inserted. When you place a view on a sheet, AutoCAD attaches the drawing with the named view as an xref. The view is listed as a paper space view on the **View List** tab. From the **View List** tab, you can add a view number to the paper space view. Numbers added to the paper space view are updated when the drawing is regenerated.

 Figure 20-13

Sheet Set Manager - Resource Drawings tab with the Shortcut menu to place a view

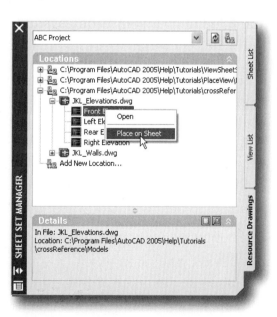

Instead of using the OPEN command to open a drawing, you can also open the drawing using the Sheet Set Manager. Double-click the name of the drawing in the **Resource Drawings** tab, and it will open in a new window. When the drawing is open, it will be locked automatically and no other user can open it at the same time. The lock status is indicated in the details section of the selected item. When you close the drawing, the status will be changed to Accessible.

View List

The **View List** tab (see Figure 20–14) displays all named views (also called sheet views) on the layouts in your sheet set. You can use the view list to keep track of all sheet views in the sheet set. You can navigate to any sheet view in the sheet set. You can also link sheet views together for coordination across the sheet set.

Figure 20–14

Sheet Set Manager with View List tab

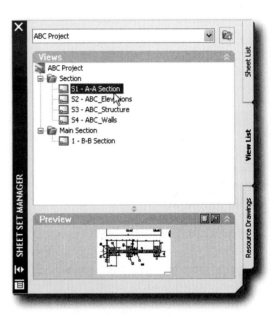

Sheet views can be created in two ways: You can create a sheet view by creating a named view in paper space on any layout by invoking the VIEW command. In addition, AutoCAD also automatically creates a sheet view whenever you place a view of a drawing on a sheet (for details on placing a view, see the section "Placing a View on a Sheet").

With the Sheet Set Manager, you can apply label and callout blocks to views on sheets. A callout block refers to other views in the sheet set. It is a symbol that shows, for example, a cross-reference to an elevation, a detail, a section and so on (see Figure 20–15). With the Sheet Set Manager, you can automatically update the information in your label and callout blocks when the reference information changes. For example, when you renumber or rename a view, information on the label and callout blocks is updated when the drawing regenerates. To place a callout block for the selected view, select Place Callout Block, and choose the type of callout block you want to use (see Figure 20–16). Click anywhere on the sheet to place the callout block. This callout block contains information about the sheet number of the drawing in which the view is saved and the view number (see Figure 20–15).

Figure 20–15

Example callout symbols

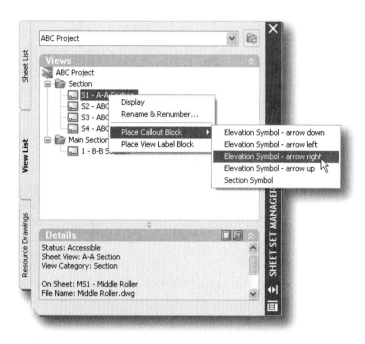

> Figure 20-16
>
> *Sheet Set Manager – View List tab with the Shortcut menu to place a callout block*

If necessary, you can rename and renumber a view. To rename or renumber the selected view, choose Rename & Renumber from the Shortcut menu. AutoCAD displays Rename & Renumber Sheet dialog box. Specify a view number and view title for the selected view. The **Number** specifies the view number of the selected view. The **View title** specifies the view title of the selected view. The **Next** option loads the next view into this dialog box.

Instead of using the OPEN command to open a drawing, you can also open the drawing using the Sheet Set Manager. Double-click the name of the drawing or choose Display from the Shortcut menu in the **View List** tab, and it will open in a new window. When the drawing is open, it will be locked automatically and no other user can open it at the same time. The lock status is indicated in the details section of the selected item. When you close the drawing, the status will be changed to Accessible.

Creating a Sheet List Table

With the Sheet Set Manager you can create a sheet list table and then update it to match changes in your sheet list. Start by opening the sheet in which you want to create the sheet list table. Select the **Sheet List** tab, select the sheet set, and from the Shortcut menu, select Insert Sheet List Table (see Figure 20–17 at left). AutoCAD displays the Insert Sheet List Table dialog box as shown at right in Figure 20–17. There are several ways to change the appearance of the table before you insert it. For example, when you select **Show Subheader**, the table includes rows displaying subheaders, or the subsets in the sheet set. After making necessary changes in the table, choose **OK**. Click anywhere on the sheet to place the sheet list table. This table lists all the sheets and subsheets in the sheet set, as shown in Figure 20–18. If you remove, add, or make any changes to the sheet number or sheet name, you can update the sheet list table by selecting the table and choosing Update Sheet List Table from the Shortcut menu. AutoCAD updates the sheet list table with the changes.

note If you modify the sheet list table manually, the changes are temporary and are lost when you update the table.

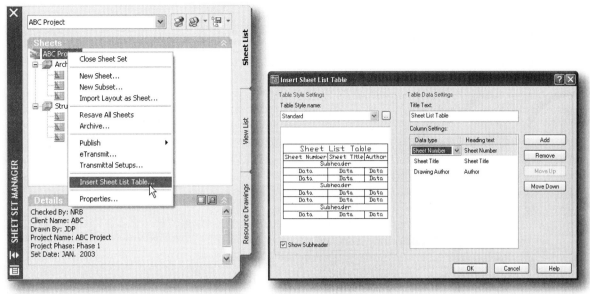

Figure 20-17

*Sheet Set Manager – Sheet List tab with the Shortcut menu to Insert Sheet List Table;
Insert Sheet List Table dialog box*

Figure 20-18

*An example table created from the Sheet
Set Manager*

Structural List of Drawings		
Sheet Number	Sheet Title	Author
Architectural		
A1	Plans	GVK
A2	Elevations	TOMS
A3	Sections	GVK
Structural		
S1	Foundation Plan	GVK
S2	Second Floor	TOMS
MS1	Middle Roller	GVK

Creating a Transmittal Package

The Sheet Set Manager allows you to eTransmit a sheet set, selected sheets, or a subset. eTransmit packages a set of files for Internet transmittal. In a transmittal package, sheet set data files, xrefs, plot configuration files, font files, and so on are automatically included. To create a transmittal package, first select the sheet set, one or more sheets, or subset you want to include in the transmittal package and select eTransmit from the Shortcut menu. Figure 20–19 at left shows the Architectural subset selection and eTransmit selected from the Shortcut menu. AutoCAD displays the Create Transmittal dialog box as shown at right in Figure 20–19. The **Sheets** tab lists all the sheets in the transmittal package for the Architectural subset. The **Files Tree** tab lists all the xref, sheet set data, and template files for the transmittal package. Choose **Transmittal**

Setups to customize the transmittal setup that will define how your transmittal is packaged. (For a detailed explanation on the transmittal setup, refer to Chapter 21.) Choose **OK** to create the eTransmit package.

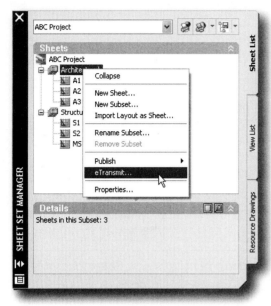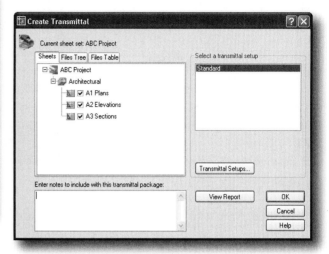

 Figure 20–19

Sheet Set Manager – Sheet List tab with the Shortcut menu for eTransmit selection; Create Transmittal dialog box

Creating an Archive of the Sheet Set

The Sheet Set Manager allows you to archive the selected sheet set. **Archive** brings together for archiving purposes the files associated with the current sheet set. To create an archive package, first select the sheet set to include in the archive package and select Archive from the Shortcut menu, as shown on the left in Figure 20–20. AutoCAD displays the **Archive a Sheet Set** dialog box as shown on the right in Figure 20–20. The **Sheets** tab lists all the sheets in the archive package for the selected sheet set. The **Files Tree** tab lists all the xref, sheet set data, and template files for the archive package. Choose **Modify Archive Setup** to customize the archive setup that will define how your archive is packaged. If necessary, enter information relative to the archive package in the **Enter notes to include with this Archive** box. The information is included in the archive report. Choose **OK** to create the archive package. Be sure that the files to be archived are not open.

---

note Be sure that the files to be archived are not open.

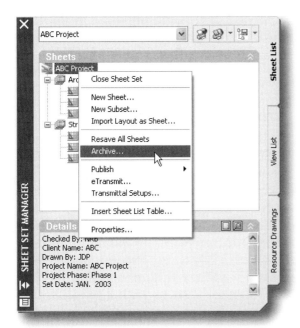
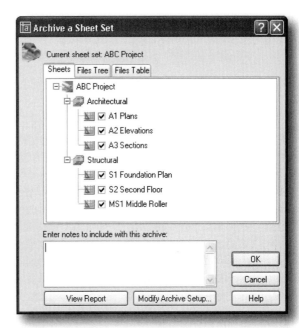

 Figure 20-20

Sheet Set Manager – Sheet List tab with the Shortcut menu for Archive selection; Archive a Sheet Set dialog box

Plotting the Sheet Set and Publishing to DWF

The Sheet Set Manager allows you to plot to the default plotter or printer or publish to specified DWF (Drawing Web Format) format a sheet set, selected sheets or subset. To plot, first select the sheet set, one or more sheets, or subset you want to include. Then select Publish from the Shortcut menu and Publish to Plotter from the submenu as shown in Figure 20–21. AutoCAD will plot the selected sheet(s) to the default plotter. (For detailed information on plotting and its related settings, refer to Chapter 16.) Similarly, to publish to specified DWF format, first select the sheet set, one or more sheets, or subset you want to include and select Publish from the Shortcut menu and Publish to DWF from the submenu. AutoCAD will create the DWF file of the selected sheet(s). (For detailed information on publishing by DWF, refer to Chapter 21.)

 Figure 20-21

Sheet Set Manager – Sheet List tab with the Shortcut menu to Publish to Plotter selection

review questions

1. Sheets in a sheet set are created from:
   a. *Adobe PDF files.*
   b. *AutoCAD drawing backup files.*
   c. *Layouts in AutoCAD drawing files.*
   d. *DWF format files.*

2. Drawings associated with a sheet set should:
   a. *All have the same prefix in their name.*
   b. *All be in the same location.*
   c. *Never exceed 1 megabyte.*
   d. *Contain external references.*
   e. *None of the above.*

3. Sheet set properties can include the following:
   a. *Name.*
   b. *Sheet set data file name.*
   c. *Callout blocks.*
   d. *Page setup template file.*
   e. *All of the above.*

4. Creating a sheet set from an example sheet set:
   a. *Is the only way to create a sheet set.*
   b. *Requires importing the sheets from an example set into the newly created sheet set.*
   c. *Does not copy any sheet in the sample set into the new set.*
   d. *Deletes the sheets in the sample set.*

5. From the Import Options dialog box of the Create Sheet Set – Choose Layouts wizard page you can:
   a. *Import music files for playing music while drawing.*
   b. *Prefix sheet titles with the file name.*
   c. *Create subsets based on folder structure.*
   d. *Both b and c above.*

6. The Sheet Set Manager has the following tab:
   a. *Sheet List.*
   b. *View List.*
   c. *Resource Drawings.*
   d. *All of the above.*

7. The Sheet List tab Details section shows:
   a. *Architectural details.*
   b. *Information about the sheet set selected in the tree view.*
   c. *Details about external references.*
   d. *Startup information about the AutoCAD program.*

busy bees and workhorses

introduction

There are many added features in AutoCAD that make it much easier to accomplish your work. The assemblage of these is referred to as utility commands and features. Many of them are found in the Tools menu.

After completing this chapter, you will be able to do the following:

- Create, customize, and use tool palettes

- Partial Load drawings

- Manage drawing properties

- Manage named objects

- Delete unused named objects

- Use the utility display commands

- Change the display order of objects

- Use object properties

- Set Security, passwords, and encryption

- Set custom settings with the Options dialog box

- Export and Import objects

- Understand CAD standards

- Access Internet utilities

TOOL PALETTES

Figure 21-1

Invoking the TOOLPALETTES command from the Standard toolbar

Tool Palettes are created to make it quicker and easier to insert blocks, draw hatch patterns and implement custom tools developed by a third party. Blocks and hatch patterns are the primary tools that are managed with tool palettes.

A special facility of a Tool Palette is the ability to preload it with objects with specific properties that, when these objects are selected and dragged into the drawing area, they take with them their properties. For instance, if a green polyline on layer 7 is selected and dragged onto a Tool Palette, then dragging it back into the drawing area invokes the pline command, and the resulting polyline will be green and on layer 7, regardless of the current layer or color control. The following objects can be dragged onto a Tool Palette:

- Blocks
- Dimensions
- External references
- Gradient fills
- Objects such as arcs, circles, lines, and polylines
- Raster images
- AutoCAD commands or strings of commands

Tool palettes are separate tabbed areas within the Tool Palettes window. This allows blocks and hatch patterns of similar usage and type to be grouped in their own tool palette. For example, one tool palette is named Architectural, with blocks representing architectural objects, furniture and fixtures as shown in Figure 21–2.

The default Tool Palettes window (see Figure 21–2) that comes with AutoCAD has other tabs attached: Annotation, Mechanical, Electrical, Civil/Structural, Hatches, and Command Tools, which contains icons representing blocks, hatch patterns, and AutoCAD commands. After you invoke the tool palettes command (see Figure 21–1), AutoCAD displays the Tool Palettes window as shown in Figure 21–2.

Figure 21-2

The Tool Palettes window in the docked position with the Architectural tab displayed

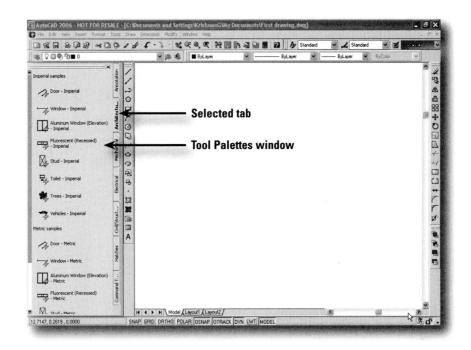

Docking and Undocking the Tool Palettes Window

The default position for the Tool Palettes window is docked on the left side of the screen. Its position can be changed by placing the cursor over the double line bar at the top of the window and either double-clicking or dragging the window into the screen area (or across to a docking position on the right side of the screen). Double-clicking causes the Tool Palettes window to become undocked and to float in the drawing area as shown in Figure 21–3. When the Tool Palettes window is undocked, it can be docked by double-clicking in the title bar (which may be on the left or right side of the window) or by placing the cursor over the title bar and dragging the window all the way to the side where you wish to dock it.

Figure 21-3

The Tool Palettes window in the floating position

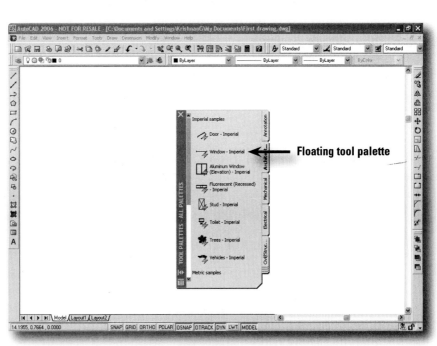

Tool Palettes Window Shortcut Menus

Four different shortcut menus are available for managing the Tool Palettes window, its palettes, and elements on the palettes, depending on the location (tab, tool palette element, open area in tool palette, and tool palette window title bar) of the cursor when you right-click.

The shortcut menu displayed when you right-click the active tab in the Tool Palettes window (see Figure 21–4) includes these options: **Move Up**, **Move Down**, **New Palette**, **Delete Palette**, **Rename Palette**, **View Options**, and **Paste** (**Delete Palette**, **View Options**, and **Paste** are not available when an inactive tab is selected).

 Figure 21–4

*The Shortcut menu when right-clicking
a Tool Palettes window tab*

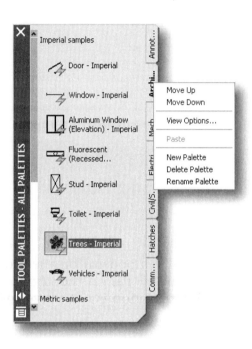

Choose **Move Up** or **Move Down** to move the selected tab up or down one place respectively in the order of tabs.

Choose **New Palette** to create a new palette. AutoCAD will create a new tool palette to which you can add tools. Enter a name or press ENTER to use the default name.

> **note** Tool palettes can be used only in the current or later version of AutoCAD in which they were created. For example, you cannot use a tool palette that was created in AutoCAD 2006 in AutoCAD 2005.

Choose **Rename Palette** to rename the current tool palette.

Choose **Delete Palette** to remove the current tool palette.

Choose **View Options** to display the View Options dialog box (see Figure 21–5) from which you can control the display of tools in the current tool palette or in all tool palettes. The **Image size** slider bar allows you to change the size of the images. The **View style** section lets you set how the tool elements are displayed. Selecting **Icon only** causes the block/hatch pattern icon to be displayed as an image only without text. Selecting **Icon with text** causes the block/hatch pattern icon to be displayed as an image with the descriptive text below it. Selecting **List view** causes the block/ hatch pattern icon to be displayed as an image with the descriptive text to its right, allowing for a more compressed listing of the symbols when used with a small image. The **Apply to** list box allows you to choose whether the changes are applied to the **Current Palette** or to **All Palettes**. To exit the View Options dialog box and accept the changes, choose **OK**. To exit without accepting the changes, choose **Cancel**.

Figure 21-5

The View Options dialog box

Choose **Paste** to paste a block/hatch pattern shortcut (that has been copied to the Clipboard) on the tool palette. If there is nothing on the Clipboard, or if whatever is on the Clipboard is not a block/hatch pattern shortcut, then the **Paste** option is not active and cannot be selected.

The shortcut menu displayed when you right-click one of the elements in a tool palettes window (see Figure 21–6) includes the options **Cut**, **Copy**, **Delete**, **Rename**, **Update tool image**, **Block Editor**, and **Properties** (**Update tool image** and **Block Editor** are not available if the selected element is a command or hatch pattern).

Figure 21-6

The Shortcut menu when right-clicking an element in the Tool Palettes window

Choose **Cut** to delete the selected element and place it on the Clipboard.

Choose **Copy** to copy the selected element and place it on the Clipboard.

Choose **Delete** to remove the selected element.

Choose **Rename** to rename the selected element.

Choose **Properties** to display the Tool Properties dialog box showing the properties of the selected element. Tool properties are explained later in this chapter.

Choose **Update tool image** to update the selected image of the tool when the definition for a block, xref, or raster image is changed. You must save the drawing before you can update the tool image. The icon for a block, xref, or raster image in a tool palette is not automatically updated if its definition changes.

Choose **Block Editor** to invoke the BEDIT command, explained in the chapter covering Dynamic Blocks.

The shortcut menu that is displayed when you right-click in an open area of the Tool Palettes window (see Figure 21–7) includes the options **Allow Docking**, **Auto-Hide**, **Transparency**, **View Options**, **Sort By**, **Paste**, **Add Text**, **Add Separator**, **New Palette**, **Delete Palette**, **Rename Palette**, and **Customize**.

 Figure 21–7

The Shortcut menu when right-clicking in an open area of the Tool Palettes window

Allow Docking, when checked, allows you to drag the Tool Palettes window to one side of the screen and dock it. When it is not checked, the window cannot be docked.

Auto-hide, when checked, causes the Tool Palettes window (only when floating) to be hidden, except for the title bar, when the cursor is not over the title bar or the window. To display the window, move the cursor over the title bar.

Transparency, when selected, causes the Transparency dialog box to be displayed as shown in Figure 21–8. When the indicator on **Less-More** is to the left

(Less), the Tool Palettes window (only when floating) is opaque. The closer the indicator is to the right (More), the more transparent the window will become. **Turn off window transparency**, when checked, prevents the Tool Palettes window from becoming transparent.

Figure 21-8

The Transparency dialog box

The **View Options** option, when selected, causes the View Options dialog box to be displayed. It is similar to the one explained earlier.

Choosing **Sort By** causes a shortcut menu to be displayed from which you can sort the elements on the tool palette by **Name** or **Type**.

The **Paste** option functions similarly to the one explained earlier.

Choose **Add Text** to add descriptive text (at the location of the cursor when right-clicked).

Choose **Add Separator** to create a separating line on the palette at the location of the cursor when right-clicked.

The **New Palette**, **Delete Palette**, and **Rename Palette** options function similarly to the ones explained earlier.

Customize causes the Customize dialog box to be displayed with the **Tool Palettes** tab selected, as shown in Figure 21–9.

Figure 21-9

The Customize dialog box with the Tool Palettes tab selected

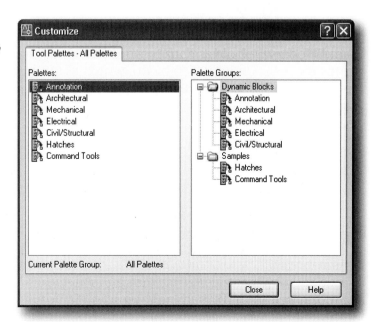

AutoCAD lists all available tool palettes. Click and drag a tool palette to move it up or down in the list. Right-click a tool palette in the list to rename, delete, or export the tool palette. (When you export a tool palette, it is saved to a file with an *.xtp* extension.) Right-click in the **Tool Palettes** area to import a tool palette or to create a new, blank one.

Inserting Blocks/Hatch Patterns from a Tool Palette

To insert a block from a tool palette, place the cursor on the block symbol in the tool palette, press the pick button, and drag the symbol into the drawing area. The block will be inserted at the point where the cursor is located when the pick button is released. This procedure is best implemented by using the appropriate OSNAP mode. Another method of inserting a block from a tool palette is to select the block symbol in the tool palette and then select a point in the drawing area for the insertion point.

Blocks dragged from a tool palette are automatically scaled according to the ratio of block-defined units to units defined in the current drawing. For example, if the current drawing uses centimeters as its units and a tool palette block uses meters, the ratio of the units is 1 centimeter/1 meter. This means the block will be dragged into the drawing at 100 times its scale.

---

note The **Source Content Units and Target Drawing Units** settings of the **User Preferences** tab of the Options dialog box are used when **Drag-and-Drop Scale** is set to Unitless, either in the target drawing or the source block.

To draw a hatch pattern that is a tool in a tool palette, place the cursor on the hatch pattern symbol in the tool palette, press the pick button, and drag the symbol into the boundary to receive the hatch pattern and release the pick button. Another method of drawing a hatch pattern that is a tool in a tool palette is to select the hatch pattern symbol in the tool palette and then select a point within a boundary in the drawing area.

Block Tool Properties

Blocks whose symbols appear in a tool palette are not, as a rule, blocks defined in the current drawing. Usually, they reside as block definitions in another drawing, or in some cases they might even be drawing files. As a tool in a tool palette, a block has tool properties. To access the block/drawing tool properties, right-click on the tool's symbol in the tool palette, and select Properties from the Shortcut menu. The Tool Properties dialog box will be displayed as shown is Figure 21–10.

Figure 21-10

The Tool Properties dialog box for block/drawing tools

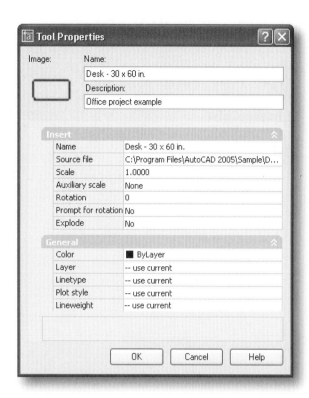

The Insert properties of a block/drawing tool in a tool palette include **Name**, **Source file**, **Scale**, **Rotation**, and **Explode**. The General properties include **Color**, **Layer**, **Linetype**, **Plot style,** and **Lineweight**.

The **Name** specifies the name of the block of the selected tool. The **Source file** property edit box lists the path to the drawing file where the block that is a tool in the tool palette resides. Or, if the tool is a drawing file, the edit box lists the path to it.

note Any change to the path/file name in the **Source file** property text box will prevent the block/drawing from being inserted.

The **Scale** factor specifies the XYZ scale factor of the block. **Auxiliary scale** overrides the regular scale setting and multiplies your current scale setting by the plot scale or the dimension scale. The **Rotation** specifies the rotation angle of the block. If the **Prompt for rotation** list box is set to **No**, the block will be inserted with the default rotation angle. If it is set to **Yes**, you will be prompted for the angle of rotation. The **Explode** list box determines whether or not the block is exploded when inserted. If the **Explode** list box is set to **No**, the block will be inserted as an unexploded block. If it is set to **Yes**, the separate parts that make up the block will be drawn as separate entities, that is, as a block that is inserted and exploded. The **Color**, **Layer**, **Linetype**, **Plot Style**, and **Lineweight** options specify the override for the color, layer, linetype, plot style, and lineweight, respectively.

Pattern Tool Properties

Hatch patterns whose symbols appear in a tool palette have tool properties. The Pattern properties (see Figure 21–11) of a hatch pattern tool in a tool palette include **Tool type**, **Type**, **Pattern name**, **Angle**, **Scale**, **Spacing**, and **ISO pen width**, and **Double**. The General properties include **Color**, **Layer**, **Linetype**, **Plot style**, and **Lineweight**.

Figure 21–11

The Tool Properties dialog box for pattern tools

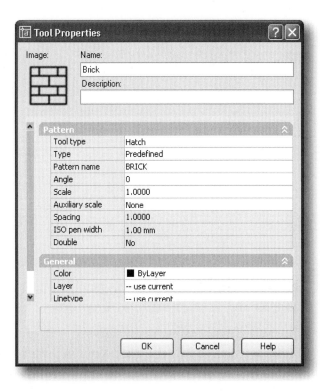

Tool type specifies whether the tool is a hatch or a gradient. **Type** specifies the pattern type of the hatch. The pattern types include **User-defined**, **Predefined** from one of the hatch patterns listed in the **Pattern** list box, or **Custom** from one of the hatch patterns in the drawing as listed in the **Custom Pattern** list box. **Pattern name** specifies the pattern name of hatch. **Angle**, **Scale**, and **Spacing** specify the angle, scale and spacing of the hatch, respectively. **Auxiliary scale** overrides the regular scale setting and multiplies your current scale setting by the plot scale or the dimension scale. **ISO pen width** specifies the ISO pen width of an ISO hatch pattern. **Double** determines whether the hatch pattern is double. **Color**, **Layer**, **Linetype**, **Plot style**, and **Lineweight** specify the override for the color, layer, linetype, plot style, and lineweight respectively.

Other tools, such as commands and tables, have their own Tool Properties dialog boxes with properties specific to the type of tool selected.

Creating and Populating Tool Palettes

A new tool palette can be created from the Tool Palette Shortcut menu by selecting the New Tool Palette option as described earlier in this section.

You can also create a new tool palette from the Tree view or content area of the DesignCenter. In the Tree view or content area, highlight an item and then right-click to display the Shortcut menu. If the selected item is a folder, one of the available options in the Shortcut menu is Create Tool Palette of Blocks. If there are no blocks in the folder, a message will be displayed stating "Folder does not contain any drawing files." If the selected item is a drawing, one of the available options in the Shortcut menu is Create Tool Palette. If the drawing does not contain any blocks, a message will be displayed stating "Drawing does not contain any block definitions."

When one of the options to create a tool palette is selected, AutoCAD creates a new tool palette, which will be populated with the drawings from the selected folder or blocks from the selected drawing.

From the content area, in addition to creating tool palettes by right-clicking a folder or drawing, you can also right-click on a block and select the Create Tool Palette option from the Shortcut menu. In this case the new tool palette will contain only the selected block. In this case you will be prompted to name the tool palette.

You can also drag and drop a block from a drawing or a drawing from the content area of the DesignCenter onto the one of the existing palettes.

As mentioned earlier, you can also drag and drop an object with properties assigned to it onto one of the existing palettes. Once you add a command to a tool palette, you can click the tool to execute the command. For example, clicking a Line tool on a tool palette invokes the line command with the settings of the pre-assigned properties.

PARTIAL OPEN

The OPEN command allows you to work with just part of a drawing by loading geometry within specific views or layers. When a drawing is partially open, specified and named objects are loaded. Named objects include blocks, layers, dimension styles, linetypes, layouts, text styles, viewport configurations, UCSs, and views. To access the Partial Open option, click on the right arrow to the right of the **Open** button.

The partiaload command allows you to load additional geometry into a current partially loaded drawing by loading geometry only from specific views or layers.

---

note   If you invoke the PARTIALOAD command in a drawing that has not been partially opened, AutoCAD will respond with the message "Command not allowed unless the drawing has been partially opened."

DRAWING PROPERTIES

Drawing properties allow you to keep track of your drawings by having properties assigned to them. The properties you assign will identify the drawing by its title, author, subject, and keywords for the model or other data. Hyperlink addresses or paths can be stored, along with ten custom properties. Invoke the Drawing Properties command from the File menu; AutoCAD displays the Drawing Properties dialog box as shown in Figure 21–12.

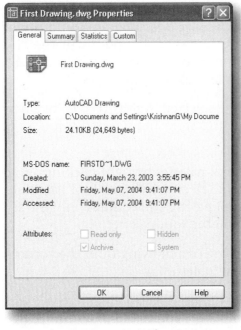

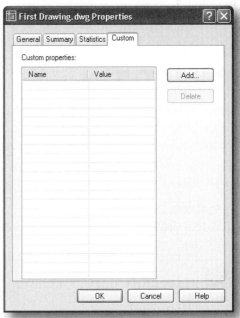

Figure 21-12

The Drawing Properties dialog box: General tab (top left), Summary tab (top right), Statistics tab (bottom left) Custom tab (bottom right)

The Drawing Properties dialog box has four tabs: **General**, **Summary**, **Statistics**, and **Custom**.

The **General** tab (see Figure 21–12 at top left) displays the drawing type, location, size, and other information. These come from the operating system and the fields are read-only. However, the attributes options are made available by the operating system if you access file properties through Windows Explorer.

The **Summary** tab (see Figure 21–12 at top right) lets you enter a title, subject, author, keywords, comments, and a hyperlink base for the drawing. Keywords for drawings sharing a common property will help in your search. For a hyperlink base, you can specify a path to a folder on a network drive or an Internet address.

The **Statistics** tab (see Figure 21–12 at bottom left) displays information such as the dates files were created and last modified. You can search for all files created at a certain time.

The **Custom** tab (see Figure 21–12 at bottom right) lets you specify up to ten custom properties. Enter the names of the custom fields in the left column, and the value for each custom field in the right column.

---

note Properties entered in the Drawing Properties dialog box are not associated with the drawing until you save the drawing.

MANAGING NAMED OBJECTS

The RENAME command allows you to change the names of blocks, dimension styles, layers, linetypes, plot styles, table styles, text styles, views, User Coordinate Systems, or viewport configurations. Invoke the RENAME command from the Format menu; AutoCAD displays the Rename dialog box, shown in Figure 21–13.

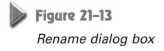

Figure 21–13

Rename dialog box

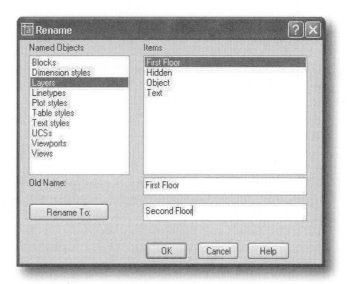

---

note Except for the layer named 0 and the linetype named Continuous, you can change the name of any of the named objects.

In **Named Objects**, select the type of object you want to change. **Items** displays the names of all objects that can be renamed. To change the object's name, select the name in the **Items** list box or type it in the **Old Name** box. Type the new name in the **Rename To** box, and select **Rename To** to update the object's name in the **Items** list box. To close the dialog box, choose **OK**.

DELETING UNUSED NAMED OBJECTS

The PURGE command is used to selectively delete any unused named objects. Invoke the PURGE command from the File menu; AutoCAD displays the Purge dialog box as shown in Figure 21– 14.

 Figure 21-14

Purge dialog box with the Items not used in drawing list displayed

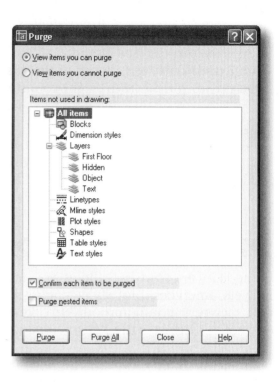

If **View items you can purge** is selected, AutoCAD lists categories of named items, under which are listed the individual named items that have been defined in the drawing but are not currently being used. For example, if a layer has been defined but has nothing drawn on it and it is not the current layer, it can be purged. Or, if the drawing contains a block definition but the block has not been inserted, it can be purged. Objects such as lines, circles, and other basic unnamed drawing elements cannot be purged.

If **Confirm each item to be purged** is checked, AutoCAD displays the Confirm Purge dialog box and asks you to reply by selecting **Yes** or **No** before continuing.

If **Purge nested items** is checked, then AutoCAD will purge nested items within any item selected. Otherwise the purge command removes only one level of reference.

Once an individual named item in the list has been selected or a category of items has been selected that has items that can be purged, **Purge** at the bottom of the dialog box becomes operational. **Purge All** causes all items that can be purged to be purged.

If **View items you cannot purge** is selected, AutoCAD lists categories of named items, under which are listed the individual named items that have been defined in the drawing and are currently being used.

Views, User Coordinate Systems, and viewport configurations cannot be purged, but the commands that manage them provide options to delete those that are not being used.

COMMAND MODIFIER—MULTIPLE

MULTIPLE is not a command, but when used with another AutoCAD command, it causes automatic recalling of that command when it is completed. You must press ESC or right-click and select Cancel from the short-cut menu to terminate this repeating process. Here is an example of using this modifier to cause automatic repeating of the ARC command:

multiple (ENTER)

Enter command name to repeat: **arc**

You can use the multiple command modifier with any of the draw, modify, and inquiry commands. plot, however, will ignore the multiple command modifier.

UTILITY DISPLAY COMMANDS

The utility display commands include VIEW, REGENAUTO, and BLIPMODE.

Saving Views

The VIEW command allows you to give a name to the display in the current viewport and have it saved as a view. When you name and save a view, settings are saved such as magnification, center point, view direction, location of the view (Model or specific layout), layer visibility, user coordinate system, 3D perspective, and clipping and view category assigned to the view (optional). You can recall a view later by using the VIEW command and responding with the name of the view desired. This is useful for moving back quickly to needed areas in the drawing without having to resort to zooming and panning. Invoke the VIEW command from the View menu; AutoCAD displays the View dialog box, shown in Figure 21–15.

 Figure 21–15

View dialog box with the Named Views tab displayed

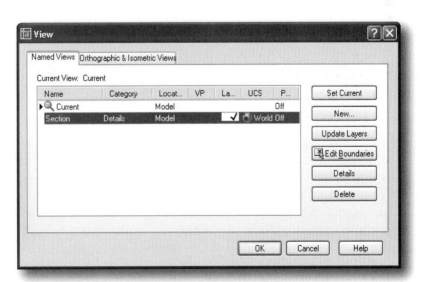

In the **Named Views** tab, AutoCAD lists any saved view(s) in the list box.

To create a new view, choose **New**. AutoCAD displays the New View dialog box. In the New View dialog box, specify a name for the new view in the **View name** box. Specify a category for the named view in the **View Category** box, for example, FloorPlans or

Sections. Select a view category from the list, enter a new category, or leave this option empty. Select one of the two options to specify the area for the new view: **Current display** or **Define window**. **Current display** selection will include the current display as the new view. **Define window** selection allows you to specify diagonally opposite corners of a window that will define the area for the new view. Select one of the available Coordinate Systems (User or World) from the **UCS name** in the **UCS Settings** section. Set **Save UCS with view** to ON to save the coordinate system that is displayed in the **USC name** with the new view. Select **OK** to close the New View dialog box.

To restore the selected named view choose **Set Current** in the View dialog box. You can also restore a named view by double-clicking its name in the list or by right-clicking its name and clicking Set Current on the Shortcut menu.

Choosing **Details** causes the View Details dialog box to be displayed, which has information about the view selected in the **Current View** box of the View dialog box. Information includes the area of the view in width and height and twist, target point coordinates, direction point coordinates, front and back offset clipping, perspective lens length, and coordinate system that the view is relative to.

To update the layer information saved with the selected named view to match the layer visibility in the current model space or layout viewport, choose **Update Layers**.

To edit boundaries of the selected view, choose **Edit Boundaries**. AutoCAD displays the selected named view centered and zoomed out and with the rest of the drawing area in a lighter color to show the boundaries of the named view. You can specify opposite corners of a new boundary repeatedly until you press enter to accept the results.

To delete the selected named view, choose **Delete**.

After making necessary changes, choose **OK** to close the dialog box and save the changes.

Controlling Regeneration

The REGENAUTO command controls automatic regeneration. When the REGENAUTO is set to ON, AutoCAD drawings regenerate automatically. When it is set to OFF, you may have to regenerate the drawing manually to see the current status of the drawing. After you invoke the REGENAUTO command from the On-screen prompt, AutoCAD prompts:

Enter mode: *(select an option)*

When you set regenauto to OFF, what you see on the screen may not always represent the current state of the drawing. When changes are made by certain commands, the display will be updated only after you invoke the regen command. But the time waiting for regeneration can be avoided as long as you are aware of the status of the display. Turning the regenauto setting back to ON will cause a regeneration. If a command should require regeneration while regenauto is set to OFF, you will be prompted:

About to regen, proceed? <Y>

Responding with **No** will abort regeneration.

Regeneration during a transparent command will be delayed until a regeneration is performed after that transparent command. The following message will appear in the text screen window:

REGEN QUEUED

Controlling the Display of Marker Blips

The BLIPMODE command controls the display of marker blips. When BLIPMODE is set to ON, a small cross mark is displayed when points on the screen are specified with the cursor or by entering their coordinates. After you edit for a while, the drawing can become cluttered with these blips. They have no effect other than visual reference and will be removed by using the REDRAW, REGEN, ZOOM, or PAN command. Any other command requiring regeneration causes the blips to be removed. When BLIPMODE is set to OFF, the blips marks are not displaced. When you invoke the BLIPMODE command, AutoCAD prompts:

Enter mode: *(select an option from the Shortcut menu)*

CHANGING THE DISPLAY ORDER OF OBJECTS

 Figure 21–16

Invoking the DRAWORDER command from the Modify II toolbar

The draworder command allows you to change the display order of objects as well as images. This will ensure proper display and plotting output when two or more objects overlap one another. For instance, when a raster image is attached over an existing object, AutoCAD obscures them from view. By using the draworder command, you can make the existing object display over the raster image. Invoke the draworder command (see Figure 21–16), AutoCAD prompts:

Select objects: *(select the objects for which you want to change the display order, and press ENTER to complete object selection)*

Enter object ordering option [Above object/Under object/Front/Back] <Back>: *(select one of the available options from the Shortcut menu)*

Select reference object: *(select the reference object for changing the order of display)*

When multiple objects are selected for reordering, the relative display order of the objects selected is maintained.

The Above object selection moves the selected object(s) in front of a specified reference object.

The Under object selection moves selected object(s) behind a specified reference object.

The Front selection moves selected object(s) to the front of the drawing order.

The Back selection moves selected object(s) to the back of the drawing order.

note The DRAWORDER command terminates when selected object(s) are reordered. The command does not continue to prompt for additional objects to reorder.

OBJECT PROPERTIES

There are three important properties that control the appearance of objects: color, line-type, and lineweight. You can specify the color, linetype, and lineweight for the objects to be drawn with the help of the LAYER command, as explained in Chapter 6. You can do the same thing by means of the COLOR, LINETYPE, and LINEWEIGHT commands.

Setting an Object's Color

 Figure 21-17

Selecting from the Color control of the Properties toolbar

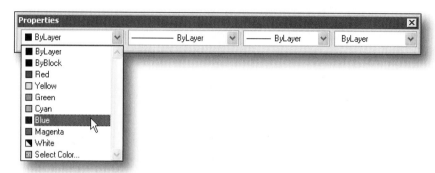

The color command allows you to specify the color for new objects. You can select ByLayer, ByBlock or use the Select Color dialog box (see Figure 21–18) to define the color of objects selecting from the 255 AutoCAD Color Index (ACI) colors, true colors, and Color Book colors.

 Figure 21-18

Select Color dialog box with Index Color tab displayed

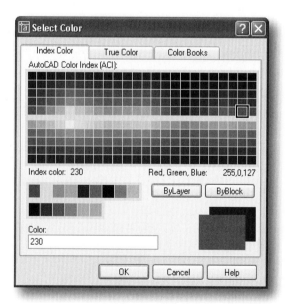

You can also select one of the colors from the chart or one of the color bars below the chart on the **Index Color** tab of the Select Color dialog box. All new objects you create are drawn with this color, regardless of the color of the current layer, until you again set the color to ByLayer or ByBlock. ByLayer (default) causes the objects drawn to assume the color of the layer on which it is drawn. ByBlock causes objects to be drawn in white until selected for inclusion in a block definition. Subsequent insertion of a block that contains objects drawn under the ByBlock option causes those objects to assume the color of the current setting of the color command.

Instead of choosing from 256 standard colors, you can also choose colors from the True Color graphic interface located on the **True Color** tab with its controls for Hue, Saturation, Luminance, and Color Model or from standard Color Books (such as Pantone) located on the **Color Books** tab. True Color and Color Books options make it easier to match colors in your drawing with colors of actual materials.

---

note   As noted in Chapter 6, the options to specify colors by both layer and the COLOR command can cause confusion in a large drawing, especially one containing blocks and nested blocks. You are advised not to mix the two methods of specifying colors in the same drawing.

Setting an Object's Linetype

▶ **Figure 21–19**

Selecting from the Linetype control of the Properties toolbar

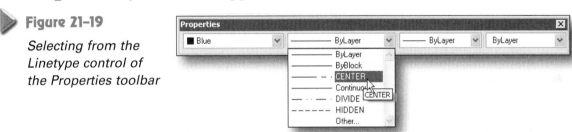

The linetype command loads, sets, and modifies linetypes. A linetype must exist in a library file and be loaded before you can apply it to an object or layer. Standard linetypes are in the library file called *acad.lin* and are not loaded with the layer command. You must load the linetype before you assign it to a specific layer.

Linetypes are combinations of dashes, dots, and spaces. Customized linetypes permit "out of line" objects in a linetype such as circles, wavy lines, blocks, and skew segments.

Lines with dashes (not all dots) usually have dashes at both ends. AutoCAD automatically adjusts the lengths of end dashes to reach the endpoints of the adjoining line. Intermediate dashes will be the lengths specified in the definition. If the overall length of the line is not long enough to permit the breaks, the line is drawn continuous.

There is no guarantee that any segments of the line fall at a particular location. For example, when placing a centerline through circle centers, you cannot be sure that the short dashes will be centered on the circle centers as most conventions call for. To achieve this effect, the short and long dashes have to be created by either drawing them individually or by breaking a continuous line to create the spaces between the dashes. This also creates multiple in-line lines instead of one line of a particular linetype. Or you can use the dimension command Center option to place the desired mark.

Individual linetype names and definitions are stored in one or more files whose extension is *.lin*. The same name may be defined differently in two different files. If you redefine a linetype, loading it with the linetype command will cause objects drawn on layers assigned to that linetype to assume the new definition.

You can select ByLayer, ByBlock, one of the linetypes listed, or invoke the linetype command by selecting Other from the Properties toolbar (See Figure 21–19). After you invoke the linetype command, AutoCAD displays the Linetype Manager dialog box, similar to Figure 21–20.

Figure 21-20

Linetype Manager dialog box

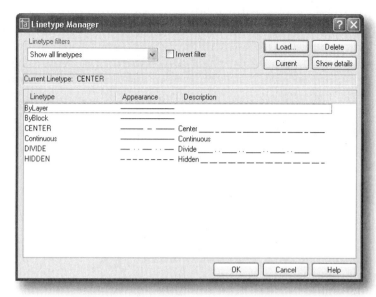

AutoCAD lists the available linetypes for the current drawing and displays the current linetype setting. By default, it is set to ByLayer. To change the current linetype setting, double-click its name in the list. All new objects you create will be drawn with the selected linetype, regardless of the layer you are working with, until you again set the linetype to ByLayer or ByBlock. ByLayer causes the object drawn to assume the linetypes of the layer on which it is drawn. ByBlock causes objects to be drawn in Continuous linetype until selected for inclusion in a block definition. Subsequent insertion of a block that contains objects drawn under the ByBlock option will cause those objects to assume the linetype of the block.

To load a linetype into your drawing, choose **Load**. AutoCAD displays the Load or Reload Linetypes dialog box, as shown in Figure 21–21.

Figure 21-21

Load or Reload Linetypes dialog box

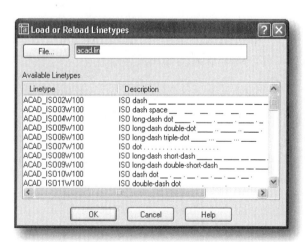

By default, AutoCAD lists the available linetypes from the *acad.lin* file. Select the linetype to load from the **Available Linetypes** list box, and choose **OK**.

If you need to load linetypes from a different file, choose **File** in the Load or Reload Linetypes dialog box. AutoCAD displays the Select Linetype File dialog box. Select the appropriate linetype file and choose **OK**. In turn, AutoCAD lists the available linetypes from the selected linetype file in the Load or Reload Linetypes dialog box. Select the appropriate linetype to load from the **Available Linetypes** list box, and choose **OK**.

To delete a linetype that is currently loaded in the drawing, first select the linetype from the list box in the Linetype Manager dialog box, and then choose **Delete**. You can delete only linetypes that are not referenced in the current drawing. You cannot delete linetype Continuous, ByLayer, or ByBlock. To display additional information about a specific linetype, first select the linetype from the **Linetype** list box in the Linetype Manager dialog box, and then choose **Show details**. AutoCAD displays an extension of the dialog box, listing additional settings.

After making the necessary changes, choose **OK** to keep the changes and close the Linetype Manager dialog box.

 note   As noted in Chapter 6, the options to specify linetypes by both the LAYER and the LINETYPE commands can cause confusion in a large drawing, especially one containing blocks and nested blocks. You are advised not to mix the two methods of specifying linetypes in the same drawing.

Setting an Object's Lineweight

 Figure 21–22

Selecting from the Lineweight control of the Properties toolbar

The Lineweight menu of the Properties toolbar (see Figure 21–22) allows you to specify a lineweight for the objects to be drawn, separate from the assigned lineweight for the layer. You can select ByLayer, ByBlock, or one of the lineweights listed, or invoke the lineweight command and select lineweight from the Lineweight Settings dialog box, as shown in Figure 21–23.

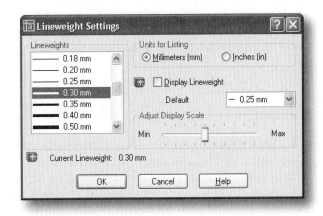

Figure 21–23

Lineweight Settings dialog box

AutoCAD displays the current lineweight at the bottom of the dialog box. To change the current lineweight, select one of the available lineweights from the **Lineweights** list box. Bylayer is the default. All new objects you create are drawn with the current lineweight, regardless of which layer is current, until you again set the lineweight to Bylayer, Byblock, or Default. Bylayer causes the objects drawn to assume the lineweight of the layer on which it is drawn. Byblock causes objects to be drawn using the default lineweight until selected for inclusion in a block definition. Subsequent insertion of a block that contains objects drawn under the Byblock option causes those objects to assume the lineweight of the current setting of the lineweight command. The **Default** selection causes the objects to be drawn to the default value as set by the lwdefault system variable, and defaults to a value of 0.01 inches or 0.25 mm. You can also set the default value from the **Default** option menu located in the right side of the dialog box. The lineweight value of 0 plots at the thinnest lineweight available on the specified plotting device and is displayed at one pixel wide in model space. You can use the properties command to change the lineweight of the existing objects.

The **Units for Listing** section specifies whether lineweights are displayed in millimeters or inches. The **Display Lineweight** check box controls whether lineweights are displayed in the current drawing. If it is set to ON, lineweights are displayed in model space and paper space. AutoCAD regeneration time increases with lineweights that are represented by more than one pixel. If it is set to OFF, AutoCAD performance improves.

The **Adjust Display Scale** slider controls the display scale of lineweights on the Model tab. Lineweights are displayed using a pixel width in proportion to the real-world unit value at which they plot. If you are using a high-resolution monitor, you can adjust the lineweight display scale to better display different lineweight widths.

The **Lineweights** list reflects the current display scale. Objects with lineweights that are displayed with a width of more than one pixel may increase AutoCAD regeneration time. If you want to optimize AutoCAD performance when working in the Model tab, set the lineweight display scale to the minimum value or turn off lineweight display altogether.

Choose **OK** to close the dialog box and keep the changes in the settings.

---

note   As noted in Chapter 6, the options to specify lineweights by both the LAYER and the LINEWEIGHT commands can cause confusion in a large drawing, especially one containing blocks and nested blocks. You are advised not to mix the two methods of specifying lineweights in the same drawing.

SECURITY, PASSWORDS, AND ENCRYPTION

Electronic drawing files, like their paper counterparts, often need to have the information they contain protected from unauthorized viewing. AutoCAD provides password and encryption capabilities to achieve this. Also, it might be necessary to determine that the person who last edited and saved the drawing is the person who was supposed to edit and save it. To accomplish this, AutoCAD allows the use of Digital Signatures.

Setting Password

AutoCAD password protection makes it possible to prevent a drawing file from being opened without first entering the assigned password. Invoke the SECURITYOPTIONS command from the Command prompt; AutoCAD displays the Security Options dialog box, similar to Figure 21–24.

Figure 21-24

Security Options dialog box with the Password tab displayed

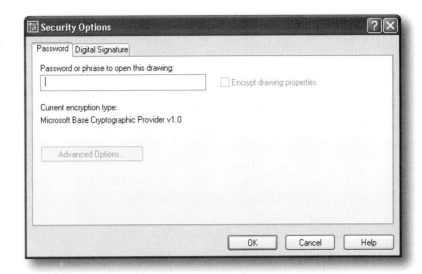

On the **Password** tab of the Security Options dialog box, enter a password in the **Password or phrase to open this drawing** box. This prevents the drawing from being opened without entering the password specified. Passwords can be a single word or a phrase and are not case-sensitive.

To view data in a password-protected drawing, open the drawing in a standard way and enter the password in the **Enter password to open drawing** box in the Password dialog box. Unless the title, author, subject, keywords, or other drawing properties were encrypted when the password was attached, you can view the properties in the Properties dialog box in Windows Explorer.

Setting Encryption

You can encrypt drawing properties, such as the title, author, subject, and keywords, thus requiring a password to view the properties and thumbnail preview of the drawing. If you decide to specify an encryption type and key length, you can select them from the ones available on your computer. On the **Password** tab of the Security Options dialog box, after you have entered a password in the **Password or phrase to open this drawing** box, set the **Encrypt drawing properties** check box to ON. Under the **Password or phrase to open this drawing** box, AutoCAD displays the current encryption type. To change the encryption type, select **Advanced Options**. AutoCAD displays the Advanced Options dialog box. The note in the upper windows warns "Note: Encryption providers vary depending on operating system and country. Before changing the encryption provider you should confirm that the intended recipient of this drawing has a computer with the encryption provider you choose." In the **Choose an encryption provider** box, you can select one of the encryption providers listed. From the **Choose a key length** box, you can choose a key length. The higher the key length, the higher the protection.

Digital Signature

AutoCAD provides a means to sign the drawing file electronically. This means that it is possible to verify that a drawing has had a distinct and unique digital signature attached to it when it was last saved. Along with this positive electronic identification, you can also apply a time stamp and comments.

To attach a digital signature to a drawing, you must first obtain a digital ID. This can be done by contacting a certificate authority through a search engine in your Internet browser, using the term "digital certificate." Once a digital ID has been established on your computer, invoke the securityoptions command. AutoCAD displays the Security Options dialog box. Select the **Digital Signature** tab, and AutoCAD displays various options available for utilizing a Digital Signature.

AutoCAD displays a list of digital IDs that you can use to sign files. It includes information about the organization or individual to whom the digital ID was issued, the digital ID vendor who issued the digital ID, and when the digital ID expires. Select one of the available digital IDs and set **Attach digital signature after saving drawing** to ON. From the **Signature information** section of the **Digital Signature** tab, you can select a time stamp to be attached with the digital ID from the **Get time stamp from** box. You can also add comments to the digital ID in the **Comment** box.

A digital ID has a name, expiration date, serial number, and certain certifying information. The certificate authority issuing the digital ID, can provide Low, Medium, and High levels of security.

From the digital signature feature, you can determine whether the file was changed since it was signed, whether the signers are who they claim to be, and if they can be traced. The digital signature is considered invalid if the file was corrupted when the digital signature was attached, if it was corrupted in transit, or if for any other reason the digital signature is no longer valid. In order to maintain validity of the digital signature, you must not add a password to the drawing or modify or save it after the digital signature has been attached.

note The digital signature status is displayed when you open a drawing if the SIGWARN system variable is set to ON. If it is set to OFF, the signature status is displayed only if the signature is invalid.

In the **Open and Save** tab of the Options dialog box, if **Display digital signature information** is checked, then when you open a drawing that has a digital signature attached, the Digital Signature Contents dialog box is displayed, providing information on the status of the drawing and the signer. In the **Other Fields** list, you can obtain information about the issuer, beginning and expiration dates, and serial number of the digital signature.

CUSTOM SETTINGS WITH THE OPTIONS DIALOG BOX

The Options dialog box allows you to customize the AutoCAD settings. AutoCAD allows you to save and restore a set of custom preferences called a profile. A profile can include preference settings that are not saved in the drawing, with the exception of pointer and printer driver settings. By default, AutoCAD stores your current settings in a profile named <<Unnamed Profile>>. Invoke the OPTIONS command from the Tools menu; AutoCAD displays the Options dialog box (see Figure 21–25). From the Options dialog box, the user can control various aspects of the AutoCAD environment. The Options dialog box has nine tabs; to make changes to any of the sections, select the corresponding tab from the top of the Options dialog box.

Figure 21–25

Options dialog box with the Files tab selected

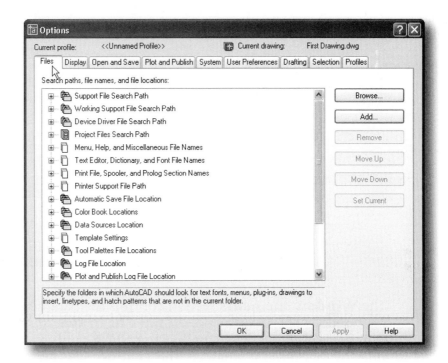

The **Files** tab of the Options dialog box, shown in Figure 21–25, specifies the directory in which AutoCAD searches for support files, driver files, project files, template drawing file location, temporary drawing file location, temporary external reference file location, and texture maps. It also specifies the location of menu, help, log, text editor, and dictionary files.

Browse displays the Browse for Folder or Select a File dialog box, depending on what you selected from the list.

Add adds a search path for the selected folder.

Remove removes the selected search path or file.

Move Up moves the selected search path above the preceding search path.

Move Down moves the selected search path below the following search path.

Set Current makes the selected project or spelling dictionary current.

The **Display** tab of the Options dialog box (see Figure 21–26), controls preferences that relate to AutoCAD performance.

Figure 21-26

Options dialog box with the Display tab selected

The **Window Elements** section controls the parameters of the AutoCAD drawing window. The **Display scroll bars in drawing window** check box (set to OFF by default) specifies whether to display scroll bars at the bottom and right sides of the drawing window. The **Display screen menu** check box (OFF by default) specifies whether to display the screen menu on the right side of the drawing window. Additional options are provided related to toolbars. Choose **Colors**, and AutoCAD displays the AutoCAD Window Colors dialog box, which can be used to set the colors for drawing area, screen menu, text window, and command line. Choose **Fonts**, and AutoCAD displays the Graphics Window Font dialog box, which can be used to specify the font AutoCAD uses for the screen menu and command line, and in the text window.

The **Display resolution** section lets you set the resolution for **Arc and circle smoothness**, **Segments in a polyline curve**, **Rendered object smoothness**, and **Contour lines per surface**.

The **Layout elements** section allows you to set the default settings related to the layout elements: **Display Layout and Model tabs**, **Display printable area**, **Display paper background**, **Display paper shadow**, **Show page setup dialog for new layouts**, and **Create viewport in new layouts**.

The **Display performance** section allows you to set the default settings related to display performance: **Pan and zoom with raster image**, **Highlight raster image frame only**, **Apply solid fill**, **Show text boundary frame only**, and **Show silhouettes in wireframe**.

The **Open And Save** tab of the Options dialog box (see Figure 21–27) lets you determine formats and parameters for drawings, external references, and ObjectARX applications as they are opened or saved.

Figure 21-27

Options dialog box with the Open And Save tab selected

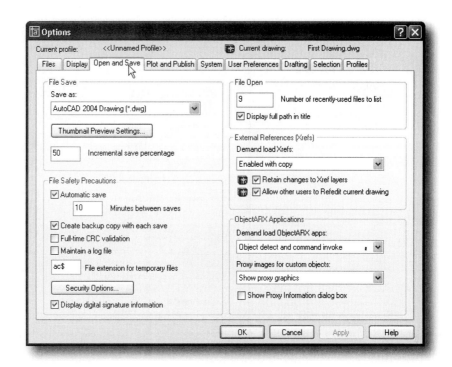

In the **File Save** section, **Save as** lets you select a default save format when you invoke the SAVEAS command.

The **File Open** section shows the number of recently used files to list and has a check box that causes AutoCAD to display the full path in the title when checked.

The **External References (Xrefs)** section controls the settings that relate to editing and loading external references.

The **File Safety Precautions** section helps detect errors and avoid losing data. **Automatic save** with the **Minutes between saves** allows you to determine if and at what intervals periodic automatic saves will be performed. **Create backup copy with each save** allows you to determine whether a backup copy is created when you save the drawing. **Full-time CRC validation** allows you to determine whether a cyclic redundancy check (CRC) is performed when an object is read into the drawing. Cyclic redundancy check is a mechanism for error checking. **Maintain a log file** allows you to determine whether the contents of the text window are written to a log file. Use the **Files** tab in the Options dialog box to specify the name and location of the log file. The **File extension for temporary files** text box allows you to specify an extension for temporary files on a network. The default extension is *.ac$*. **Security Options** provides options for a digital signature and password that are invoked when you save a file, and **Display digital signature information** presents digital signature information when a file with a valid digital signature is opened.

The **ObjectARX Applications** section controls parameters for AutoCAD Runtime Extension applications and proxy graphics.

The **Plot and Publish** tab of the Options dialog box (see Figure 21–28) lets you select the parameters for plotting your drawing.

Figure 21-28

Options dialog box with the Plot and Publish tab selected

The **Default plot settings for new drawings** section determines plotting parameters for new drawings. The **Use as default output device** option determines the default output device for new drawings. The list displays all plotter configuration files (*.PC3*) that are found in the plotter configuration search path. It also displays all system printers configured in the system. **Use last successful plot settings**, when selected, uses the settings of the last successful plot for the current settings. **Add or Configure Plotters** lets you add or configure a plotter from the Plotters program window. See the chapter on plotting for adding and configuring plotters.

The **General plot options** section lets you set general parameters such as paper size, system printer alert parameters, and OLE objects. **Keep the layout paper size if possible**, when selected, applies the paper size in the **Layout Settings** tab in the Page Setup dialog box provided the selected output device is able to plot to this paper size. If it cannot, AutoCAD displays a warning message and uses the paper size specified either in the plotter configuration file (PC3) or in the default system settings if the output device is a system printer. **Use the plot device paper size** selection applies the paper size in either the plotter configuration file (PC3) or in the default system settings if the output device is a system printer. The **System printer spool alert** list box selection determines if a warning will be displayed if the plotted drawing is spooled through a system printer because of an input or output port conflict. The **OLE plot quality** list box selection determines plotted OLE objects' quality. **Use OLE application when plotting OLE objects** selection starts the application that creates the OLE object when you plot a drawing with OLE objects. This will help optimize quality of OLE objects. The **Hide system printers** selection controls whether Windows system printers are displayed in the Plot and Page Setup dialog boxes

The **Plot to File** section specifies the default location for plot to file operations. You can enter a location or click the browse **[...]** button to specify a new location.

The **Background processing options** section specifies options for background plotting and publishing. You can use background plotting to start a job you are plotting or publishing and immediately return to work on your drawing while your job is plotted or published as you work.

The **Plot and publish log file** section controls options for saving a plot and publish log file as a comma-separated value (*.CSV*) file that can be viewed in a spreadsheet program.

The **Specify plot offset relative to** section specifies whether the offset of the plot area is from the lower-left corner of the printable area or from the edge of the paper.

The **Plot Stamp Settings** button opens the Plot Stamp Settings dialog box, which allows you to specify information for the plot stamp.

The **Plot Style Table Settings** button opens the Plot Style Table Settings dialog box, which allows you to specify settings for plot style tables.

The **System** tab of the Options dialog box (see Figure 21–29), has sections for managing the 3D graphics display, pointing devices, dbConnect options, and general options.

Figure 21-29

Options dialog box with the System tab selected

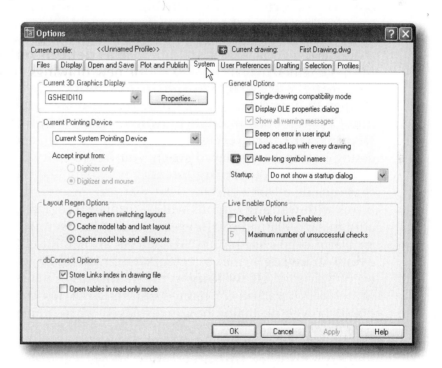

The **Current 3D Graphics Display** section controls settings that relate to system properties and configuration of the 3D graphics display system.

Choosing **Properties** displays a 3D Graphics System Configuration dialog box for the current 3D graphics display system. In the 3D Graphics System Configuration dialog box, you set options that affect the way objects are displayed and system resources are used in the 3D Orbit view. The options you set also affect the way objects are shaded with SHADEMODE.

The **Current Pointing Device** section determines the parameter for the pointing device(s) being used. The list box lists available pointing device drivers from which to choose. The **Accept input from** selection specifies whether AutoCAD accepts input from both a mouse and a digitizer or ignores mouse input when a digitizer is set.

The **Layout Regen Options** section lets you specify how AutoCAD updates the display list in the Model and layout tabs. The display list for each tab is updated either by regenerating the drawing when you switch to that tab or by saving the display list to memory and regenerating only the modified objects when you switch to that tab. **Regen when switching layouts**, when selected, causes AutoCAD to regenerate the drawing each time you switch tabs. **Cache model tab and last layout**, when selected, saves the display list to memory for the Model tab and the last layout made current. It suppresses regenerations when you switch between the two tabs. Regenerations for all other layouts still occur when you switch to those tabs. **Cache model tab and all layouts**, when selected, causes AutoCAD to regenerate the drawing the first time you switch to each tab. For the remainder of the drawing session, when you switch to those tabs, the display list is saved to memory, and regenerations are suppressed.

The **dbConnect Options** section allows you to manage the options associated with database connectivity. The **Store links index in drawing file** check box, when selected, causes AutoCAD to store the database index within the drawing file. The **Open tables in read-only mode** check box, when selected, causes AutoCAD to open database tables in read-only mode within the drawing file.

The **General Options** section controls general options that relate to system settings. **Single-drawing compatibility mode**, when selected, causes AutoCAD to open only one drawing at a time (Single-drawing Interface or SDI). Otherwise AutoCAD can open multiple drawing sessions (Multi-drawing Interface or MDI). **Display OLE properties dialog**, when selected, causes the OLE Properties dialog box to be displayed when you insert OLE objects into AutoCAD drawings. **Show all warning messages**, when selected, causes all dialog boxes that include a Don't Display This Warning Again option to be displayed. Dialog boxes with warning options will be displayed regardless of previous settings specific to each dialog box. **Beep on error in user input**, when selected, causes an alarm beep when AutoCAD detects an invalid entry. **Load acad.lsp with every drawing**, when selected, causes AutoCAD to load the *acad.lsp* file into every drawing. **Allow long symbol names**, when selected, allows you to use up to 255 characters for named objects. **Startup** controls whether the Startup dialog box or no dialog box is displayed when starting AutoCAD or creating a new drawing.

The **Live Enabler Options** section allows you to specify how AutoCAD checks for Object Enablers. Using Object Enablers, you can display and use custom objects in AutoCAD drawings even when the ObjectARX application that created them is unavailable. **Check Web for Live Enablers**, when selected, causes AutoCAD to check for Object Enablers on the Autodesk Web site. The **Maximum number of unsuccessful checks** box allows you to specify the number of times AutoCAD will continue to check for Object Enablers after unsuccessful attempts.

The **User Preferences** tab of the Options dialog box (see Figure 21–30) controls options that optimize the way you work in AutoCAD.

Figure 21-30

Options dialog box with the User Preferences tab selected

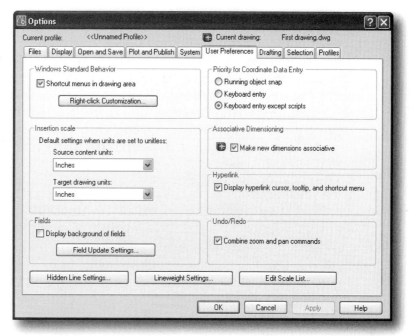

The **Windows Standard Behavior** section lets you apply Windows techniques and methods in AutoCAD. Selecting **Shortcut menus in drawing area** causes shortcut menus to be displayed when the pointing device is right-clicked. Otherwise, right-clicking is the same as pressing ENTER. Choosing **Right-click customization** causes the Right-Click Customization dialog box to be displayed. The Right-Click Customization dialog box determines whether right-clicking in the drawing area displays a shortcut menu or is the same as pressing ENTER. This will allow you to have a right-click invoke ENTER while a command is active. You can also disable the following Command shortcut menu options: Choosing **Turn on time-sensitive right-click** controls right-click behavior. A quick click is the same as pressing ENTER. A longer click displays a shortcut menu. You can set the duration of the longer click in milliseconds. The **Default Mode** section of the Right-Click Customization dialog box determines the effect of right-clicking when no objects are selected. Selecting **Repeat Last Command** causes right-clicking to be the same as pressing ENTER. Selecting **Shortcut Menu** causes right-clicking to display a shortcut menu when applicable. The **Edit Mode** section of the Right-Click Customization dialog box determines the effect of right-clicking when one or more objects are selected. Selecting **Repeat Last Command** causes right-clicking to be the same as pressing ENTER. Selecting **Shortcut Menu** causes right-clicking to display the Edit shortcut menu. The **Command Mode** section of the Right-Click Customization dialog box determines the effect of right-clicking when a command is in progress. Selecting ENTER causes right-clicking to be the same as pressing ENTER when a command is in progress. Selecting **Shortcut Menu: always enabled** causes right-clicking to display the Command shortcut menu. Selecting **Shortcut Menu: enabled when command options are present** causes right-clicking to display the Command shortcut menu to be displayed only when options are currently available from the command line. Otherwise, rightclicking is the same as pressing ENTER.

The **Insertion scale** section of the **User Preferences** tab allows you to control the default scale for dragging objects into a drawing using i-drop or DesignCenter. The **Source content units** list box allows you to set the units AutoCAD uses for an object being inserted into the current drawing when no insert units are specified with the INSUNITS system variable. The **Target drawing units** list box allows you to set the units AutoCAD uses in the current drawing when no insert units are specified with the INSUNITS system variable. (This is stored in the INSUNITS-DEFTARGET system variable.) Each of these allow you to choose from the following units: **Inches**, **Feet**, **Miles**, **Millimeters**, **Centimeters**, **Meters**, **Kilometers**, **Microinches**, **Mills**, **Yards**, **Angstroms**, **Nanometers**, **Microns**, **Decimeters**, **Decameters**, **Hectometers**, **Gigameters**, **Astronomical Units**, **Light Years**, and **Parsecs**. If **Unspecified-Unitless** is selected, the object is not scaled when inserted.

The **Fields** section of the **User Preferences** tab sets preferences related to fields. Selecting **Display background of fields** displays fields with a light gray background that is not plotted. When this option is cleared, fields are displayed with the same background as any text. Choosing **Field Update Settings** displays the Field Update Settings dialog box, which allows you to set the fields that will be updated automatically.

The **Priority for Coordinate Data Entry** section of the **User Preferences** tab determines how input of coordinate data affects AutoCAD's actions. Selecting **Running object snap** causes running object snaps to be used at all times instead of specific coordinates. Selecting **Keyboard entry** causes the coordinates that you enter to be used at all times and overrides running object snaps. Selecting **Keyboard entry except scripts** causes the specific coordinates that you enter to be used rather than running object snaps, except in scripts.

The **Associative Dimensioning** section of the **User Preferences** tab controls whether new dimensions are associative. Selecting **Make new dimensions associative** causes new dimensions to be drawn as associative dimensions and will be associated with the objects being dimensioned.

The **Hyperlink** section of the **User Preferences** tab determines display property settings of hyperlinks. Selecting **Display hyperlink cursor**, **tooltip**, and **shortcut menu** causes the hyperlink cursor, tooltip, and shortcut menu to be displayed when the cursor is over an object that contains a hyperlink.

The **Undo/Redo** section of the **User Preferences** tab controls UNDO and REDO for ZOOM and PAN. Selecting **Combine zoom and pan commands** causes these two commands to act together.

Choosing **Hidden Line Settings** causes the Hidden Line Settings dialog box to be displayed (see Figure 21–31). This allows you to change the display properties of hidden lines. These settings are in effect only when the HIDE command is used or when the Hidden option of the SHADEMODE command is used.

Figure 21-31

Hidden Line Settings dialog box

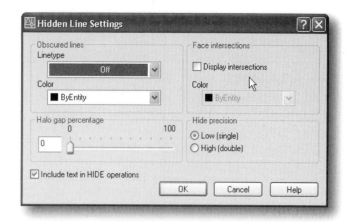

The **Obscured lines** section of the Hidden Line Settings dialog box allows you to specify the linetype and color of obscured lines, which are lines that are made visible by changing its color and linetype. The **Linetype** list box allows you to select from a list of linetypes or select **Off**.

The **Color** list box allows you to select from available colors.

The **Halo gap percentage** section lets you specify the distance to shorten a haloed line at the point where it will be hidden. Moving the slider bar specifies the distance as a percentage of one inch. It is not affected by the zoom level.

Selecting **Include text in HIDE operations** causes text objects created by the TEXT, DTEXT, or MTEXT command to be included during a HIDE command.

The **Face intersections** section controls the display of intersections. Selecting **Display intersections** causes intersection polylines to be displayed. The **Color** drop down allows you specify the color of intersection polylines.

The **Hide precision** section controls the accuracy in the method of creating hides and shades. Selecting **Low (single)** results in low accuracy (and low memory usage). Selecting **High (double)** results in high accuracy.

Choosing **Lineweight Settings** causes the Lineweight Settings dialog box to be displayed. This allows you to set lineweight options. The **Lineweights** section lets you select lineweights in the list box. Options include **ByLayer**, **ByBlock**, **Default**, and a list of varying lineweights. The **Units for Listing** lets you choose between **Millimeters** and **Inches**. Selecting **Display Lineweight** causes lineweights to be displayed in model space and paper space. The **Adjust Display Scale** slider bar lets you adjust the how wide the selected linewidth will be displayed.

Choosing **Edit Scale List** causes the Edit Scale List dialog box to be displayed, which controls the list of scales available for layout viewports, page layouts, and plotting. Select a scale from the **Scale List** list box and then choose one of the options, which include **Add**, **Edit**, **Move Up**, **Move Down**, **Delete**, and **Reset**.

The **Drafting** tab of the Options dialog box (see Figure 21-32), specifies a number of general editing options.

Figure 21-32

Options dialog box with the Drafting tab selected

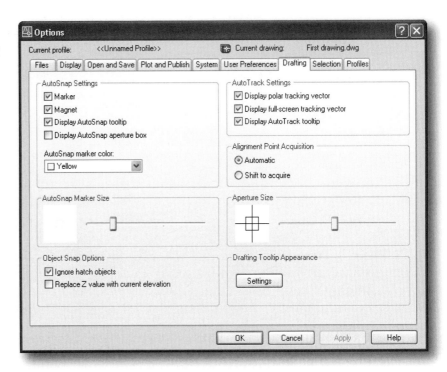

The **AutoSnap Settings** section controls settings that relate to the AutoSnaps, which are displayed when you use object snaps.

The **AutoSnap Marker Size** section sets the display size for the AutoSnap marker.

The **AutoTrack Settings** section controls the settings that relate to AutoTrack™ behavior, which is available when polar tracking or object snap tracking is set to ON.

The **Alignment Point Acquisition** section controls the method of displaying alignment vectors in a drawing.

The **Aperture Size** section sets the display size for the AutoSnap aperture.

The **Object Snap Options** section has the **Ignore hatch objects** check box that, when selected, causes AutoCAD to ignore hatch objects when employing an Object Snap mode.

The **Selection** tab of the Options dialog box (see Figure 21–33), lets you customize selection options in AutoCAD.

Figure 21-33

Options dialog box with the Selection tab selected

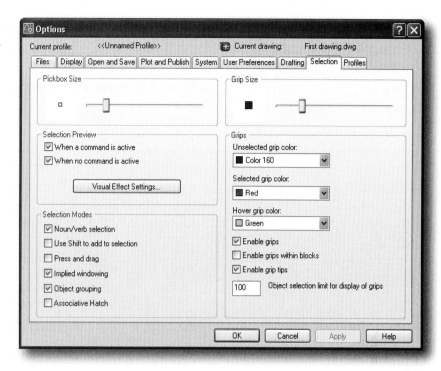

The **Pickbox Size** slider controls the display size of the AutoCAD pickbox. The pickbox is the format that the cursor assumes for object selection in editing commands.

The **Selection Preview** section controls whether there is a preview **When a command is active** or **When no command is active**. Choosing **Visual Effect Settings** causes the Visual Effect Settings dialog box to be displayed where you can control how objects appear when selected.

The **Selection Modes** section controls settings that relate to object selection methods. Refer to Chapter 11 for a detailed explanation about selection modes.

The **Grips** section controls the settings that relate to grips. Grips are small squares displayed on an object after it has been selected. For a detailed explanation about grips, refer to Chapter 11.

The **Profiles** tab of the Options dialog box (see Figure 21-34), lets you manage profiles. A profile is a named and saved group of environment settings. This profile can be restored as a group when desired. AutoCAD stores your current options in a profile named <<Unnamed Profile>>. AutoCAD displays the current profile name, as well as the current drawing name, in the Options dialog box. The profile data is saved in the system registry and can be written to a text file (an *.arg* file). AutoCAD organizes essential data and maintains changes in the registry as necessary.

Figure 21-34

*Options dialog box with
the Profiles tab selected*

A profile can be exported to or imported from different locations. If changes have been made to your current profile during an AutoCAD session and you want to save them in the *.arg* file, the profile must be exported. After the profile with the current profile name has been exported, AutoCAD updates the *.arg* file with the new settings. Then the profile can be imported again into AutoCAD, thus updating your profile settings.

The **Set Current** makes the profile that is highlighted in the **Available profiles** list the current profile. Choosing **Add to List** lets you name and save the current environment settings as a profile. **Rename** lets you rename the highlighted profile. **Delete** lets you delete the highlighted profile. **Export** causes the Export Profiles dialog box to be displayed. This is a file manager dialog box in which the highlighted profile can be saved to the path you specify. **Import** causes the Import Profiles dialog box to be displayed. This is a file manager dialog box in which you can select a profile from a saved path to be imported. **Reset** causes the highlighted profile to be reset.

After making all the necessary changes, choose **Apply** to apply the changes. Then choose **OK** to close the Options dialog box.

SAVING OBJECTS IN VARIOUS FILE FORMATS

The EXPORT command allows you to save selected object(s) in other file formats, such as *.bmp, .dxf, .dwf, .sat, .3ds,* and *.wmf.* Invoke the EXPORT command from the File menu; AutoCAD displays the Export Data dialog box.

In the **Files of type** list box, select the format type in which you wish to export objects. Enter the file name in the **File name** box. Select **Save**, and AutoCAD prompts:

Select objects: *(select the objects to export, and press* ENTER *to complete object selection)*

AutoCAD exports the selected objects in the specified file format using the specified file name.

IMPORTING VARIOUS FILE FORMATS

The IMPORT command allows you to import various file formats, such as *.3ds, .dxf, .eps, .sat,* and *wmf,* into AutoCAD. Invoke the IMPORT command, and AutoCAD displays the Import File dialog box.

In the **Files of Type** list box, select the format type you wish to import into AutoCAD. Select the file from the appropriate directory in the list box and click **Open**. AutoCAD imports the file into the AutoCAD drawing.

STANDARDS

AutoCAD has a feature that allows you to verify that the layers, dimension styles, linetypes, and text styles of the drawing you are working in conform to an accepted standard, such as a company, trade, or client standard. To utilize this feature, your drawing should be associated too a standards drawing.

You can create a standards file from an existing drawing, or you can create a new drawing and save it as a standards file with an extension of *.dws.* Open an existing drawing from which you wish to create a standards file, invoke the saveas command, and enter a name for the standards file in the Save Drawing As dialog box. Select AutoCAD Drawing Standards (*.*dws*) from the **Files of type** list and then choose **Save**. You can also create a new drawing and set appropriate standards for layers, text styles, dimension styles and linetypes and save as a standards file.

The standards command lets you obtain information about the standards file that is associated with the current drawing. Invoke the standards command from the Tools menu; AutoCAD displays the Configure Standards dialog box as shown in Figure 21–35.

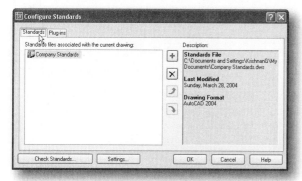 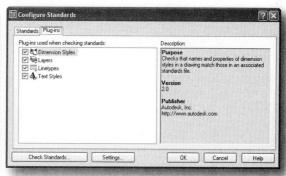

 Figure 21-35

Configure Standards dialog box with the Standards tab (left) and Plug-ins tab (right) selected

If a drawing has an associated standards file, there will be an Associated Standards File(s) icon on the status bar tray at the bottom right corner of the drawing area.

On the **Standards** tab (see Figure 21–35 at left), the **Standards files associated with the current drawing** section lists all Standards (*.dws*) files that are associated with the current drawing. To add a Standards file, choose **Add Standards File** (+ sign) or press f3. AutoCAD displays the Select Standards File dialog box. Select a Standards file from an appropriate folder. To remove a Standards file from the current drawing, choose **Remove Standards File** (x sign) or press delete on your keyboard. If conflicts arise between multiple standards in the list (for example, if two standards specify layers of the same name but with different properties), the standard that appears first in the list takes precedence. To change the position of a Standards file in the list, select it and choose **Move Up** or **Move Down**.

In the **Plug-ins** tab of the Configure Standards dialog box (see Figure 21–35 at right), the **Plug-ins used when checking standards** section lists the standards plug-ins that are installed on the current system. For the CAD Standards Extension, a standards plug-in is installed for each of the named objects for which standards can be defined (layers, dimension styles, linetypes, and text styles). The **Description** section has descriptions of the Purpose, Version, and Publisher of the plug-in that is highlighted in the **Plug-ins used when checking standards** section.

Choosing **Check Standards** analyzes the current drawing for standards violations. AutoCAD displays the Check Standards dialog box as shown in Figure 21–36.

Figure 21–36

Check Standards dialog box

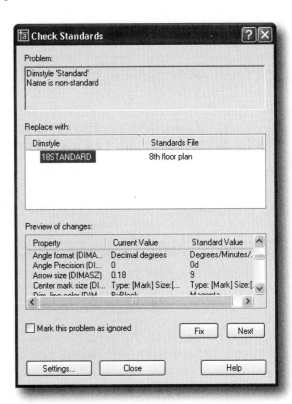

The Check Standards dialog box has sections titled **Problem**, **Replace with**, and **Preview of changes**.

> The **Problem** section provides a description of a nonstandard object in the current drawing. To fix a problem, select a replacement from the **Replace with** list, and then choose **Fix** or press F4.

> The **Replace with** section lists possible replacements for the current standards violation. If a recommended fix is available, it is preceded by a check mark. If a recommended fix is not available, no items are highlighted in the **Replace with** list.

> The **Preview of changes** section indicates the properties of the nonstandard AutoCAD object that will be changed if the fix currently selected in the **Replace with** list is applied.

Selecting **Mark this problem as ignored** causes AutoCAD to ignore deviations from standards by flagging the current problem as ignored. If the **Show Ignored Problems** option is turned off in the CAD Standards Settings dialog box, problems flagged as ignored are not displayed the next time the drawing is checked.

Choosing **Fix** fixes the nonstandard AutoCAD object using the item currently selected in the **Replace with** list, and advances to the next nonstandard object in the current drawing. This button is unavailable if a recommended fix does not exist or if an item is not highlighted in the **Replace with** list.

Choosing **Next** causes AutoCAD to advance to the next nonstandard AutoCAD object in the current drawing without fixing the current one.

Choosing **Settings** displays the CAD Standards Settings dialog box, which specifies additional settings for the Check Standards dialog box and the Configure Standards dialog box.

Choose **Close** to close the Check Standards dialog box without applying a fix to the standards violation currently displayed in the Problem section.

In AutoCAD you can use CAD Standards tools to check for violations as you work. You are immediately alerted whenever you create a non-standard named object.

INTERNET UTILITIES

The Internet is the most important way to convey digital information around the world. You are probably already familiar with the best-known uses of the Internet: e-mail (electronic mail) and surfing the Web (short for "World Wide Web"). E-mail lets users exchange messages and data at very low cost. The Web brings together text, graphics, audio, and video in an easy-to-use format. Other uses of the Internet include FTP (file transfer protocol, for effortless binary-file transfer), Gopher (presents data in a structured, subdirectory-like format), and Usenet, a collection of more than 100,000 news groups.

AutoCAD allows you to interact with the Internet in several ways. You can launch a Web browser from within AutoCAD. AutoCAD can create .*dwf* (short for "Design Web Format") files for viewing drawings in two-dimensional format on Web pages. AutoCAD can open and insert drawings from, and save drawings to the Internet. AutoCAD transforms into the Internet design platform, delivering enhanced Internet collaboration and communication capabilities to users.

Launching the Default Web Browser

 Figure 21-37

Invoking the BROWSER *command from the Web toolbar*

The browser command lets you start a Web browser from within AutoCAD. By default, the browser command uses the Web browser program that is registered in your computer's Windows operating system. AutoCAD prompts for the URL (uniform resource locator). The URL is the Web site address, such as *http://www.autodesk.com*. After you invoke the browser command (see Figure 21–37), AutoCAD prompts:

> Enter Web location (URL) <default location>: *(specify a new location, or press* ENTER *to accept the default location)*

You can specify a default URL in the Options dialog box. AutoCAD launches the Web browser, which contacts the Web site. Figure 21–38 shows Internet Explorer and the Autodesk Web site.

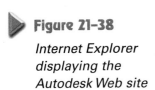

Figure 21–38

Internet Explorer displaying the Autodesk Web site

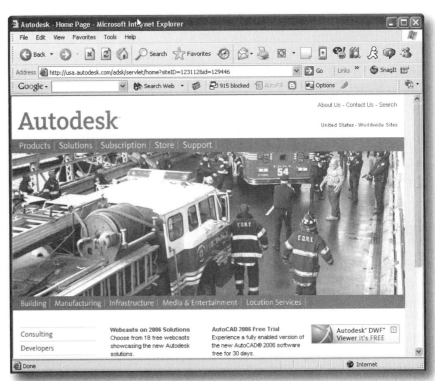

Communication Center

The Communication Center resides in the Tray section (lower-right corner of the AutoCAD window) and can be customized to offer you the desired amount of information. Here you can choose to be notified about such things as maintenance patches, product support information, subscription information and extension announcements, articles, and tips. Maintenance patches include any program updates and fixes to the existing product. Product support information includes ground-breaking news from the Product Support team at Autodesk. Subscription information and extension announcements include announcements and subscription program news if you are an Autodesk subscription member. If new information becomes available, a bubble announcement is displayed.

The Communication Center is an interactive feature that must be connected to the Internet to deliver content and information. Each time the Communication Center is connected, it sends information to Autodesk so that the correct information can be returned. All information is sent anonymously to maintain your privacy. The information that will be sent to Autodesk includes Product Name, Product Release Number, Product Language, and Country.

Opening and Saving Drawings from the Internet

AutoCAD allows you to open and save drawing files from the Internet or an intranet. You can also attach externally referenced drawings stored on the Internet/intranet to drawings stored locally on your system. Whenever you open a drawing file from an Internet or intranet location with the OPEN command, it is first downloaded into your computer and opened in the AutoCAD drawing area. Then you can edit the drawing and save it, either locally or back to the Internet or intranet location for which you have appropriate access privileges.

Working with Hyperlinks

AutoCAD allows you to create hyperlinks that provide jumps to associated files. Hyperlinks provide a simple and powerful way to quickly associate a variety of documents with an AutoCAD drawing. For example, you can create a hyperlink that opens another drawing file from the local drive or network drive, or from an Internet Web site. You can also specify a named location to jump to within a file, such as a view name in an AutoCAD drawing, or a bookmark in a word processing program. You can also attach a URL to jump to a specific Web site. You can attach hyperlinks to any graphical object in an AutoCAD drawing.

AutoCAD allows you to create both *absolute* and *relative* hyperlinks in your AutoCAD drawings. Absolute hyperlinks store the full path to a file location, whereas relative hyperlinks store a partial path to a file location, relative to a default URL or name of the directory you specify using the hyperlinkbase system variable. You can also specify the relative path for a drawing in the Drawing Properties dialog box (**Summary** tab).

Whenever you attach a hyperlink to an object, AutoCAD provides cursor feedback when you position the cursor over the object. To activate the hyperlink, first select the object (make sure the system variable is set to 1). Right-click to display the Shortcut menu and activate the link from the Hyperlink submenu.

When you create a hyperlink that points to an AutoCAD drawing template (*.dwt*) file, AutoCAD creates a new drawing file based on the selected template instead of opening the actual template when you activate the hyperlink. With this method there is no risk of accidentally overwriting the original template.

When you create a hyperlink that points to an AutoCAD named view and activate the hyperlink, the named view that was created in model space is restored in the Model tab, and the named view that was created in paper space is restored in the Layout tab.

Creating a Hyperlink

To create a hyperlink, invoke the HYPERLINK command from the Insert menu, and AutoCAD prompts:

> Select objects: *(select one or more objects to which the hyperlink will be attached and press* ENTER*)*

AutoCAD displays the Insert Hyperlink dialog box, as shown in Figure 21–39.

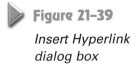

Figure 21-39

Insert Hyperlink dialog box

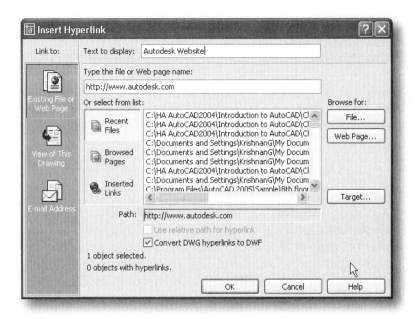

Specify a description for the hyperlink in the **Text to display** box. This is useful when the file name or URL is not helpful in identifying the contents of the linked file. Specify the URL or path with the name of the file that you wish to have associated with the selected objects in the **Type the file or Web page name** box. Or choose one of the **Browse for** buttons: **File**, **Web Page**, or **Target**. Choosing **File** opens the Browse the Web – Select Hyperlink dialog box (standard file selection dialog box). Use the dialog box to navigate to the file that you want associated with the hyperlink. Choosing **Web Page** opens the AutoCAD browser. Use the browser to navigate to a Web page to associate with the hyperlink. Choosing **Target** opens the Select Place in Document dialog box, in which you specify a link to a named location in a drawing. The named location that you select is the initial view that is restored when the hyperlink is executed. You can also select the path, with the name of the file or URL, from the list box categorized from Recent Files, Browsed Pages, and Inserted Links.

Path displays the path to the file associated with the hyperlink. If **Use relative path for hyperlink** is selected, only the file name is listed. If **Use relative path for hyperlink** is cleared, the full path and the file name are listed. Choosing **Convert DWG Hyperlinks to DWF** specifies that the DWG hyperlink will convert to a DWF file hyperlink when you publish or plot the drawing to a DWF file

Choose **OK** to create the hyperlink to the selected objects and close the Insert Hyperlink dialog box. Whenever you attach a hyperlink to an object, AutoCAD provides cursor feedback when you position the cursor over the object. To activate the hyperlink, first select the object (make sure the pickfirst system variable is set to 1). Right-click to display the Shortcut menu and activate the hyperlink associated with the currently selected object from the Hyperlink submenu.

AutoCAD allows you to attach hyperlinks to blocks, including nested objects contained within blocks. If the blocks contain any relative hyperlinks, the relative hyperlinks adopt the relative base path of the current drawing when you insert them.

Editing and Removing the Hyperlink

To edit a hyperlink, invoke the HYPERLINK command and select the object that has a hyperlink. AutoCAD displays the Edit Hyperlink dialog box. Make the necessary changes in the **Text to display** box, and/or the **Type the file or Web page name** text box. To remove the existing hyperlink, choose **Remove Link**. AutoCAD removes the hyperlink. Choose **OK** to accept the changes and close the Edit Hyperlink dialog box.

Design Web Format

Design Web Format™ (DWF) is an open, secure file format developed by Autodesk for the transfer of drawings over networks, including the Internet. The *.dwf* files are highly compressed, so they are much smaller (less than half the size of a *.dwg* file) and faster to transmit, enabling the communication of rich design data, without the overhead associated with typical larger CAD drawings. The *.dwf* files are not a replacement for native CAD formats such as *.dwg* and don't allow editing of the data within the file. The sole purpose of *.dwf* is to allow designers, engineers, developers, and their colleagues to communicate design information and intent to anyone needing to view, review, or print design information.

The latest release of DWF is DWF 6. DWF 6 has been re-designed to enable users to build a complex set of design documents, pages, or layouts in one *.dwf* file. The latest release of Autodesk DWF Viewer supports viewing and printing of 3D models published from nearly every Autodesk design application. You can also take advantage of new features like print preview, viewing new DWF MapBooks, and accessing block attribute data published from AutoCAD 2006.To view DWF files, you can use Autodesk DWF Viewer, free, downloadable application (you can download from www.autodesk.com) or Autodesk DWF Composer which enables complete round-tripping of markups, annotations, and other changes back into Autodesk Revit® products and the AutoCAD 2006 family of software products, so you never have to reenter information.

To view *.dwf* files, you need Autodesk DWF Viewer (formerly Autodesk Express Viewer), which is a free, lightweight, high-performance application. Autodesk DWF Viewer enables users to view *.dwf* files using the advanced graphics systems found in many of the Autodesk design products. Autodesk DWF Viewer has a very flexible printing system that enables users to print to scale, fit to page, tile, or selectively print a variety of layouts from the new DWF 6 publishing format. Autodesk DWF Viewer is available as both a stand-alone application or as a control that is used within the Microsoft Internet Explorer browser, providing a simple, easy to use, and common Windows interface for even the novice user to master viewing and printing of design data. Check the *www.autodesk.com* Web site for the latest version of Autodesk DWF Viewer. Always use the latest version of the application to view the *.dwf* files.

Using Design Publisher to create DWF files

Design Publisher allows you to assemble a collection of drawings and plot directly to paper or publish to a DWF (Design Web Format) file. AutoCAD allows you to publish your drawing sets as either a single multi-sheet DWF format file or multiple single-sheet DWF format files, or to plot to the designated plotter in the page setup. You can publish to devices (plotters or files) specified in the page setups for each layout. With Design Publisher, you have the flexibility to create electronic or paper drawing sets for distribution. The recipients can then view or plot your drawing sets.

You can customize your drawing set for a specific user, and you can add and remove sheets in a drawing set as a project evolves. Design Publisher allows you to publish directly to paper or to an intermediate electronic format that can be distributed using email, FTP sites, project Web sites, or CD. You can open DWF files with Autodesk DWF Viewer.

Figure 21-40

Invoking the PUBLISH *command from the Standard toolbar*

Invoke the publish command from the Standard toolbar (see Figure 21–40); AutoCAD displays the Publish dialog box similar to Figure 21–41.

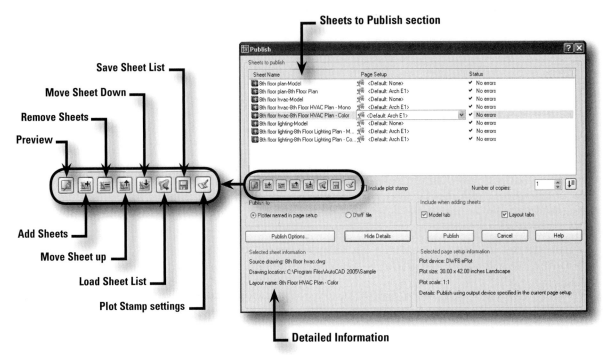

Figure 21-41

Publish Drawing Sheets dialog box

The **Sheets to publish** section lists the drawing sheets to be included for publishing.

> The **Sheet Name** column displays the combined name of the drawing and the layout name separated with a dash (-). If necessary, you can rename from the Rename option available from the Shortcut menu. Drawing sheet names must be unique within a single *.dwf* file.

> The **Page Setup** column displays the named page setup for the sheet. You can change the page setup by clicking the page setup name and selecting another page setup from the list. Select Import to import page setups from another *.dwg* file through the Import Page Setups for Publishing dialog box. Only Model tab page setups can be applied to Model tab sheets, and only paper space page setups can be applied to paper space layouts.

> The **Status** column displays the status of the sheet when it is loaded to the list of sheets.

To add sheets to the existing selection, choose **Add Sheets**; AutoCAD displays a standard file selection dialog box, where you can add sheets to the list of drawing sheets. The layout names from those files are extracted, and one sheet is added to the list of drawing sheets for each layout. New drawing sheets are always appended to the end of the current list.

To remove sheets, choose **Remove Sheets**, and AutoCAD deletes the currently selected drawing sheet from the list of sheets.

To move the selected drawing sheet up one position in the list, choose **Move Sheet Up**, and to move the selected drawing sheet down one position in the list choose **Move Sheet Down**.

The **Include when adding sheets** section specifies whether the model and layouts contained in a drawing are added to the sheet list when you add sheets. At least one option must be selected.

> The **Model tab** selection specifies whether the model is included when drawing sheets are added.

> The **Layout tabs** selection specifies whether all layouts are included when drawing sheets are added.

To save the current selection of the drawing sheets, choose **Save Sheet List**, and AutoCAD displays the Save List dialog box, where you can save the current list of drawings as a *.dsd* (Drawing Set Descriptions) file. These *.dsd* files are used to describe lists of drawing files and selected lists of layouts within those drawing files.

To load a saved list to the current selection, choose **Load Sheet List**, and AutoCAD displays a standard file selection dialog box. You can select a *.dsd* file or a *.bp3* (Batch Plot) file to load. AutoCAD displays the Replace or Append dialog box if a list of drawing sheets is present in the Publish Drawing Sheets dialog box. You can either replace the existing list of drawing sheets with the new sheets or append the new sheets to the current list.

To include a plot stamp on a specified corner of each drawing and log it to a file, set the **Include plot stamp** check box to ON. To customize the plot stamp settings, open the Plot Stamp dialog box by choosing **Plot Stamp Settings**. AutoCAD displays the Plot Stamp dialog box, in which you can specify the information, such as drawing name and plot scale, that you want applied to the plot stamp.

The **Publish to** section defines how to publish the list of sheets. You can publish to either a multisheet *.dwf* file (an electronic drawing set) or to the plotter specified in the page setup (a paper drawing set or a set of plot files).

> Select **Plotter named in page setup** to plot to output devices for each drawing sheet page setups.

> Select **DWF file** to publish as a *.dwf* file.

Specify the number of copies to publish in the **Number of copies** box. If the **DWF file** option is selected in the **Publish to** section, the **Number of copies** setting defaults to 1 and cannot be changed.

Choose **Preview** to display the drawing as it will appear when plotted on paper by executing the preview command.

To customize publishing, choose **Publish Options**, and AutoCAD opens the Publish Options dialog box (see Figure 21–42), in which you can specify options for publishing.

Figure 21–42

Publish Options dialog box

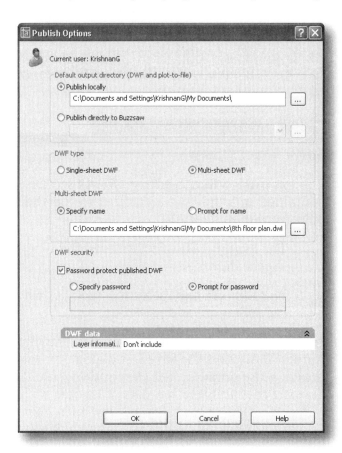

The name of the current user or current sheet set is displayed. When the name of the current user is shown, changes made in the dialog box are saved in the current user's profile. When the name of the current sheet set is shown, changes made in the dialog box are saved with the sheet set.

Specify where *.dwf* and plot files are saved when you publish drawing sheets in the **Default output directory (DWF and plot-to-file)** section.

Specify whether single sheet *.dwf* or a single multisheet *.dwf* is created in the **DWF type** section.

The **Multi-sheet DWF** section allows you to specify file naming options for creating multisheet *.dwf*. Select **Specify name** and specify the name and location in the name field to the multisheet *.dwf* file. Instead select **Prompt for name**, and name and location will be provided when it is prompted for published *.dwf* files.

The **DWF security** section specifies options for protecting *.dwf* files with passwords. Set **Password protect published DWF** to ON to specify that published DWF files will have a password applied to them. A recipient of a *.dwf* file that has a password applied to it must have the password to open the DWF file. Select **Specify password** to specify that the password in the password field is applied to the DWF file and specify the password to apply to the *.dwf* file. DWF passwords are case sensitive. The password or phrase can be made up of letters, numbers, punctuation, or non-ASCII characters. Instead select **Prompt for password** to specify that the DWF Password dialog box is displayed when you click **Publish**.

note If you lose or forget the password, it cannot be recovered. Keep a list of passwords and their corresponding *.dwf* file names in a safe place.

The **DWF data** section lists and allows you to specify the data that can optionally be included in the *.dwf* file.

Choose **OK** to save the changes and close the Publish Options dialog box.

Choose **Show Details** in the Publish dialog box to display the **Selected sheet information** and **Selected page setup information** sections (see Figure 21–41).

Choose **Publish** to publish the selected layouts. AutoCAD begins the publishing operation, creating one or more single-sheet *.dwf* files or a single multisheet *.dwf* file, or plotting to a device or file, depending on the option selected in the **Publish to** section and the options selected in the Publish Options dialog box.

If a drawing sheet fails to plot, Design Publisher continues plotting the remaining sheets in the drawing set. A log file is created that contains detailed information, including any errors or warnings encountered during the publishing process. You can stop publishing after a sheet has finished plotting. If you stop publishing a multi-sheet *.dwf* file before it is complete, no output file is generated. After publishing is complete, the **Status** field is updated to show the results.

Viewing DWF Files

AutoCAD itself cannot display *.dwf* files, nor can *.dwf* files be converted back to *.dwg* format without using file translation software from a third-party vendor. In order to view a *.dwf* file, you need Autodesk DWF Viewer.

Figure 21–43 shows an example of *dwf* displayed in Autodesk DWF Viewer.

Figure 21–43

Autodesk DWF Viewer displaying DWF file

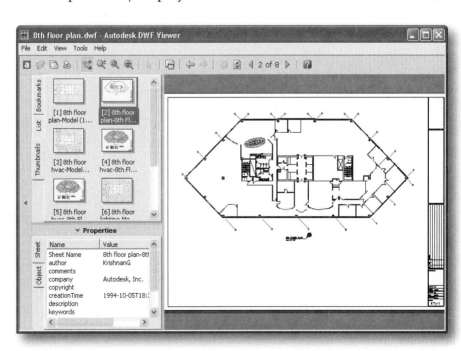

Autodesk DWF Viewer makes it easier to navigate multisheet drawing sets, access intelligent design data, and customize the look and feel of the user interface. The navigation pane next to the image includes thumbnail images to help you browse a multisheet *.dwf* faster and more easily. The index list in the navigation pane helps make browsing a multisheet *.dwf* faster and easier. Autodesk DWF Viewer shows you rich design data that is not available in *.pdf*, including hyperlinks, named views, and properties of objects in the design, such as door size, height, and materials. In addition, you can view detailed information for all sheets in the *.dwf* file–including who created the document, and when. You can customize flexible, high-fidelity printing for whatever orientation and size you need. Hyperlinks and named views make it easy to guide users to the design data they need.

eTransmit Utility

The eTransmit utility allows you to select and bundle together the drawing file and its related files. You can create a transmittal set of files as a compressed self-extracting executable file, as a compressed *.zip* file, or as a set of uncompressed files in a new or existing folder. You can include all the reference files attached to the drawing file, Word files, spreadsheet, etc. to be part of the bundle. It is easier to transmit by e-mail one single compressed file consisting of a drawing file and several related files.

To create a transmittal set of a drawing and related files, invoke the etransmit command from the File menu. AutoCAD displays the Create Transmittal dialog box, shown in Figure 21–44.

Figure 21-44

Create Transmittal dialog box

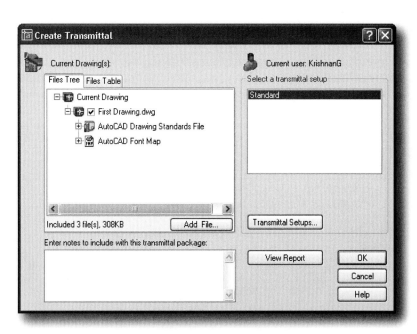

The **Files Tree** tab lists the files to be included in the transmittal package in a hierarchical tree format. By default, all files associated with the current drawing (such as related xrefs, plot styles, and fonts) are listed. You can add files to the transmittal package or remove existing files.

To add files to the transmittal package, choose **Add File**. AutoCAD opens a standard file selection dialog box, in which you can select an additional file to include in the transmittal package.

Enter notes related to a transmittal package in the space under **Enter notes to be included with this transmittal package**. The notes are included in the transmittal report. You can specify a template of default notes to be included with all your transmittal packages by creating an ASCII text file called *etransmit.txt*. This file must be saved to a location specified by the **Support File Search Path** option on the **Files** tab in the Options dialog box.

Select the transmittal setup from **Select a transmittal setup** list. AutoCAD lists previously saved transmittal setups. The default transmittal setup is named STANDARD.

To create, modify, and delete transmittal setups, choose **Transmittal Setups**. AutoCAD displays the Transmittal Setup dialog box. You can create or modify an existing transmittal setup that specifies the type of transmittal package created.

To view the report that is included with the transmittal package, choose **View Report**. AutoCAD displays report information that includes any transmittal notes that you entered and distribution notes automatically generated by AutoCAD that detail what steps must be taken for the transmittal package to work properly.

Choose **OK** to create the transmittal set and close the Create Transmittal dialog box.

review questions

1. All of the following can be renamed using the RENAME command, except:
   a. *A current drawing name.*
   b. *Named views within the current drawing.*
   c. *Block names within the current drawing.*
   d. *Text style names within the current drawing.*

2. The PURGE command can be used:
   a. *After an editing session.*
   b. *At the beginning of the editing session.*
   c. *At any time during the editing session.*
   d. *A and b only.*

3. The most common setting for the current drawing color is:
   a. *Red.*   d. *White.*
   b. *Bylayer.*   e. *None of the above.*
   c. *Byblock.*

4. Which of the following are not valid color names in AutoCAD?
   a. *Brown*   d. *Magenta*
   b. *Red*   e. *None of the above (all are valid)*
   c. *Yellow*

5. To change the background color for the graphics drawing area, you should use:
   a. *COLOR.*   d. *SETTINGS.*
   b. *BGCOLOR.*   e. *CONFIG.*
   c. *OPTIONS.*

6. AutoCAD release 14 can be used to edit a drawing saved as an AutoCAD 2005 drawing.
   a. *True*
   b. *False*

7. AutoCAD 2005 can be used to edit a drawing that is saved in an AutoCAD release 14 drawing.
   a. *True*
   b. *False*

8. All the following items can be purged from a drawing file, except:
   a. *Text styles.*   d. *Views.*
   b. *Blocks.*   e. *Linetypes.*
   c. *System variables.*

9. The following can be deleted with the PURGE command, except:
   a. *Blocks not referenced in the current drawing.*
   b. *Linetypes that are not being used in the current drawing.*
   c. *Layer 0.*
   d. *None of the above (all can be purged).*

10. The VIEW command:

    a. *Serves a purpose similar to the PAN command.*
    b. *Will restore previously saved views of your drawing.*
    c. *Is normally used on very small drawings.*
    d. *None of the above.*

11. Which command permits a user to work with just a portion of a drawing by loading geometry from specific views or layer?

    a. *PLOAD*   **c.** *OPEN*
    b. *PARTOPN*   **d.** *PLTOPN*

12. What command permits additional geometry to be loaded into the current partially loaded drawing?

    a. *PLOAD*   **c.** *PARTADD*
    b. *PARTIALOAD* **d.** *ADDGEO*

13. Can the PURGE command be used in partially open drawings?

    a. *Yes*
    b. *No*

14. What term can be added to draw, modify, or inquiry commands that will cause them to repeat?

    a. *Redo*   **d.** *Multiple*
    b. *Repeat*   **e.** *No such option*
    c. *Return*

15. When one of more objects overlay each other, which command is used to control their order of display?

    a. *DWGORDER*   **c.** *VIEWORDER*
    b. *DRAWORDER*   **d.** *ARRANGE*

16. DWF is short for:

    a. *DraWing Format.*
    b. *Design Web Format.*
    c. *DXF Web Format.*

17. The purpose of DWF files is to view:

    a. *2D drawings on the Internet.*
    b. *3D drawings on the Internet.*
    c. *3D drawings in another CAD system.*
    d. *All of the above.*

18. Compression in the DWF file causes it to take _____ time to transmit over the Internet.

    a. *More*
    b. *Less*
    c. *Same*
    d. *None of the above*

19. Which AutoCAD command is used to open an AutoCAD drawing file from an Internet site?

    a. *LAUNCH*

    b. *OPEN*

    c. *START*

    d. *LAUNCHFILE*

20. Once an Internet AutoCAD drawing file has been modified, it can only be saved back to the Internet site.

    a. *True*   **b.** *False*

21. AutoCAD creates hyperlinks as being _____.

    a. *Absolute*

    b. *Relative*

    c. *Polar*

    d. *Only a or b*

22. Hyperlinks can be attached only to text objects within a drawing.

    a. *True*   **b.** *False*

23. DWF files are compressed to as much as _____ of the original .DWG file size.

    a. *1/8*

    b. *1/4*

    c. *1/2*

    d. *Equal to size*

Hardware Requirements

> **note** Open the Appendix A *PDF* file on the accompanying CD.

Listing of AutoCAD Commands (Menu and Command Prompt)

note Open the Appendix B *PDF* file on the accompanying CD.

AutoCAD Toolbars

note Open the Appendix C *PDF* file on the accompanying CD.

System Variables

 Open the Appendix D *PDF* file on the accompanying CD.

appendix e

Hatch and Fill Patterns

 Open the Appendix E *PDF* file on the accompanying CD.

appendix f

Fonts

note Open the Appendix F *PDF* file on the accompanying CD.

Linetypes and Lineweights

 Open the Appendix G *PDF* file on the accompanying CD.

appendix h

Command Aliases

 Open the Appendix H *PDF* file on the accompanying CD.

appendix i

Express Tools

> **note** Open the Appendix I *PDF* file on the accompanying CD.

Note: AutoCAD keyed commands appear in SMALL CAPS.

T

IMPORTANT-READ CAREFULLY: This End User License Agreement ("Agreement") sets forth the conditions by which Delmar Learning, a division of Thomson Learning Inc. ("Thomson") will make electronic access to the Thomson Delmar Learning-owned licensed content and associated media, software, documentation, printed materials and electronic documentation contained in this package and/or made available to you via this product (the "Licensed Content"), available to you (the "End User"). BY CLICKING THE "I ACCEPT" BUTTON AND/OR OPENING THIS PACKAGE, YOU ACKNOWLEDGE THAT YOU HAVE READ ALL OF THE TERMS AND CONDITIONS, AND THAT YOU AGREE TO BE BOUND BY ITS TERMS CONDITIONS AND ALL APPLICABLE LAWS AND REGULATIONS GOVERNING THE USE OF THE LICENSED CONTENT.

1.0 SCOPE OF LICENSE

1.1   Licensed Content. The Licensed Content may contain portions of modifiable content ("Modifiable Content") and content which may not be modified or otherwise altered by the End User ("Non-Modifiable Content"). For purposes of this Agreement, Modifiable Content and Non-Modifiable Content may be collectively referred to herein as the "Licensed Content." All Licensed Content shall be considered Non-Modifiable Content, unless such Licensed Content is   ted.

1.2   Subject to the End User's compliance with the terms and conditions of this Agreement, Thomson Delmar Learning hereby grants the End User, a nontransferable, non-exclusive, limited right to access and view a single copy of the Licensed Content on a single personal computer system for noncommercial, internal, personal use only. The End User shall not (i) reproduce, copy, modify (except in the case of Modifiable Content), distribute, display, transfer, sublicense, prepare derivative work(s) based on, sell, exchange, barter or transfer, rent, lease, loan, resell, or in any other manner exploit the Licensed Content; (ii) remove, obscure or alter any notice of Thomson Delmar Learning's intellectual property rights present on or in the License Conte   se engineer or otherwise reduce the Licensed Content.

2.0 TERMINATION

2.1   Thomson Delmar Learning may at any time (without prejudice to its other rights or remedies) immediately terminate this Agreement and/or suspend access to some or all of the Licensed Content, in the event that the End User does not comply with any of the terms and conditions of this Agreement. In the event of such termination by Thomson Delmar Learning, the End User shall immediately return any and all copies of the Licensed Content to Thomson Delmar Learning.

3.0 PROPRIETARY RIGHTS

3.1   The End User acknowledges that Thomson Delmar Learning owns all right, title and interest, including, but not limited to all copyright rights therein, in and to the Licensed Content, and that the End User shall not take any action inconsistent with such ownership. The Licensed Content is protected by U.S., Canadian and other applicable copyright laws and by international treaties, including the Berne Convention and the Universal Copyright Convention. Nothing contained in this Agreement shall be construed as granting the End User any ownership rights in or to the Licensed Content.

3.2   Thomson Delmar Learning reserves the right at any time to withdraw from the Licensed Content any item or part of an item for which it no longer retains the right to publish, or which it has reasonable grounds to believe infringes copyright or is defamatory, unlawful or otherwise objectionable.

4.0 PROTECTION AND SECURITY

4.1   The End User shall use its best efforts and take all reasonable steps to safeguard its copy of the Licensed Content to ensure that no unauthorized reproduction, publication, disclosure, modification or distribution of the Licensed Content, in whole or in part, is made. To the extent that the End User becomes aware of any such unauthorized use of the Licensed Content, the End User shall immediately notify Delmar Learning. Notification of such violations may be made by sending an Email to delmarhelp@thomson.com.

5.0 MISUSE OF THE LICENSED PRODUCT

5.1   In the event that the End User uses the Licensed Content in violation of this Agreement, Thomson Delmar Learning shall have the option of electing liquidated damages, which shall include all profits generated by the End User's use of the Licensed Content plus interest computed at the maximum rate permitted by law and all legal fees and other expenses incurred by Thomson Delmar Learning in enforcing its rights, plus penalties.

6.0 FEDERAL GOVERNMENT CLIENTS

6.1   Except as expressly authorized by Delmar Learning, Federal Government clients obtain only the rights specified in this Agreement and no other rights. The Government acknowledges that (i) all software and related documentation incorporated in the Licensed Content is existing commercial computer software within the meaning of FAR 27.405(b)(2); and (2) all other data delivered in whatever form, is limited rights data within the meaning of FAR 27.401. The restrictions in this section are acceptable as consistent with the Government's need for software and other data under this Agreement.

7.0 DISCLAIMER OF WARRANTIES AND LIABILITIES

7.1   Although Thomson Delmar Learning believes the Licensed Content to be reliable, Thomson Delmar Learning does not guarantee or warrant (i) any information or materials contained in or produced by the Licensed Content, (ii) the accuracy, completeness or reliability of the Licensed Content, or (iii) that the Licensed Content is free from errors or other material defects. THE LICENSED PRODUCT IS PROVIDED "AS IS," WITHOUT ANY WARRANTY OF ANY KIND AND THOMSON DELMAR LEARNING DISCLAIMS ANY AND ALL WARRANTIES, EXPRESSED OR IMPLIED, INCLUDING, WITHOUT LIMITATION, WARRANTIES OF MERCHANTABILITY OR FITNESS OR A PARTICULAR PURPOSE. IN NO EVENT SHALL THOMSON DELMAR LEARNING BE LIABLE FOR: INDIRECT, SPECIAL, PUNITIVE OR CONSEQUENTIAL DAMAGES INCLUDING FOR LOST PROFITS, LOST DATA, OR OTHERWISE. IN NO EVENT SHALL DELMAR LEARNING'S AGGREGATE LIABILITY HEREUNDER, WHETHER ARISING IN CONTRACT, TORT, STRICT LIABILITY OR OTHERWISE, EXCEED THE AMOUNT OF FEES PAID BY THE END USER HEREUNDER FOR THE LICENSE OF THE LICENSED CONTENT.

8.0 GENERAL

8.1   Entire Agreement. This Agreement shall constitute the entire Agreement between the Parties and supercedes all prior Agreements and understandings oral or written relating to the subject matter hereof.

8.2   Enhancements/Modifications of Licensed Content. From time to time, and in Delmar Learning's sole discretion, Thomson Thomson Delmar Learning may advise the End User of updates, upgrades, enhancements and/or improvements to the Licensed Content, and may permit the End User to access and use, subject to the terms and conditions of this Agreement, such modifications, upon payment of prices as may be established by Delmar Learning.

8.3   No Export. The End User shall use the Licensed Content solely in the United States and shall not transfer or export, directly or indirectly, the Licensed Content outside the United States.

8.4   Severability. If any provision of this Agreement is invalid, illegal, or unenforceable under any applicable statute or rule of law, the provision shall be deemed omitted to the extent that it is invalid, illegal, or unenforceable. In such a case, the remainder of the Agreement shall be construed in a manner as to give greatest effect to the original intention of the parties hereto.

8.5   Waiver. The waiver of any right or failure of either party to exercise in any respect any right provided in this Agreement in any instance shall not be deemed to be a waiver of such right in the future or a waiver of any other right under this Agreement.

8.6   Choice of Law/Venue. This Agreement shall be interpreted, construed, and governed by and in accordance with the laws of the State of New York, applicable to contracts executed and to be wholly preformed therein, without regard to its principles governing conflicts of law. Each party agrees that any proceeding arising out of or relating to this Agreement or the breach or threatened breach of this Agreement may be commenced and prosecuted in a court in the State and County of New York. Each party consents and submits to the non-exclusive personal jurisdiction of any court in the State and County of New York in respect of any such proceeding.

8.7   Acknowledgment. By opening this package and/or by accessing the Licensed Content on this Website, THE END USER ACKNOWLEDGES THAT IT HAS READ THIS AGREEMENT, UNDERSTANDS IT, AND AGREES TO BE BOUND BY ITS TERMS AND CONDITIONS. IF YOU DO NOT ACCEPT THESE TERMS AND CONDITIONS, YOU MUST NOT ACCESS THE LICENSED CONTENT AND RETURN THE LICENSED PRODUCT TO THOMSON DELMAR LEARNING (WITHIN 30 CALENDAR DAYS OF THE END USER'S PURCHASE) WITH PROOF OF PAYMENT ACCEPTABLE TO DELMAR LEARNING, FOR A CREDIT OR A REFUND. Should the End User have any questions/comments regarding this Agreement, please contact Thomson Delmar Learning at delmarhelp@thomson.com.